# Protein Arrays, Biochips, and Proteomics

# Protein Arrays, Biochips, and Proteomics

## The Next Phase of Genomic Discovery

*edited by*

## Joanna S. Albala
*Lawrence Livermore National Laboratory*
*Livermore, California, U.S.A.*

## Ian Humphery-Smith
*University of Utrecht*
*Utrecht, The Netherlands*

**CRC Press**
Taylor & Francis Group
Boca Raton   London   New York

CRC Press is an imprint of the
Taylor & Francis Group, an **informa** business

CRC Press
Taylor & Francis Group
6000 Broken Sound Parkway NW, Suite 300
Boca Raton, FL 33487-2742

First issued in paperback 2019

ISBN-13: 978-0-8247-4312-3 (hbk)
ISBN-13: 978-0-367-39507-0 (pbk)

**Library of Congress Cataloging-in-Publication Data**
A catalog record for this book is available from the Library of Congress.

**Visit the Taylor & Francis Web site at**
**http://www.taylorandfrancis.com**

**and the CRC Press Web site at**
**http://www.crcpress.com**

# Foreword

During the lean years of proteomics, the field was largely dominated by techniques such as two-dimensional gel electrophoresis. More recently, the spectacular innovations in mass spectrometry have given proteomics a shot in the arm and transformed the discipline. The complete sequencing of the human genome and that of other model organisms has further boosted proteomics in many ways, not least by providing a sequence-based framework for mining the human and other proteomes. Clearly, however, to make a substantial impact in biomedicine, from disease-marker identification to accelerating drug development, proteomics has to evolve much further in the direction of providing high-throughput, high-sensitivity, proteome-scale profiling. Unlike genomic-type profiling, which tends to be unidimensional, as exemplified by DNA microarrays that allow RNA abundance to be measured, there is a need at the protein level to capture a multitude of protein attributes. There is also a need to determine in a cell and tissue context not just the abundance of protein constituents but also their posttranslational modifications, as well as their functional states and their interactions with other proteins and molecules, all with requisite high-throughput and high-sensitivity. The emerging field of protein biochips and microarrays is intended to address such needs and will likely mark yet another evolution in proteomics. The stakes are high and the challenges are enormous.

The milestones in any emerging field sooner or later include the publication of books that review progress and provide both critical and forward-looking perspectives. This is the case for this timely book with the catchy title *Protein Arrays, Biochips, and Proteomics: The Next Phase of Genomic Discovery*. The editors have all the desired credentials and are well-suited for the task of assembling contributing authors who are experts in the field. The editors have devoted much effort in their careers to activities that define the current status of protein chips and microarrays. They are very well connected and are prominently featured in meetings devoted to the subject.

Commensurate with the need to assay a wide range of protein attributes, an equally wide range of chip types have become available that are reviewed in this book with respect to their merits and limitations. Innovative technologies in this field have been developed by academics and by biotechnology companies, thus contributing creative solutions to challenging problems. However, the most challenging problem of all—delivering content on a proteome scale—is beyond the reach of both academics and most biotech companies, simply because of the very high costs involved in producing the tens—and more likely hundreds—of thousands of proteins encoded just in the human genome, or to produce capture agents directed against these proteins and their various epitopes. A consortium approach not unlike that put together for sequencing the genome or for cataloging genome-wide single-nucleotide polymorphisms may need to be implemented to meet this challenge. Strategic considerations such as these are being pursued, for example, by the Human Proteome Organization with its proteome-scale antibody initiative.

So what is in this book for the reader? Obviously, not all applications of protein chips need to be on a proteome scale. Much could be accomplished, particularly by academic investigators, through focused approaches that target a family of proteins, a specific signaling pathway, or a particular posttranslational modification. This book contains a wealth of information that brings the reader up to date in the field of proteomics, protein biochips, and array-based protein strategies, from the theoretical to the practical aspects, with topics ranging from functionalized chip surfaces and the performance of ultrasensitive ligand assays using microarrays to strategies for expressing proteins. There is even a chapter that reviews the proteomics market in its various aspects. The text is easy to read, as are the numerous figures and charts befitting a book on chips and microarrays.

It is rather gratifying to see that the field of proteomics now encompasses chemical engineers, analytical chemists, biochemists, cell and molecular biologists, clinical scientists, and bioinformaticians, just to list a few of the subspecialties. I am confident that people in the field of proteomics or those who are contemplating using proteomics, however varied their interests, will derive valuable knowledge from reading this book.

*Sam Hanash*
*President, Human Proteome Organization*
*Professor of Pediatrics*
*University of Michigan*
*Ann Arbor, Michigan, U.S.A.*

# Preface

Wasinger and colleagues (*Electrophoresis*, 1995, 16: 1090–1094) first defined the term *proteome* as: "the total protein complement able to be encoded by a given genome." It is important to note that this encoded complement can vary significantly, temporally, and with respect to cell and tissue type, while the temporal variation can occur over very short time intervals. In an immunological context it is this antigenic diversity (temporal, cellular, and tissue-specific) that constitutes *self*. A central tenet of modern immunology is that healthy individuals with developing lymphocytes must be exposed to most of *self*, so as to avoid the dysfunctional state of autoimmunity. Thus, on a daily basis, the human body is faced with—and presumably succeeds at—the task of teaching developing lymphocytes the nature of *self* antigens, i.e., the human proteome in its innumerable iterations. Currently, however, experimental proteomics is far from achieving similar analytical success; the task of accessing and detecting all elements within an entire mammalian proteome looms as an almost insurmountable charge, due mostly to the predominance of low-abundance gene products that continue to defy detection. A proteome of a living cell or organism is a highly dynamic entity, and following its many facets in health and disease constitutes a major challenge to the biomedical and scientific community as we collectively attempt to build upon the wealth of understanding afforded by completion of the Human Genome Project. A variety of technologies will be required to come to grips with this technological challenge.

Herein, we have attempted to bring together authors at the forefront of their discipline to provide an overview of current and emerging trends and their applications to the study of proteomics, particularly array-based procedures that offer the promise of "near-to-total" proteomic screening in a high-throughput microenvironmment, including analysis of complex mammalian proteomes, in a manner similar to that achieved for entire genomes and transcriptomes. Of noteworthy importance, however, are the associated financial and infrastructural resources likely to be required. They are no less daunting than was the initiation of the Human Genome Project more than a decade ago; the Human Proteome Project will require equally grandiose means on a global scale, if success is to be forthcoming over the next decade. For both the pharmaceutical industry and academics, the stimulus to proceed remains paramount in that it is the proteins, and not the nucleic acids, that are the molecular workhorses of the cell, that is, the physical players that decide physiological fates in action-packed scenarios with multiple possible endpoints more complex and perverse than the greatest suspense thriller of Alfred Hitchcock or Agatha Christie. Whether the knives and forks are employed for a banquet or a massacre depends on the ordered permutations of protein isoforms, all of which await deciphering within the infinite world of the multidimensional complexity associated with intracellular molecular interactions.

The study of proteomics combines biochemistry, genetics, genomics, and molecular biology to explore cellular networks in a parallelized, high-throughput, global format. Proteomics has its roots in protein profiling by two-dimensional gel electrophoresis and yet appears to some as a newcomer on the scientific scene, a logical next phase in genomic research. Because the nature of science is dynamic, this textbook attempts to address proteomics past, present, and future. The aim is to present a variety of technologies and applications for proteomics research that will have broad application for the individual researcher and that should assist in the introduction of important concepts to newcomers.

The first five chapters focus on the emerging technology of protein arrays and biochips in proteomic research and advances in their application to protein diagnostics and therapeutics. Chapters 1 and 2 provide a global overview of the emerging protein array field as well as a thorough historical perspective. Chapters 3–5 expand on the details of generating and developing protein array technologies.

Chapters 6 and 7 explore array-based proteomics focusing on the use of resources from genomic strategies, particularly ESTs (expressed sequence tags), cDNA databases, and robotics for generating protein content through high-throughput recombinant expression techniques. The chapter that follows examines second-generation proteomics and describes methods that integrate protein profiling by mass spectrometry with protein biochips. Chapter 9 describes shotgun proteomics applications using several mass spectrometry techniques.

Chapters 10 and 11 examine analysis of protein function, specifically protein–protein interaction assays, and explore unique applications in proteomics relating various species, moving through the phylogenetic tree, exemplifying how proteomics can be exploited in model organisms for application to more complex biological systems. Chapter 12 explores advances in structural proteomics aimed at providing a greater understanding of protein biochemistry and cellular function. Then, reflecting an age in which we are inundated with information, Chapter 13 focuses on the integration of genomics and proteomics information. Finally, Chapter 14 provides an educated insight into the growing proteomics market and its emerging biotech sector.

This text aims to be the first to present a variety of genomic-based, high-throughput strategies for the study of proteins by the scientists who are defining proteomics. It provides a foundation from which to examine the field of proteomics as it evolves, to broaden our collective scientific outlook on the future direction of biological research.

*Joanna S. Albala*
*Ian Humphery-Smith*

# Contents

*Foreword*     Sam Hanash                                                    *iii*

*Preface*                                                                       *v*

*Contributors*                                                                 *xi*

1. Protein Biochips and Array-Based Proteomics                                  1
   *Ian Humphery-Smith*

2. Ultrasensitive Microarray-Based Ligand Assay Technology                     81
   *Roger Ekins and Frederick Chu*

3. Practical Approaches to Protein Microarrays                                127
   *Brian Haab*

4. Protein Biochips: Powerful New Tools to Unravel the
   Complexity of Proteomics?                                                  145
   *Steffen Nock and Peter Wagner*

5. Functionalized Surfaces for Protein Microarrays: State of the
   Art, Challenges, and Perspectives                                         159
   *Erik Wischerhoff*

6. High-Throughput Protein Expression, Purification, and
   Characterization Technologies                                             173
   *Stefan R. Schmidt*

7.  Miniaturized Protein Production for Proteomics                    203
    *Michele Gilbert, Todd C. Edwards, Christa Prange, Mike
    Malfatti, Ian R. McConnell, and Joanna S. Albala*

8.  Protein Profiling: Proteomes and Subproteomes                     217
    *Eric T. Fung and Enrique A. Dalmasso*

9.  Shotgun Proteomics and Its Applications to the Yeast Proteome     233
    *Anita Saraf and John R. Yates III*

10. Forward and Reverse Proteomics: It Takes Two (or More) to
    Tango                                                             255
    *David E. Hill, Nicolas Bertin, and Marc Vidal*

11. Dynamic Visualization of Expressed Gene Networks                  277
    *Ingrid Remy and Stephen W. Michnick*

12. High-Throughput Structural Biology and Proteomics                 299
    *Wuxian Shi, David A. Ostrov, Sue Ellen Gerchman, Jadwiga H.
    Kycia, F. William Studier, William Edstrom, Anne Bresnick, Joel
    Ehrlich, John S. Blanchard, Steven C. Almo, and Mark R.
    Chance*

13. Integration of Proteomic, Genechip, and DNA Sequence Data         325
    *Leah B. Shaw, Vassily Hatzimanikatis, Amit Mehra, and Kelvin
    H. Lee*

14. The Proteomics Market                                             337
    *Steven Bodovitz, Julianne Dunphy, and Felicia M. Gentile*

*Index*                                                               *351*

# Contributors

**Joanna S. Albala**  Biology and Biotechnology Research Program, Lawrence Livermore National Laboratory, Livermore, California, U.S.A.

**Steven C. Almo**  Department of Biochemistry, Albert Einstein College of Medicine, Bronx, New York, U.S.A.

**Nicolas Bertin**  Cancer Biology, Dana-Farber Cancer Institute and Harvard Medical School, Boston, Massachusetts, U.S.A.

**John S. Blanchard**  Department of Biochemistry, Albert Einstein College of Medicine, Bronx, New York, U.S.A.

**Steven Bodovitz**  BioInsights, San Francisco, California, U.S.A.

**Anne Bresnick**  Department of Biochemistry, Albert Einstein College of Medicine, Bronx, New York, U.S.A.

**Mark R. Chance**  Department of Physiology and Biophysics and Department of Biochemistry, Albert Einstein College of Medicine, Bronx, New York, U.S.A.

**Frederick Chu**  Molecular Endocrinology, University College London Medical School, London, England

**Enrique A. Dalmasso**  Biomarker Discovery Center, Ciphergen Biosystems, Inc., Fremont, California, U.S.A.

**Julianne Dunphy**   BioInsights, Redwood City, California, U.S.A.

**William Edstrom**   Department of Physiology and Biophysics, Albert Einstein College of Medicine, Bronx, New York, U.S.A.

**Todd C. Edwards**   Biology and Biotechnology Research Program, Lawrence Livermore National Laboratory, Livermore, California, U.S.A.

**Joel Ehrlich**   Department of Biochemistry, Albert Einstein College of Medicine, Bronx, New York, U.S.A.

**Roger Ekins**   Molecular Endocrinology, University College London Medical School, London, England

**Eric T. Fung**   Biomarker Discovery Center, Ciphergen Biosystems, Inc., Fremont, California, U.S.A.

**Felicia M. Gentile**   BioInsights, Cupertino, California, U.S.A.

**Sue Ellen Gerchman**   Department of Biology, Brookhaven National Laboratory, Upton, New York, U.S.A.

**Michele Gilbert**   Biology and Biotechnology Research Program, Lawrence Livermore National Laboratory, Livermore, California, U.S.A.

**Brian Haab**   Van Andel Research Institute, Grand Rapids, Michigan, U.S.A.

**Vassily Hatzimanikatis**   Department of Chemical Engineering, Northwestern University, Evanston, Illinois, U.S.A.

**David E. Hill**   Cancer Biology, Dana-Farber Cancer Institute and Harvard Medical School, Boston, Massachusetts, U.S.A.

**Ian Humphery-Smith**   Department of Pharmaceutical Proteomics, University of Utrecht, Utrecht, The Netherlands

**Jadwiga H. Kycia**   Department of Biology, Brookhaven National Laboratory, Upton, New York, U.S.A.

**Kelvin H. Lee**   Department of Chemical and Biomolecular Engineering, Cornell University, Ithaca, New York, U.S.A.

**Mike Malfatti**   Biology and Biotechnology Research Program, Lawrence Livermore National Laboratory, Livermore, California, U.S.A.

**Ian R. McConnell**   Biology and Biotechnology Research Program, Lawrence Livermore National Laboratory, Livermore, California, U.S.A.

**Amit Mehra**   Department of Chemical Engineering, Northwestern University, Evanston, Illinois, U.S.A.

**Stephen W. Michnick**   Department of Biochemistry, University of Montreal, Montreal, Quebec, Canada

**Steffen Nock**   Zyomyx, Inc., Hayward, California, U.S.A.

**David A. Ostrov**   Department of Biochemistry, Albert Einstein College of Medicine, Bronx, New York, U.S.A.

**Christa Prange**   Biology and Biotechnology Research Program, Lawrence Livermore National Laboratory, Livermore, California, U.S.A.

**Ingrid Remy**   Department of Biochemistry, University of Montreal, Montreal, Quebec, Canada

**Anita Saraf**   Department of Cell Biology, The Scripps Research Institute, La Jolla, California, U.S.A.

**Stefan R. Schmidt**   Biotech Laboratory, AstraZeneca, Södertälje, Sweden

**Leah B. Shaw**   Department of Physics, Cornell University, Ithaca, New York, U.S.A.

**Wuxian Shi**   Department of Biochemistry, Albert Einstein College of Medicine, Bronx, New York, U.S.A.

**F. William Studier**   Department of Biology, Brookhaven National Laboratory, Upton, New York, U.S.A.

**Marc Vidal**   Department of Genetics, Dana-Farber Cancer Institute and Harvard Medical School, Boston, Massachusetts, U.S.A.

**Peter Wagner**   Zyomyx, Inc., Hayward, California, U.S.A.

**Erik Wischerhoff**   Utrecht University, Utrecht, The Netherlands

**John R. Yates III**   Department of Cell Biology, The Scripps Research Institute, La Jolla, California, U.S.A.

# 1

# Protein Biochips and Array-Based Proteomics

IAN HUMPHERY-SMITH
*University of Utrecht*
*Utrecht, The Netherlands*

## I. INTRODUCTION

The discipline of proteomics has evolved around the core separation technologies of two-dimensional gel electrophoresis (2DGE), advanced image analysis, chromatography, capillary electrophoresis, and mass spectrometry. Ward and Humphery-Smith [1] have reviewed the methodologies and bioinformatic procedures employed within the field for protein characterization. There are numerous shortcomings associated with these procedures (see later, pg. 7); however, 2DGE currently remains unsurpassed in its ability to resolve complex mixtures of proteins (for examples, see Figs. 1 and 2). The question remains, however, as to whether or not these very same technologies (traditional proteomics) or variants thereof are capable of scaling to allow meaningful analyses of human tissues in health and disease across multiple organ systems and for large patient cohorts. Based on lessons learned with what until recently was the most complete proteome [2], namely traditional proteome analysis of the smallest living organism, the bacterium *Mycoplasma genitalium*, the answer is clearly no. The difficulties encountered for such a small project simply do not scale to the analysis of numerous human proteomes. Thus, the above technologies need to be complemented by alternate array-based or second-generation approaches (i.e., analytical procedures conducted independently of the separation sciences) (cf. Ref 3, for definition). Array-based procedures are most likely to become the tool of choice for initial target discovery, whereby large sets of patient material will need to be examined so as to acquire the necessary statistical significance necessary for the understanding of multigenic phenomena (Fig. 3). The latter are expected to represent the greater part ($> 95\%$) of all human aliments, as opposed to monogenic disorders (e.g. the catalog of Mendelian inheritance in man) [4]. Nonethe-

1

less, rather than becoming obsolete, the need for traditional proteomics is expected to become increasingly important in defining the nature and location of co-translational and posttranslational modifications found on molecules in health and disease. Over recent years, protein characterization has become increasingly rapid and reliable, but has yet to be practiced on a scale akin to the throughputs achievable in genetic analysis of either DNA or mRNA. This is particularly relevant when one considers the enormity of the task at hand (i.e., the multitude of protein isoforms likely to be encountered within the human proteome). To date, little of the human proteome has either been observed or characterized, if one considers an estimated 300,000 to 500,000 expected elements awaiting discovery. This number is based on the gene content of the human genome lying somewhere between 30,000 and 50,000 open reading frames (ORFs) [5,6] and the observations of Langen (personal communication), whereby an average of 10 isoforms were observed per protein following matrix-assisted laser desorption/ionization–time-of-flight (MALDI-TOF) mass-spectrometric analysis of approximately 150,000 high-abundance human proteins derived from 2D gels. Notably, scientists from Oxford GlycoScience (Ch. Rohlff, personal communication) have suggested the number may only represent a multiple of five times the number of human ORFs based on their large-scale studies of human proteins. It is likely that most, if not all, human protein gene products will possess one to several cotranslational and/or posttranslational modifications (PTMs). Apart from PTMs, differential splicing and protein cleavage contribute to the variety of protein gene products able to exist as isoforms, be they amidated, glycosylated, phosphorylated, myristolated, acetylated, palmitoylated, and so forth. Humphery-Smith and Ward [1] have summarized the more commonly occurring PTMs seen in mammalian systems. Extremes include the potential to produce dozens of different protein isoforms from individual exon-rich ORFs as a result of differential splicing. Extremes here include the titin gene [7].

Here, we will review current progress with respect to protein, peptide, and antibody arrays and attempt to clarify their relevance to the Human Proteome Project. Numerous authors have now reviewed the field of protein chips and array-based proteomics [8–48]. The variance inherent within biological systems (combined with variance derived from both sample preparation and signal detection) dictates that one must replicate experiments on numerous occasions before being able to draw meaningful conclusions with high statistical significance. As seen in the area of cDNA biochips, microarrays offer the potential for reproducibility achieved through a combination of parallel (both interarray and intraarray) and miniaturized assays. Regrettably, biochips are employed too often in experiments containing too few replicates. The latter combined with large numbers of variables (i.e., elements on a particular array) prevent chance occurrences from being outnumbered by statistically validated findings. Nonetheless, when employed correctly, large numbers of observations can be exploited to detect subtle differences in population variance between two or more populations. This

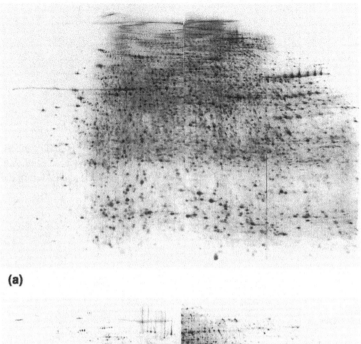

(a)

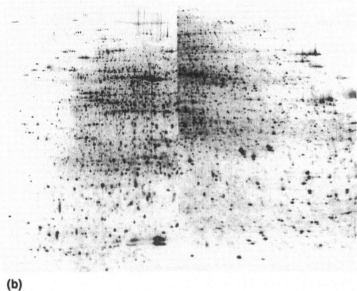

(b)

**Figure 1**    Silver-stained two-dimensional gels of whole organism lysates from (a) *Caeno-rhabditis elegans* and (b) *Arabidopsis thaliana* containing approximately 6200 and 9000 distinct protein spots, respectively. Each image is a composite of a left side generated by isoelectric focusing/polyacrylamide gel electrophoresis (IEF/PAGE) and a right side generated by nonequilibrium pH-gradient electrophoresis (NEPGHE)/PAGE.

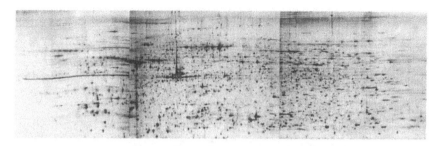

**Figure 2**   Silver-stained two-dimensional gel of a tissue lysate derived from Balb/c mouse lung containing approximately 8000 distinct protein spots. The image is a composite generated by three custom-built 18-cm immobilized pH gradients (IPG) and PAGE in the second dimension.

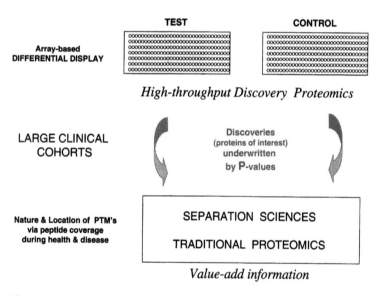

**Figure 3**   Schematic overview of high-throughput discovery proteomics underwritten initially by large numbers of observations and a high degree of proteomic coverage obtained by biochip experiments. This is then followed up by more detailed, nonparallel analysis on proteins of particular interest designed to furnish peptide coverage on protein isoforms and a detailed knowledge of their cotranslational and posttranslational modifications seen in test and control or healthy and disease study groups. The schema intends to effectively combine the strengths of both traditional and array-based proteomics in a complementary fashion.

has not always been the case with 2DGE for which the variance within both the control and test groups prevents conclusions being made for the vast majority of molecular elements resolved [49,50].

## II. NEED FOR ARRAY-BASED PROTEOMICS

Since the completion of the initial blueprint (26 June 2000) and the working draft (12 February 2001) of the human genome, proteomics has been generally hailed as the next phase of genomics (Fig. 4). This is a commonsense message, as it is, indeed, the proteins that conduct work in living systems. However, many technical obstacles await a solution if proteomics is to become the mainstay of functional genomics. It is far from clear whether proteomics will be capable of scaling to allow meaningful and reproducible analyses of human tissues in health and disease across multiple organ systems and for large patient cohorts. Credibility for proteomics in the genomic sciences will be intimately linked to its ability to deliver near-to-total coverage of the entire human and other proteomes, as has been witnessed for both genomic DNA and mRNA transcription (e.g., more than 90 genomes for which total DNA sequence is available) [cf. Institute for Genome Research website containing an overview of global DNA sequencing completed and ongoing (www.tigr.org/tdb/)]. Currently, as the complexity of an organism increases, the extent of expected proteomic coverage decreases dramatically from the 73% observed and the 32% characterized in *Mycoplasma genitalium* [2] to a point at which no more than 5% of the human proteome has yet to be observed and far less characterized by mass spectrometry (Fig. 5). Furthermore, this situation is exacerbated whereby as protein abundance decreases, the sample processing time increases at the expense not only of throughput but also peptide coverage and analytical reproducibility. In bacterial systems, 10% of genes consistently encode more than 50% of the protein bulk found in living cells [51]. In eukaryotic cells, as much as 90% of the proteome has been estimated to be contributed by just 10% of the proteins [52], and the situation is even more extreme in body fluids, such as serum, whereby albumin, transferrin, haptoglobulin, and immunoglobulin make up an estimated 90% of the protein content. As a result, the majority of proteins, including those with "housekeeping" functions, usually occur at very low intracellular abundance. Most of these are beyond the resolution and analytical capacity of traditional proteomics in complex metazoans. A knowledge of these very same proteins is nonetheless essential to our understanding of disease genesis. Researchers from both industry and academia are consistently confronted with the problem of seeing over and over again the *same high-abundance proteins* in different cells and tissues (e.g., structural proteins from mitochrondria, nucleus, cell wall and endoplasmic reticulum and enolase, ATPase, ribosomal proteins, etc.) and *without the ability* to look deeper into the protein constituents of living

## The Discovery Chain in Genomics

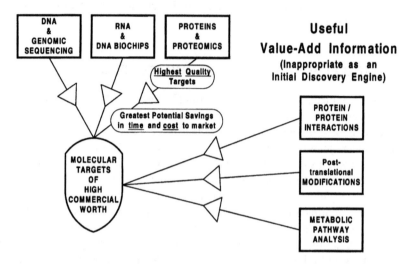

**Figure 4** The discovery chain within the genomic sciences wherein proteomics is thought to offer the greatest potential relevance to the discovery of therapeutic targets because it is the proteins that are the molecular workhorses intracellularly.

tissues and cells. Reproducible assays for multiple-target molecule detection in parallel by affinity ligands in an array-based setting offer potential solutions here.

Recent publications to go beyond 2DGE, but maintaining a dependence on mass spectrometry [53] have been able to characterize approximately 20% of the expected yeast proteome, whereas others [54] have highlighted the inability of 2DGE to display sufficient elements from within the same proteome. Indeed, recent proteomic studies of full-sequenced micro-organisms would vindicate this view [55–60]. The work of Lipton et al. [61] may represent an extension beyond this one-third barrier in the bacterium *Deinococcus radiodurans*, but it is difficult to interpret with respect to the total expected proteome, as isoforms cannot be revealed using peptide analysis by Fourier transform ion cyclotron resonance mass spectrometry (FTICR-MS) in a 2DGE-independent platform. Nonetheless, an increase in the number of proteomic studies is currently being witnessed due to the increased user-friendliness with respect to previous iterations of both the display and analysis technologies of respectively 2DGE and MS. However, no group has yet to deliver anything close to complete protein resolution even for a micro-organism at a given time point in the highly dynamic world of intracellular protein expression and degradation.

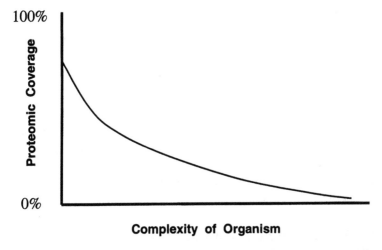

**Figure 5**   Plot of proteomic coverage versus the complexity of the organism being studied. The starting point on the vertical axis is intended to represent the simplest known living organism, *M. genitalium*, and the 73% of the expected proteome (the number of ORFs plus the addition of protein isoforms observed on a significant slice of the total proteome), as observed by Wasinger, Pollack, and Humphery-Smith [2].

In summary, the following are responsible for hampering our ability to provide highly reproducible analyses of the proteome of complex organisms by traditional proteomics based on the separation sciences:

- Inability to display the entire proteome, particularly low abundance, small, highly hydrophobic and large highly basic proteins
- Absence of a polymerase chain reaction (PCR) equivalent
- Need for highly skilled operators for both 2DGE and MS
- Nonparallel analyses (MS) incompatible with screening large sample populations (e.g., 10,000 patient biopsies from multiple tissues or body fluids)
- Experiment-to-experiment variance still too high and not reproducible on a global scale
- Image analysis of 2DGE still requires too much manual editing
- Loss of analyte during protein extraction from sieving matrix
- Lack of run-to-run reproducibility for large-scale multidimensional chromatography
- Poor statistical confidence in results obtained

Furthermore, the low abundance of the greater part of the protein content found within living cells and tissues means that reproducibility and analytical throughput decreases once the first 10–15% of a given proteome has been characterized. Thus, proteomics is currently suffering from an inability to reproducibly display the greater part of the proteome. The variances associated with biological phenomena, sample preparation techniques, and signal detection all combine to render most results of limited value (i.e., if one is to contemplate a holistic view of the workings of cellular molecular biology within living systems at the level of proteins). As such, array-based proteomics is increasingly being hailed as the path forward wherein miniaturization, parallelization, and automation can be implemented in a proteomics context. Additional benefits include reduced cost of manufacture, high level of reproducibility, low level of operator expertise required for analysis, speed of fabrication, ease of distribution, reduction in analyte volume, and sensitivity of detection [8].

## III. IMPORTANCE OF ANTIBODIES AND OTHER AFFINITY LIGANDS

As protein biochips attempt to follow from where cDNA biochips left off, there remains an unmet need for affinity ligands able to specifically recognize the proteins produced by each ORF in the human genome either as individual isoforms or collectively as a family of protein isoforms. In the latter, detection is based on a special class of linear epitopes, "signature peptides," acting as a common denominator [62–65] or conserved conformational epitopes or "mimotopes" [66–69] that are not cleaved during protein processing, common to all splice variants and unencumbered by PTMs. In the absence of a PCR or reverse transcriptase-PCR (RT-PCR), used respectively for the amplification of DNA and mRNA, protein science must call upon analyte enrichment and/or signal-amplification strategies.

The task at hand is daunting, but perhaps not more so than the task of sequencing the entire human genome as perceived a decade ago. In short, without access to large numbers of high-specificity, high-affinity ligands, the Human Proteome Project, and the analysis of its respective elements in health and disease, will remain inaccessible. In turn, the availability of these affinity ligands render the separation sciences more efficient through effective affinity enrichment of the target molecules awaiting analysis, either one at a time or in parallel using array-based technologies. Hayhurst and Georgiou [70] expressed the situation thus: "To define the proteome, there is a need for robust and reproducible methods for the quantitative detection of all the polypeptides in a cell. The ability to isolate and produce antibodies *en masse* to large numbers of targets is critical." As mentioned earlier, in even the small proteomes of micro-organisms, the greater portion of the expected proteome (number of ORFs plus additional protein isoforms) is likely to go undetected at the current detection threshold of mass spectrometry. Without affinity enrichment, high levels of peptide coverage across a given polypeptide and thus a knowledge of adducts linked to health and disease also remain unlikely deliverables. To overcome these technical hurdles, affinity ligands are seen as a means to allow proteins to be examined in large numbers both in clinical and research settings. The initial dilemma is, of course, the generation of large numbers of recombinant antigens or synthetic peptides. This will be discussed later. Although Hayhurst and Georgiou [70] mention "antibodies" specifically, a number of ligand classes are equally attractive for the production of high-affinity, high-specificity target binders. These include the following:

- Polyclonals antibodies
- Monoclonal antibodies [71–73]
- Phage display antibodies [10,74–81]
- Receptins: affibodies and antibody mimics [82–92]
- Aptamers [93–97]
- Peptide and combinatorial libraries

Each of the above mentioned classes has its respective merits and technological challenges for large-scale implementation. These are referred to briefly here, but the reader is directed to the literature cited for a more in-depth discussion of the issues at hand. Notably, polyclonal antibodies afford multiepitope recognition, including denatured proteins. They are relatively inexpensive to produce and can be employed across species boundaries and in association with histochemistry, tissue arrays, Western blots, and cell sorting. This is not so easily achieved with monoclonal antibodies. Monospecific polyclonal antibodies may be plausible for large-scale applications if generated against low-homology 100–150-amino-acid domains devoid of transmembrane-spanning regions (M. Uhlen, personal communication).

On the other hand, a monoclonal antibody offers an unlimited resource of a less cross-reactive ligand. Importantly, however, monoclonal antibodies are not necessarily of high specificity if directed against a highly conserved epitope within the human proteome (Fig. 6). The latter can number in the several thousands for co-occurring linear epitopes as detected within known genomic sequence. These conserved epitopes, be they linear or conformational, highlight the need for screening to check for the absence of cross-reactivity even among the highest-affinity target binders.

Phage display technologies allow for ligands with monoclonal properties without continued recourse to living animals, either as a result of cloning of the variable region diversity from immunized or naïve mice to produce populations of light- and heavy-chain fragments from which binders with desirable properties can be isolated. This is usually achieved by repeated biopanning of phage or phagmids against individual antigens or vice versa.

Receptins have the advantage of relative stability across a wide range of denaturants and extremes of pH. In the latter, diverse binding moieties are linked to a low-molecular-mass backbone derived from antibodies or other proteins.

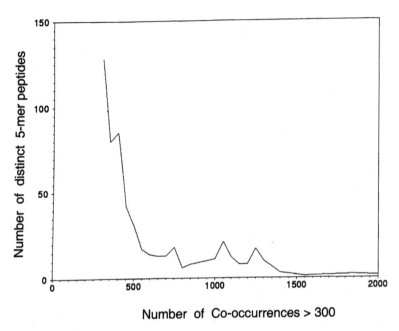

**Figure 6** The number of linear 5-mer epitopes or drug-binding sites observed among 46,461 human SwissProt entries as of March 2002. It is important to note that these linear epitopes are outnumbered in the ratio 2 : 1 by conformational epitopes. Frequency was grouped at intervals of 50 commencing at 300 or more occurrences.

The nonprotein nature of aptamers provide them with perhaps a unique virtue in that the absence of proteins on chips can allow protein-specific staining of bound proteins to an array [94], provided nonspecific binding to substrate is minimized. A healthy proportion of high-affinity binders is considered less likely among peptide and combinatorial libraries, but if these can be isolated within large populations of molecules, they, too, will find applications in proteomics.

## IV. PARALLEL GENERATION AND SCREENING OF ANTIBODIES

Traditional monoclonal antibodies derived from any number of mammalian systems are likely to produce large numbers of high-affinity ligands because of their inherent advantage of exploiting the mammalian immune system as an efficient sieving mechanism for detecting nonself molecules and through time affinity maturing the molecules (antibodies) that it employs to bind nonself elements within the body. Competing technologies that rely upon in vitro biopanning, as opposed to in vivo immunological biopanning, encounter the logistics of examining each antigen one at a time against huge libraries of $10^{10}$ out to $10^{15}$ distinct possible binders. That said, any class of affinity binder is likely to prove useful to the protein sciences and should not be excluded *a priori* form the outset. Indeed, the next decade is likely to hold the secret to this technological conundrum. The task here is to render the production of affinity binders genomically cost and time relevant, again through the implementation of parallelized procedures both for antigen generation, ligand generation (chemical or biological, in vivo or in vitro), and antibody screening. Our good health as human beings is largely dependent on the efficiency with which our bodies raise in parallel many hundreds of thousands distinct antibodies per day against large numbers of different antigens. Currently, the production of monoclonal antibodies is focused on one or two antigens of interest in a given experimental animal. If large numbers of hybridomas or immortalised B-cells derived form a well-immunized mice can be housed effectively in an automated setting, protein arrays offer the potential to screen large numbers of immunogens in parallel from individual animals, thereby parallelizing and reducing the cost of monoclonal production by dissecting the appropriate ligand-recognition patterns from within multitudes (Fig. 7). Similarly, protein arrays can be exploited to screen synthetic binder libraries or any number of affinity ligands for target selectivity. An individual monoclonal antibody must take a similar length of time to generate in vivo, but if this process can be adapted to accommodate hundreds of antigens, then parallelization can afford the means to produce many during a similar period, thereby reducing the overall cost. For both naturally occurring and synthetically derived ligands, each must be screened individually. The dilemma is to initially increase the abundance of high-affinity, high-specificity ligands with respect to cross-reactive and low-affinity binders by prior enrichment in vivo or in vitro.

A number of steps await the successful implementation for large-scale production of monoclonal antibodies, the current gold standard for affinity binders. These include the immunization protocol, adjuvant technologies, and immunogenicity enhancement of each recombinant antigen inoculated in the antigen "soup". Recent results have produced up to 95 successful immunizations (polyclonal response) in Balb/c mice following immunization with some 102 human recombinant antigens. Successful immunization was assessed by at least three on-array replicates of protein spots responding above background with respect to preimmunization fluorescent intensity of labeled mouse serum (Fig. 8). Similar results were obtained in 1999 using 12 different antigens purchased from a catalog (Fig. 9). The presence of a good polyclonal response means that the next phase of this technological challenge must encompass large-scale automated culture of immortalised mammalian B-cells or traditional hybridomas (Fig. 10). Here, the anticipated infrastructure costs are significant. However, the challenge lies very much

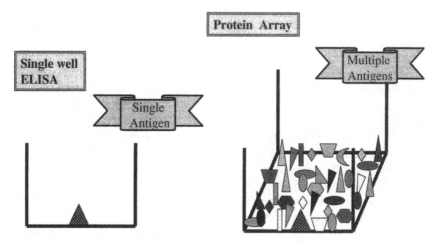

**Figure 7** Protein arrays offer the means to screen affinity ligands simultaneously for target recognition, lack of cross-reactivity, and a qualitative measure of affinity. This can significantly compress otherwise lengthy and costly screening procedures needing to be conducted on numerous binding agents one at a time.

with the need to look deeper into the B-cell population within well-immunized animals. In fact, as more and more B-cells are examined, the likelihood of encountering desirable high-affinity, high specificity binders is increased concomitantly. The process adopted is identical to the methods employed for traditional monoclonal antibody production, except that immunization is parallelized and screening is conducted against multiple replicates of the recombinant antigens used for parallel immunization and spotted onto arrays. Figure 11 is an example of a robotic system designed to process 288 protein chips every 3.5 h using fluorescent dyes conjugated to anticlass antibodies to detect simultaneously target recognition, lack of cross-reactivity with respect to the other antigens placed on array, and a qualitative measure of affinity afforded by fluorescent intensity (i.e., low affinity will not result in strong signal). A responder mouse by standard protocols for monoclonal antibody production will yield several dozen ligands able to recognize a particular target. The same can be said for phage display technologies, where it is a rare antigen that is not recognized by at least a few dozen binders. By both procedures, the next phase of ligand selection can take many months, whereby one at a time, each affinity ligand is assessed for antibody class, target specificity (lack of cross-reactivity), and level of affinity for target (Figure 12). Streamlining and parallelization of this lengthy procedure is likely to render ligand generation and selection more genomically relevant. The same process can be employed to speed considerably the selection of any number of affinity binders

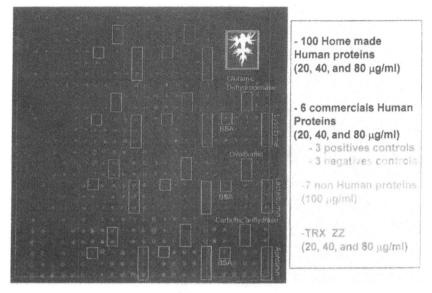

**Figure 8** The results of on-array screening of the polyclonal response obtained following parallel immunization in a single mouse with 102 recombinant human antigens. Of these, 95 produced at least three positive responses as indicated by the fluorescent signal above the background on a protein array comprising up to 10 replicates of each of the 102 parallel immunogens. Highlighted in pink and yellow are the positive and negative human recombinant antigens not produced "in-house," in blue the nonhuman proteins, and in red the absence of signal due to fusion elements employed during immunogenicity enhancement of each recombinant fusion protein. See also Fig. 23 for more details. (See the color plate.)

or to better assess target selectivity of existing off-catalog binders. The latter will not possess the same level of "specificity" as the number of distinct recombinant protein targets exposed on array is increased. Results to date have shown a monoclonal antibody manifesting good specificity when exposed to 131 different human recombinant proteins, but cross-reactivity when exposed to 361. A scenario encompassing exposure to 4000 or 40,000 distinct recombinant antigens will no doubt cause the definition of specificity to be redefined and/or necessitate accurate diagnosis of a "specific" response based on the use of an ensemble of monoclonal antibodies. Nonetheless, the level of specificity of ligands employed will need to improve dramatically if meaningful in-roads are to be made into understanding the function of the protein complement encoded by the human genome. Indeed, the existence of undesirable levels of ligand cross-reactivity and screening to avoid such are currently holding back progress in this field more than any other factors.

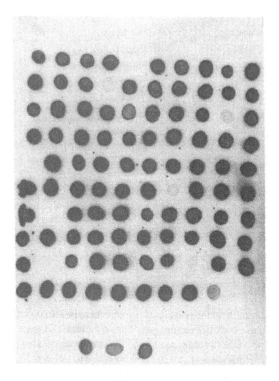

**Figure 9**  A primitive "dot-blot" protein array showing multiple hybridoma supernatants recognizing each of 12 different antigens inoculated in parallel into a single mouse.

In summary, if initially a common denominator approach is adopted, affinity ligands need only to be raised against conserved portions of the proteins encoded by each of some 40,000 genes found in the human genome (Fig. 13). In turn, these will enable the greater part of the human proteome to be followed individually or in parallel (i.e., an estimated 200,000 to 400,000 proteins, not including the diversity of immunoglobulins engendered). As the Human Proteome Project looks for a definable beginning and end within a highly dynamic entity, none is more fitting than the task of generating ligands to facilitate the detailed analysis of the output of each and every gene with the human genome.

## V.  ORIGINS AND APPLICATIONS OF PROTEIN BIOCHIP TECHNOLOGY

Protein biochips have their origins as a logical extension of dot-blot hybridization of immobilized DNA (i.e., protein dot blots of isolated proteins onto membranes

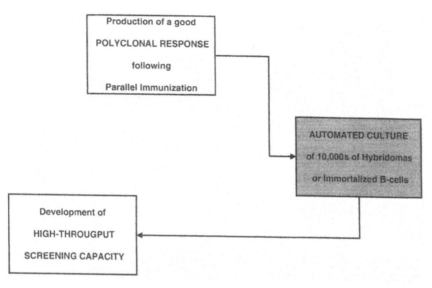

**Figure 10**   Schematic view of the parallelization of traditional monoclonal antibody production and the elements reduced to practice, namely a good polyclonal response following parallel immunization and the development of high-throughput screening robotics designed to process large numbers of biochips containing arrays of immunogens. The shaded box shows the missing link, namely fully automated culture of tens of thousands of mammalian cells, be they distinct hybridoma cultures or immortalized B-cells.

or electrophoretically separated proteins transferred to membranes for further analysis). Passive transfer of proteins out of a seiving gel onto a membrane was ineffective and, thus, Towbin, et al. [98] instigated electrotransfer to improve transfer efficiency. This procedure became known as "Western" blotting in 1981 [99,100]. There is little practical difference between a Western blot of a 2D electrophoresis gel with its random distribution of proteins across the $x$ and $y$ axes of the substrate, and a substrate on which the proteins have been arranged in ordered rows and columns. Certainly by the late 1980s, Western blotting of 2D electrophoresis gels and exposure of the resulting membranes to sera was commonplace in many research settings, whereas Western blots of 1D electrophoresis gels were already part of nationally registered disease diagnosis protocols [e.g., a confirmatory assay for human immunodeficiency virus (HIV) infection].

   In 1983, Chang [101] demonstrated the use of immobilized antibody arrays for the capture of cells, namely mouse thymocytes and human mononuclearcytes, and shortly thereafter, Ekins and colleagues [102–109] demonstrated the utility of antibody "microspot" arrays for the immunodiagnosis of proteins occurring

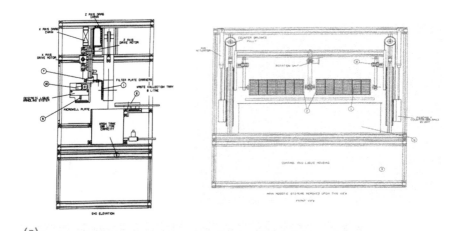

(a)

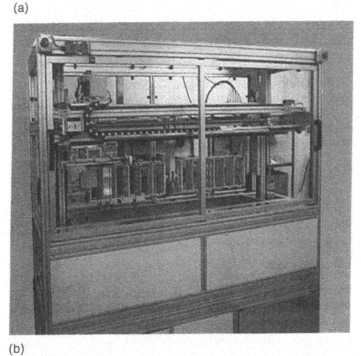

(b)

**Figure 11**  High-throughput screening robotics: (a) engineering blueprint and (b) completed instrument, designed to process 288 biochips in parallel every 3.5 h. The number of enzyme-linked immunosorbent assay (ELISA)-equivalents is then determined by the number of elements contained on each biochip, the number of robots, and the number of cycles achieved per day.

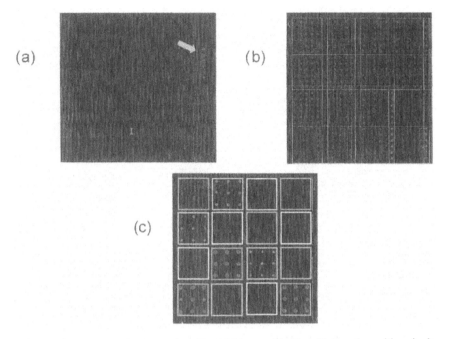

**Figure 12** Biochips demonstrating (a) a highly specific ligand interaction with a single binding partner on an array containing 130 nontarget elements, (b) a highly cross-reactive ligand recognizing most of 131 elements on array, and (c) exposure of 2 affinity ligands to 9 different targets showing antigen recognition and low-background signal in the absence of a blocking step. (See the color plate.)

in plasma. (See also patents published by Chang; and Chin and Wang about the same time, namely patents US5,486,452 and US6,197,599 respectively.) During this period, Geysen, Meloen, and Barteling [110] were synthesizing *in situ* hundreds of peptides on pins to assess antibody binding properties and, in particular, epitope mapping via mimiotope technology [111,112]. These pin-based methods were then expanded to mixtures of hundreds of thousands of octapeptides as combinatorial libraries which were later able to incorporate peptide and protein diversity on the surface of a filamentous phage [113,114]. Immobilized libraries were able to be assayed for binding to soluble receptors such as antibodies, but the substrate had by now evolved to "one-bead/one peptide" [115]. These latter approaches lead Fodor et al. [116] in 1991 to go a step further and employ light-directed, spatially addressable, parallel chemical synthesis to produce an array of 1024 peptides and demonstrate their interaction with a monoclonal antibody. The methods developed were to facilitate high-density, rapid *in situ* synthesis of oligonucleotide arrays [117].

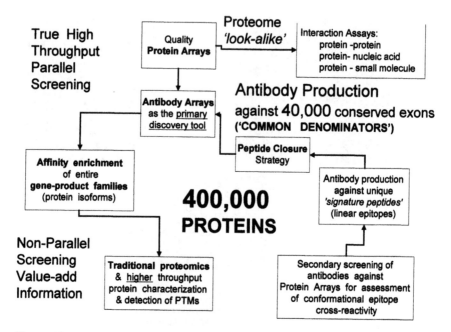

**Figure 13** A "common denominator" strategy designed to achieve coverage of the greater part of the entire human or other proteomes. These "common denominators" are elements likely to be conserved within all protein isoforms generated by a given ORF, thereby reducing the need for a separate affinity ligand for each protein isoform to be studied. Access to these affinity ligands then allows traditional proteomics and its dependence on the separation sciences to be conducted with greater efficiency and result in higher levels of peptide coverage, for example.

Ekins demonstrated that "microspot" technology had the potential to increase sensitivity with respect to that achieved by traditional enzyme-linked immunosorbent assay (ELISA) approaches. The counterintuitive process of determining local ambient analyte concentration (LAAC) was distinct from that of immune precipitation and ELISA [34,42,44,102–104]. The latter is governed by the valence of the antigen (number of epitopes per antigen) and the concentration of both the antigen and the antibody. Cross-linking of antigen and antibody molecules is maximal near the equivalence point. On the other hand, LAAC samples only a minor portion of the analyte is in solution in a noncompetitive manner, whereby sensitivity and dynamic range are enhanced by a reduction in spot size. (*Note*: The dynamic range may become compromised when the number of possible binders immobilized are too few in the presence of a high concentration binders and/or unlimited time.) The recent explosion in the use of cDNA and oligonucleotide arrays for differential transcription analysis in the postgenomic era has re-awakened the biomedical research community's interest in these technologies (i.e., the protein equivalent of a cDNA biochip). Thus, a generation later, the task at hand is to transform genomic knowledge into antibody and protein arrays for discovery and diagnosis.

Another commonly employed method in molecular biology for many years has been the technique of "colony lifts." The latter is traditionally employed to verify the presence of a particular mutant in a library and/or successful cloning of one gene into another micro-organism. This is conducted by laying down a nylon or nitrocellulose membrane over plated bacterial colonies during growth. This can then be exploited as a large-scale "dot blot" for target DNA hybridization based on the DNA derived from colonies sticking to or becoming embodied within the solid substrate of the membrane. A logical extension of this technique was to include an inducer, such as IPTG or salts, in the growth medium (usually solid agar plates) with appropriately diluted bacterial suspensions spread across the surface to produce isolated colonies. Under these circumstances, the induction is concomitant with colony growth and then either nitrocellulose or nylon membranes can be exploited for Western blotting to confirm, for example, recombinant protein expression. Like the Western blots of 2D electrophoresis gels referred to earlier, here the bacterial colonies are randomly displaced across the surface of the membrane. Here, too, it became a logical extension to order bacterial colonies in rows and columns as a high-throughput method of examining expression of recombinant proteins and antibodies and interaction mapping with respect to each of the immobilized colonies on a solid substrate. The protein source being targeted is adsorbed to the substrate in the absence of any protein-specific surface chemistries following colony lysis [118–125].

To summarize, proteins have now been arrayed onto solid supports as large or small dot/blot on membranes or biochips or into microwells. The proteins themselves include affinity-purified proteins and antibodies, lysates of expression vector host cells grown off-line or *in situ*, and tissue extracts or body fluid (e.g., "reverse" arrays). Overall, antibody and protein arrays have found a wide variety of applications in biomedical research. These include the following:

- Antibody arrays interacting with antigens in solution [13,19,126–143]
- Protein arrays for monitoring
  - Protein–protein interactions [11,48,130,144–147]
  - Protein–nucleic acid interactions [144]
  - Protein–small molecule or drug interactions [11,48,146]
  - Autoimmunity [23,148,149]
- Membrane-bound receptors [150]
- Domain screening [151,152]
- Enzymatic function [11,146,153–158]

To these applications should be added the following:

- *Tissue microarrays* provide a means of conducting highly parallelized immunohistochemistry to assess the level and location of expression of both recombinant and naturally occurring antigens [159–164].
- *Peptide arrays* have been employed for many years for epitope mapping of antibodies, cellular epitopes, and other chemical binders [110–112,152,165–168]. Synthetic peptides can be manufactured on-array or spotted down onto a substrate.
- *Reverse arrays* are based on tissue lysates or fractions thereof spotted onto solid supports. A major advantage is the use of naturally occurring protein isoforms and PTMs, but problems are evident in association with batch-to-batch reproducibility and the inherent disparity in antigen abundance and accessibility [169,170]. These arrays can be exposed directly to potential binders or interfaced with MALDI-TOF MS [171–173].
- *Cellular arrays* are used to conduct multiplexed assays on recombinant proteins expressed in vitro in mammalian cells. The strength of this approach is that membrane-associated proteins are expressed in conjunction with a cell membrane of a living cell [174–177]. Caveats of this approach will be discussed later.
- *Chromatography affinity capture arrays* (e.g., Ciphergen Protein Chips™) allow low-resolution, but highly user-friendly differential profiling of tissues and body fluids [178–189].

An *interesting nuance* is the use of a DNA array as a detection strategy for monitoring intermolecular interaction events involving DNA peptide constructs and protein activity or inhibition in solution prior to exposure to the array [190].

## VI.  SURFACE CHEMISTRIES FOR PROTEIN BIOCHIPS

Groups entering the field of peptide, protein, and/or antibody arrays will rapidly be confronted by the critically important nature of developing appropriate solutions for immobilization surface chemistries. Technologies developed for nucleic acids are probably inadequate when applied to proteins and protein arrays. Simply put: "Protein is not DNA." Indeed, proteins stick to one another and also extremely well to most substrates. Workable solutions derived from nucleic acid technology on membranes and solid biochips may produce adequate results, but they will rapidly encounter problems similar to traditional proteomics; namely, results will be forthcoming only with respect to high-abundance proteins. In the short term, the advantages of parallelization and miniaturization will nonetheless provide a stimulus for the use of array-based technologies, even in the absence of dedicated surface chemistry solutions. However, if not coupled with specific solutions for the reduction of nonspecific binding (NSB) to the substrate, sensitivity will flounder and much of the proteome will defy detection. Both sensitivity and dynamic range are severely compromised if this NSB is not reduced to a minimum (Fig. 14).

During 1996 and 1997, efforts at the Center for Proteomics Research and Gene-Product Mapping at the National Innovation Centre in Sydney were the first to take up the challenge of fully automating excision of protein spots from 2D electrophoresis gels or PVDF membranes derived from Western blots of 2D electrophoresis gels and sample preparation for both high-performance liquid chromatography (HPLC) and mass spectrometry [191]. This robotic solution (Fig. 15) met its mechanical specifications in November 1997. These specifications included a 5-m $\times$ 2-m $\times$ 2-m enclosure fed sterile air, a $CO_2$ impact laser for spot excision, 25-point contour mapping of spots, a high-precision X/Y transport table linked to a vacuum, an X/Y/Z Cartesian robot for liquid handling and spot aspiration, parallel processing of 12, 96-well plates for protein digestion and peptide elution or 294 HPLC vials destined for acid hydrolysis, an orientated-lid delivery system linked to infrared position detection, a capping station, and, finally, MALDI-TOF target loading. The project represented a prototyping chal-

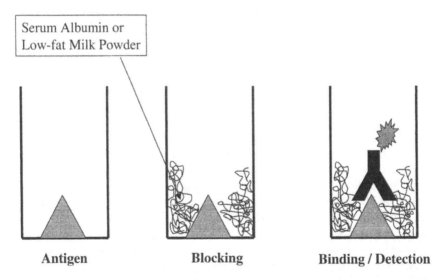

Antigen          Blocking          Binding / Detection

**Figure 14** Schematic view of the steps involved in a traditional ELISA. Much sensitivity is foregone by the use of very sticky blocking agents. Specific surface chemistry solutions are required to reduce nonspecific analyte binding to an absolute minimum during high-throughput screening.

**Figure 15**  Photograph of a high-throughput protein excision and processing robot developed at the Centre for Proteome Research and Gene-Product Mapping in Australia and designed to automatically process two-dimensional gels or Western blots.

lenge in that elements for the system were derived from Ottawa in Canada, Springfield and Boston in the United States, Newcastle in the United Kingdom, and Sydney in Australia. The sensitivity of detection for the protein spots processed in the system was appalling due to loss of sample bound to the walls of plastic wells long before analysis by mass spectrometry. A tradition Eppendorf™ tube employed in the molecular laboratory has the capacity to bind some 6 μg of protein. Needless to say, when confronted with low-abundance analytes barely visible on 2D electrophoresis gels or the remnants thereof following electrotransfer to membranes, little sample remained prior to final analysis. For many years now, mass spectrometry has been performing high-sensitivity analysis, but the manner in which samples are manipulated prior to analysis has been holding back progress in whole-proteome screening. Having expended much energy in robotic design only to learn the critical importance of surface chemistry for protein handling, it became very obvious that if a transition was to be made to array-based proteomics, one would have to pay particular attention to surface chemistry in order to ensure analytical success. In summary, appropriate surface chemistry solutions for reduction of NSB remain critical to both traditional and array-based proteomics. Elements considered important for protein and antibody arrays include the following:

- Minimal NSB, particularly when exposed to blood or serum
- Avoidance of blocking steps such as absorption with bovine serum albumin (BSA) (which itself sticks to approximately one-third of the visible proteome of bacteria, unpublished result) and/or milk powder
- Minimization of surface defects
- Covalent bonding of surface chemistries to substrate so as to afford increased robustness of the surface layers at extremes of pH
- Covalent bonding of recombinant proteins or antibodies to surface chemistry assemblages
- Compatibility with a wide variety of surfaces from noble metals, to plastics, glass, and semiconductors
- Reproducibility of fabrication in a dust-free environment
- Stability and robustness
- Biocompatibility and maintenance of molecular activity (high water content can be an asset here)
- Maximal site occupancy per unit area
- Extended shelf life
- Minimal steric hindrance of binding sites
- Maintenance of structural integrity following immobilization
- Homogeneity of substrate across array

In an effort to optimize the above, glass microscope slides destined as protein or antibody chips were subjected to plasma or piranha treatment to remove all surface-bound impurities (respectively highly caustic cleaning procedure conducted under vacuum or boiling in the presence of sulfuric acid and hydrogen peroxide). Before the surface was reexposed to air, the first chemical layer was plasma deposited on the cleaned surface. This step resulted in a covalently bound polymer layer. From here, slides were subjected to a multistep procedure designed to place down multiple polymer layers with a view to rendering the surface defect-free (Fig. 16). Finally, a 3D hydrophilic hydrogel matrix was deposited and activated esters were then used to bind amine groups of proteins (a variety of chemistries are possible here for binding amino, carboxy, and thiol groups). Subsequent to protein arraying, residual esters were then deactivated by ethanol-amine following protein gridding. This procedure allowed NSB to be reduced to approximately 0.3 ng/cm$^2$ of BSA when exposed to 4 mg/mL of BSA (Fig. 17). This process formed the basis of a European patent application (patent 00203767.9) with a priority date of 26 October 2000.

A multilayer approach to surface chemistry was able to demonstrate further a reduction of NSB binding, whereby the same hydrogel coating was placed down

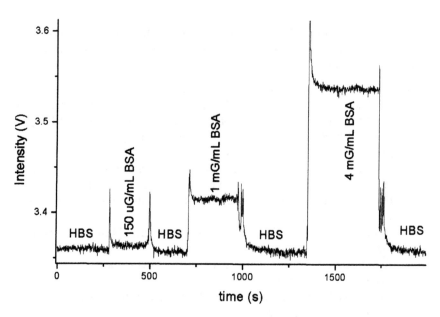

**Figure 16** Real-time plot obtained by surface plasmon resonance of substrate binding observed with three different concentrations of BSA.

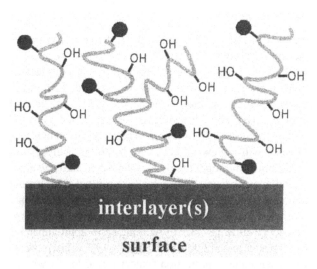

**Figure 17** Sketch of hydrogels atop multiple polymer layers designed to minimize non-specific substrate binding. Covalently bound recombinant proteins or antibodies are represented by the dark balls.

as a monolayer and was shown to underperform with respect to itself atop a multilayered surface assemblage (Fig. 18). This finding was interpreted as resulting from a further reduction of surface defects.

The initial corrosive steps involved in biochip manufacture meant that sample labeling had to be conducted in the midplane of the glass slide via laser etching. However, without such measures the use of blocking agents can severely compromise the signal-to-noise ratio obtained on biochips. Fig. 19 shows BSA binding to every one of 131 different human recombinant proteins in a manner detectable the above the off-spot background. This is a visible indication of the loss of signal-to-noise ratio due to blocking, as represented schematically in Fig. 20. Up to 12 on-array replicates of such findings allowed the protein–protein interactions to be reliably ranked—something that is impossible without minimization of NSB [48]. Biologists must realize that it is not merely a matter of placing proteins

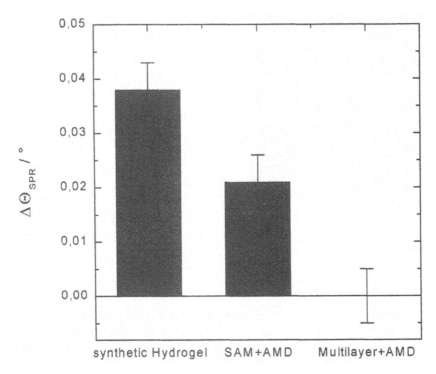

**Figure 18** Observations obtained by surface plasmon resonance of different hydrogel constructs showing the extent of nonspecific substrate binding obtained, from left to right, by a synthetic hydrogel, a patent-protected hydrogel employed as a monolayer, and the latter on top of multiple polymer layers designed to minimize surface defects.

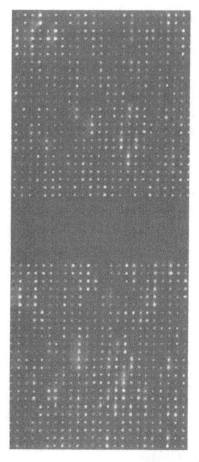

**Figure 19**   Bovine serum albumin binding with all recombinant proteins on array at levels significantly above background. See also Fig. 43 for a numerical plot of these results.

down on a substrate by absorptive processes, but, rather, there is a need for specific surface chemistry solutions designed to reduce nonspecific analyte binding to an absolute minimum during high-throughput screening. Otherwise, array-based proteomics will suffer the same fate as traditional proteomics and be limited to data obtained only from the high-abundance proteins. In the absence of specific capture strategies for polypeptides, nondesirable elements will be captured with equal alacrity as the desired ligand–target interactions being monitored. As outlined earlier, blocking agents are considered inadequate to achieve the desired

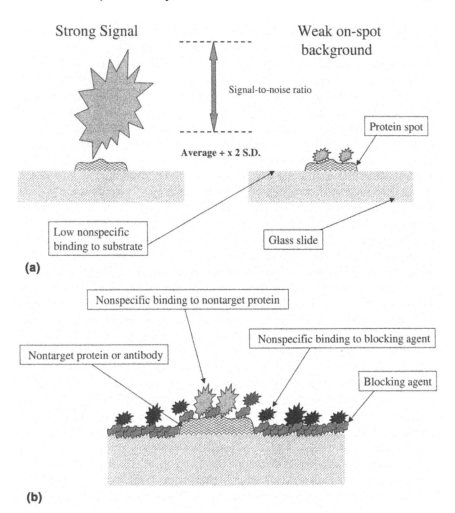

**Figure 20** Schematic representation of (a) the desired outcome of the signal obtained following exposure of a ligand to a protein target on a biochip, whereby the signal obtained is not severely compromised by either the use of blocking agents, as in Fig. 19, or nonspecific substrate binding, and (b) the reduction in the signal-to-noise ratio obtained in the absence of specific surface chemistry solutions for protein biochips.

signal-to-noise ratio necessary to screen against the low-abundance constituents of the proteome (i.e., the greater part of all intracellular proteins). Indeed, orders of magnitude in sensitivity are probably lost in association with this step alone.

A population of target molecules immobilized on a substrate cannot behave similarly during their intermolecular docking carried out in solution. Hydrogels minimize such caveats, but must, in turn, be quality assured for the type of biomolecular interactions being monitored. For example, the porosity of a three-dimensional matrix will behave differently for a small molecule–protein binding assay, as opposed to a sandwich ELISA and the introduction on three large biomolecules. Molecules must enter into the matrix and be given time and a passage to exit from the matrix or, again, NSB can be enhanced by entrapment. The need for assessing biomolecular binding kinetics within a three-dimensional matrix prompted Zacher and Wischerhoff [192] to develop dual-wavelength surface plasmon resonance (SPR) as a tool for real-time monitoring and quality assurance of the surface chemistries being employed (Fig. 21). This device permitted the measurement of kinetics of biomolecule binding to the substrate and the interactions between immobilized biomolecules and their binding partners at 10-nm intervals within a 70-nm surface. Without such quality assurance procedures and surface chemistry optimization, it is clear that a single surface chemistry solution cannot function optimally for the wide variety of biomolecules to be assessed on-array. Real-time SPR cannot be conducted on glass biochips, but these methodologies can provide the means to assess surface chemistry performance on the definitive substrate. Hydrogels afford increased site occupancy with respect to monolayers, but analyte accessibility and steric hindrance may become important issues as molecular mass of analytes and/or the thickness of the 3D matrix is increased. A more detailed review of solutions able to be adopted and their respective strengths and weaknesses is presented in Chapters 4 and 5. As biologists grapple with the challenge of delivering improved sensitivity and reproducibility for a wide variety of applications, equally challenging is the issue of immobilization itself and the risk of impeding potential binding sites [193]. For this reason, ordered one-sided linking of a population of biomolecules should be avoided, except in association with antibodies upon which the epitope binding sites are consistently located on one aspect of the molecule. Noteworthy is the fact that not all antibody classes abide by this rule. A random immobilization strategy ensures that all aspects of a biomolecule are presented to potential binders. However, the downside is that effective site occupancy is significantly reduced due to the percentage of a population of target molecules that cannot function as a result of steric hindrance, be that on a monolayer substrate or in the immediate vicinity within a 3D environment.

The question is often posed: ''How sensitive is this array-based assay system?'' A bad question cannot be given a good answer no matter how many times it is

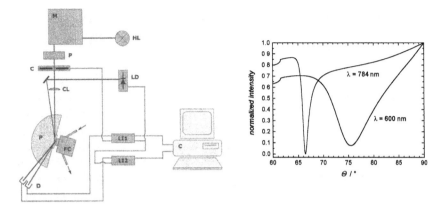

**Figure 21** The dual-wavelength surface plasmon resonance apparatus constructed by Zacher and Wischerhoff and employed to assess the dynamics of biomolecule binding to substrate and biomolecular interactions within 10-nm slices of surface chemistry assemblages as a means of optimizing surface chemistry to assay type. (From Ref. 192.)

asked by biotech analysts or however many times biotech companies make press releases using biotin/streptavidin assays to attract the attention of potential investors. In fact, the sensitivity of competing detection technologies must first take into account (1) time, (2) concentration, and (3) the affinity of the ligand for its target. For example, a radioisotope exposed to a photographic plate for 6 weeks can produce remarkable sensitivity. A recommended approach would include each of the following in solution binders being exposed for identical time at identical concentrations. A range of affinities expressed here as dissociation constants ($K_d$ values) for their respective immobilized targets should then be examined by each competing detection platform for the following:

- A small molecule at $10^{-5}$ to $10^{-6}$
- A poor antibody at $10^{-8}$
- A good antibody at $10^{-11}$ to $10^{-12}$
- Biotin/streptavidin at $10^{-14}$

The results obtained across this spectrum of affinities can then be compared in a meaningful manner. Other "one liners" should be ignored. For real-time assay procedures such as SPR or atomic force microscopy (AFM), capable of calculating on- and off-rates, the above experiments should be conducted over a fixed period of, for example, 10–15 mins at identical analyte concentrations for each of the above four categories—otherwise only meaningless comparisons are possible.

## VII.   HIGH-THROUGHPUT RECOMBINANT PROTEIN
##        PRODUCTION FOR ANTIBODY PRODUCTION
##        AND BIOCHIP CONTENT

Biochip content in the form of both recombinant proteins and affinity ligands generated against the former will pose the greatest challenge to significant progress and comprehensive coverage of the human proteome. Without these elements, analysis and diagnosis of low-abundance proteins linked to health and disease will remain beyond the sensitivity of our analytical procedures. With access to these elements, analyte concentration and signal amplification fall within the routine of protein science as conducted over the last few decades and, more importantly, analytical procedures are able to be parallelized. Thus, the high-throughput production of recombinant proteins is paramount to the success of both structural genomics [194,195] and proteomics initiatives. Molecular issues and challenges facing high-throughput generation of recombinant proteins have been addressed by several authors [9,17,23,46,48,118–125,196–205; (see also Chapter 6).]

The concept of producing recombinant proteins derived from each ORF in a genome is equally daunting as the sequencing of an entire organism. The first group to demonstrate the feasibility of such was the biotech firm Acacia Biosciences, as exposed on the front page of *Genetic Engineering News* (see Fig. 22) in 1997 [206]. This effort later culminated (prior to the end of 1998) in some 5800 distinct living clones from the brewer's yeast each with a one-gene reporter system linked to green fluorescent protein (GFP). These clones were then able to be placed on a solid support so as to report in parallel on the abundance of each protein in vivo as a function of drug administration and/or change in physiological parameters. This provided a tool for following entire biochemical pathways in a near-to-total proteome. Using a similar approach, batches of purified GST-fusion proteins were employed to follow protein function and activity of *Saccharomyces cerevisiae* in late 1999 [207]. Grayhack and Phizicky [208] termed this experimental approach of pooled analyses as ''biochemical genomics'' while others had been working to achieve global coverage with protein baits [123,209]. Eventually, the work of Zhu et al. [146] in 2001 demonstrated functionality of protein arrays for most of the ORFs of *S. cerevisiae* with respect to protein–protein and protein–phospholipid interactions. Functional enzymatic screening was demonstarted by the same group [152].

Thus, feasibility had been provided for the initial step in the development of antibody arrays, namely the construction of protein arrays designed to emulate the antigenicity encoded by an entire genome, including that of humans. The individual recombinant proteins contained can be employed to generate affinity ligands and/or screen the ligands produced for target recognition (Fig. 23). To achieve this goal for higher organisms, it would be desirable for each ORF to be

**Figure 22** Front page of *Genetic Engineering News*, September 15, 1997. These prototype arrays went on to report on the near-to-total proteome of the brewer's yeast by employing live arrays of single-gene fusions with green fluorescent protein. (From Ref. 206.)

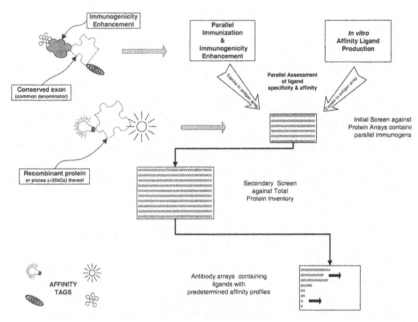

**Figure 23** Schematic overview of parallel generation and screening of affinity ligands, noting that the recombinant fusion proteins employed for generation of affinity ligands are distinct from those employed for on-array screening. The secondary screen against protein inventory to select affinity ligands with desirable levels of target specificity should also include expression of recombinant antigens in multiple expression vector hosts (cf. Fig. 24). This process is currently covered by international patent WO99/39210 and is non-trivial if one considers the work involved to facilitate analysis of the near-to-total human proteome.

expressed in bacterial, yeast, baculovirus, and mammalian cell lines. Multiple expression vector hosts are necessary in order to provide an effective emulation of the PTMs and antigenic forms likely to be encountered in an entire proteome (Fig. 24). The latter are most critical during parallel antibody screening of linear and conformational epitopes. PTMs are highly conserved in eukaryotic systems and, thus, although not all PTMs will be present on any given recombinant protein, most PTMs are likely to appear on-array as part of a sufficiently large population of recombinant proteins. As such, arrays provide a means of excluding affinity ligands that recognized such nonspecific epitopes (i.e., commonly occurring PTMs). Furthermore, any antibodies or affinity ligands recognizing a phosphory-lated, glycosylated, succinylated, and the like, epitopes will be seen to be cross-reacting with numerous proteins presented on a protein chip. Rather than a caveat

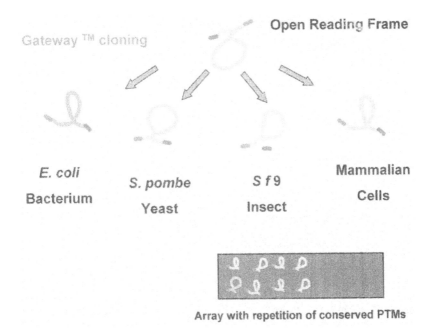

Array with repetition of conserved PTMs

**Figure 24** Expression of a particular ORF in multiple expression vector hosts allows the greater part of the expected repertoire of cotranslational and posttranslational modifications to be represented on array, but not necessarily on each recombinant protein. Nonetheless, affinity binders recognizing these often highly conserved modifications will appear as nonspecific binders during high-throughput screening.

associated with high-throughput recombinant protein production, cotranslational and posttranslational modifications represent a potential strength of the above procedure, particularly with respect to the manner in which affinity ligands can be screened for specificity in the absence of cross-reactivity.

Gateway™ vectors obtained from Invitrogen were specifically modified to meet the need for high-purity recombinant protein production following dual-affinity enrichment (Fig. 25). Affinity ligands were created from the following fusion elements placed on both sides of the cloning site, namely His (low immunogenicity, good compatibility with high-molar-urea solutions for extraction from inclusions bodies)/TRX/ORF/Strep tag or GST/ORF/Strep tag fusion proteins. The Gateway vectors allowed in vitro recombination following nested PCR reactions for high-fidelity, low-efficiency recombination and are potentially able to be processed in the absence of colony picking. Efficient cloning is not simply a process of placing a cDNA library into a protein expression vector wherein only one clone in three is expected to contain the correct reading frame [118,125].

## ORDERED DIRECTIONAL CLONING

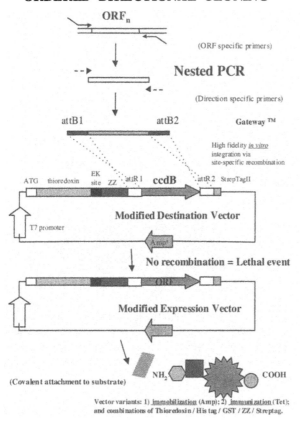

**Figure 25**   Vector design and modification of the Gateway vectors employed during synthesis of recombinant antigens destined for parallel immunization.

To avoid this problem, reamplification of target genomic DNA or cDNA fragments is desirable, as is fusion protein tags on both the 5′ and 3′ sides of the cloning site so as to avoid "read-through" even in the correct reading frame and facilitate dual-affinity purification to sufficiently enrich recombinant elements from within complex cellular mixtures. Specific induction of protein synthesis is able to maximally upregulate protein expression some four orders of magnitude, but this induction is not necessarily sufficient to render the recombinant protein among the most prevalent cellular constituents. Extremely low-abundance proteins may remain lost as low-abundance or poorly soluble elements and/or be sequestered into inclusion bodies. Therefore, multiple enrichment procedures may be required to recover a sufficient end product of sufficient purity [210,211].

Although manual colony picking has been an essential element of optimizing procedures for high-throughput cloning, this process can be replaced in the long term by a quality control step of, for example, 1000 sequencing reactions of cloned inserts to attest the overall accuracy and efficiency rate of cloning, or automated colony picking, as is common in genomic sequencing laboratories. An added advantage of the Gateway system is the readiness with which any cloned insert can be transferred from entry vectors into a variety of expression vectors compatible with different expression vector hosts. The InFusion™ PCR and Creator™ vector system from BD Biosciences is likely to offer similar efficiency for large-scale cloning and subcloning experiments. Recombinant proteins can then be incorporated on protein arrays. The process for generating large numbers of recombinant proteins has been automated in association with the elements outlined schematically in Fig. 26 and in reality in Fig. 27. The importance of an efficient Laboratory Information Management System (LIMS) as a means of following and assuring the quality of synthesis products cannot be overestimated. Descriptor files for all processes conducted during both content synthesis and chip manufacture must be generated and, in turn, these must be linked to "Go/No Go" triggers as part of Standard Operating Procedures designed to ensure end-product quality (Fig. 28). All robotic units must be interfaced with LIMS to afford systemwide compatibility and operator accessibility to all points at all times. Equally important is the need to adopt an international standard for data acquisition, processing, and publishing for protein chips, similar to that adopted for cDNA biochips (i.e., MIAME) [212,213].

The optimization of protein recovery and quality assurance of recombinant proteins is critical to the quality of results produced in association with protein biochips. This is equally important to the specificity of affinity ligands likely to be generated from a particular protein extract; that is, multiple ingredients and impurities will generate or bind selectively with multiple affinity ligands and thereby further exacerbate the search for highly specific affinity ligands for use in proteomics. There is a well-known adage in analytical chemistry: "Garbage in equals garbage out." Nowhere is this likely to be more true than in efforts to clarify target specificity in the absence of cross-reactivity. Cross-reactivity can be due to impurities within target samples or to conservation of a particular binding site within the human proteome found on protein isoforms derived from the same ORF or as a result of sequence and/or structural similarity. Based on earlier work dealing with unique "signature peptides" [62], predictions have shown numerous linear epitopes to be present on hundreds and even thousands of occasions within the human proteome, not to mention those containing highly conserved PTMs such as phosphorylation, glycosylation, myristylation, palmitoylation, and so forth. In addition, one must note that conformational epitopes are thought to outnumber linear binding sites by 2 : 1 [67,68]. However, most critical is the ability to produce high-purity, quality-assured recombinant human proteins. This is a nontrivial exercise. Current practice involves a long list of quality control

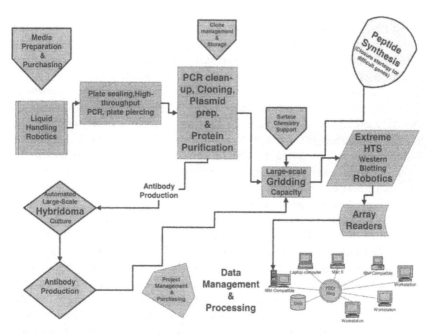

**Figure 26** Schematic representation of the infrastructure required for automation of recombinant protein production and monoclonal antibody generation.

steps on recombinant proteins to assure the purity and fidelity of product. Noteworthy is that at every step there is an attrition rate. An example of end-product purity as determined by silver staining is shown in Fig. 29. Similar results obtained with Western blotting using a specific antibody or less sensitive staining procedures can be highly misleading as to the quality of recombinant proteins.

Laboratory-based in vitro molecular biology is far more error-prone than similar processes occurring in living cells. Molecular biologists know that results must be confirmed on agarose gels at every step of a cloning procedure; yet, errors persist and these must be discarded by methodical screening. Without this attention to protein purity, the results obtained for intermolecular interactions on-chip are rendered totally uninterpretable. Although induction can significantly upregulate the abundance of recombinant protein expression, other cellular constituents significantly contaminate signals obtained during cross-reactivity assessment (i.e., binding to impurities in the protein sample placed on the array). Herein lies the need for routine dual-affinity enrichment of recombinant proteins. The following steps are involved in quality assurance of recombinant proteins placed on arrays, namely verification of the following:

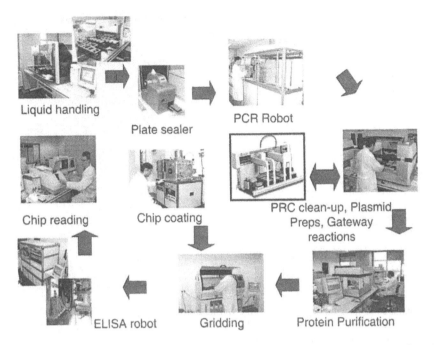

Liquid handling

Plate sealer

PCR Robot

Chip reading

Chip coating

PRC clean-up, Plasmid Preps, Gateway reactions

ELISA robot

Gridding

Protein Purification

**Figure 27** Automation infrastructure employed for the high-throughput generation of recombinant proteins, starting from the top left and proceeding in a clockwise fashion.

- PCR product—on gel (Fig. 30)
- Entry clone—on gel (Fig. 31)
- Expression clone—on gel
- Vector design to ensure only recovery of proteins in the correct reading frame and the absence of any read-through phenomenon
- Dual-affinity enrichment for enhanced protein purity
- DNA sequencing of cloned insert
- Absence of 5′ and 3′ UTRs (untranslated regions)
- Protein purity and $M_r$—on gel (Fig. 29)
- Concentration—level of expression and standardization thereof
- MALDI-TOF MS total mass spectrometry (Fig. 32)
- MALDI-TOF MS peptide mass fingerprinting, PMF (Fig. 33)
- MALDI-TOF MS peptide mass fingerprinting (Fig. 33)
- Electrospray ionization: ESI-MS-MS sequencing tagging (HTS [high-throughput screening] implementation is currently problematic due to high cost and low throughput—may change in the not too distant future); see Fig. 34.

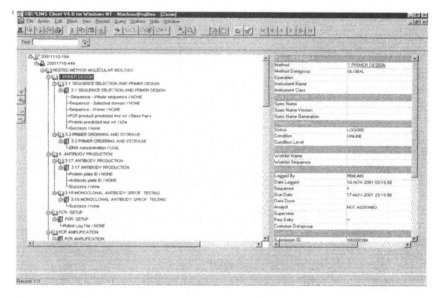

**Figure 28** The need for a Laboratory Information Management System designed to manage the production process and quality control in each step cannot be overemphasized. Presented is an example user interface.

Natsume et al. [214] have recently demonstrated similar quality assurance procedures employing MALDI-TOF MS and the ability to apply this technique to many hundreds and even thousands of recombinant proteins.

This preoccupation with quality control must also be linked to a significant throughput of production, as potentially every user will possess different requirements with respect to the protein content and/or near-target space (NTS) during applications in lead validation and optimization (see later, pgs. 47–52). The protein inventory associated with noncandidate space can be increased through time, but one cannot afford to wait many years for the synthesis of a protein repertoire required for a specific array-based application during lead optimization, for example. Thus rapid, high-quality synthesis of numerous proteins is an obligatory prerequisite for the synthesis of NTS to multiple targets and/or the components of a particular biochemical, cancer, or signal transduction pathway or series of toxicological end points, including the near-molecular relatives of the latter groups. As such, expression of sequence homologs, domain homologs and tertiary structural homologs detected by threading algorithms all become potential components for protein biochips. Some 200–300 such proteins are likely for any target molecules, particularly when the NTS is expanded by splice variants and the numerous potential PTMs afforded by expression in multiple expression vector hosts, such as bacte-

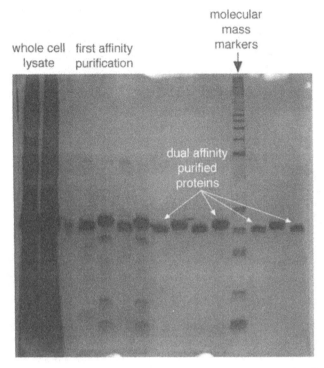

**Figure 29** Optimized recovery of recombinant proteins visualized on a silver-stained polyacrylamide gel following dual-affinity enrichment. Similar samples of single-band purity have produced clean spectrographs by both MALDI-TOF MS (Fig. 32) and ESI-MS-MS (Fig. 34). Western blots and Coomassie-stained gels can provide highly misleading indications of protein purity.

rial, yeast, insect, and mammalian systems. In mid-2002, our production capacity in *Escherichia coli* was approximately 1000 recombinant proteins from any 1500 randomly chosen human ORFs within 6–8 weeks following primer design, whether the starting material was genomic sequence alone (i.e., *in silico* detected ORFs) or cDNA clones. Both have been reduced to practice, but the latter was associated with less attrition, particularly as a result of less undesirable PCR products. In each of the above listed quality control measures listed, one must expect attrition due to errors or low efficiencies obtained during amplification, cloning, transcription, translation, and affinity enrichment. Successful production of an intended recombinant protein for chip-based applications is currently assessed as recovery of at least 100–200 μg of protein. For other applications, upscaling of production is always possible using the same expression vector and

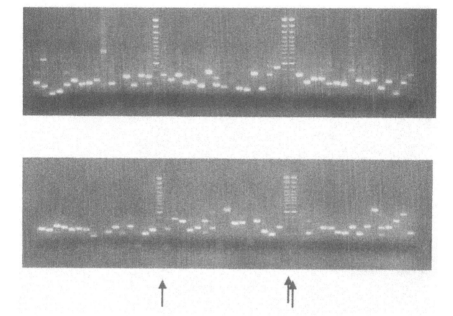

**Figure 30**  High-throughput quality assurance on agarose gels of products derived from PCRs.

host systems, provided access is available to appropriate large-volume production and recovery infrastructure. Last but not least is the quality of protein chips produced (Fig. 35).

## VIII.  PEPTIDE CLOSURE STRATEGY FOR PROBLEMATIC PROTEINS

For a number of years, my colleagues and I have been investigating the information content contained within genes and the potential for this information to do work, as defined by the extent to which nature has "fallen in love" with particular peptide strings and used them over and over again. These commonly occurring amino acid strings have been found to be associated with statistical significance measured below $p < 1.0 \times 10^{-7}$ and presumably do either intermolecular (functional) or intramolecular (structural) work in biology. To detect these much loved elements, database mining tools and word-building algorithms were employed. The work and an overview of what we currently know about the origin of information content in genes has been reviewed recently [62]. This work has now evolved

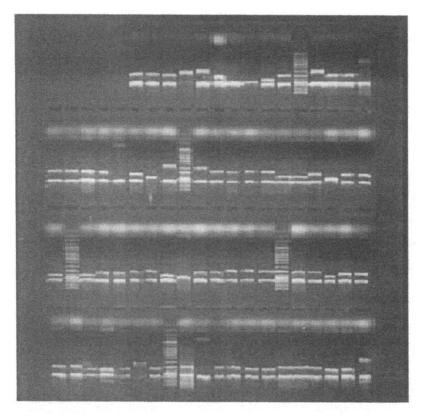

**Figure 31** High-throughput quality assurance on agarose gels of cloned inserts in entry vectors.

to a point whereby it is accompanied by an advanced graphic user interface able to link peptide strings of varying length to an available tertiary structure (Fig. 36). Most importantly for peptide closure of proteomes is the need to build on this knowledge and discard all well-liked information (peptide strings common in biology) as a means of rapidly identifying regions likely to be ''signature'' strings/peptides.

Having done so, the remaining regions are processed for the following attributes:

- Unique intragenomic and intergenomic protein sequence
- A minimum unique sting of 8–12 amino acids contained within a 20-amino-acid string

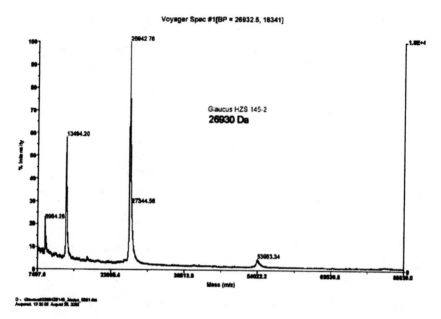

**Figure 32** Quality assurance of total molecular mass of a human recombinant protein using MALDI-TOF mass spectrometry.

- Surface exposed as predicted by Kyte–Doolittle hydrophobicity plots
- Surface exposed as predicted by existing crystal or nuclear magnetic resonance (NMR) structure
- Not encumbered by known cotranslational and posttranslational modifications as defined by ProSite
- Common peptide strings color coded for probability of occurrence in SwissProt with respect to fully randomised coding regions within a given genome
- Hyperactive cursor to assess the phylogenetic similarity of the sequence peptide strings highlighted—one notes significantly disparate affiliations between a whole protein and the peptide strings contained within a given protein

For those genes not easily cloned or expressed or which turn out to be poorly immunogenic, synthetic peptides provide a means of constructing "signature peptides" or conformational "mimotopes" following *in silico* analysis of protein sequence and structure. Antibodies or other affinity ligands can then be raised against these peptides and then screened against protein arrays to assess target selectivity. Some five to six "signature peptides" are thought to be necessary per gene to ensure a successful outcome for genes that are known to be difficult

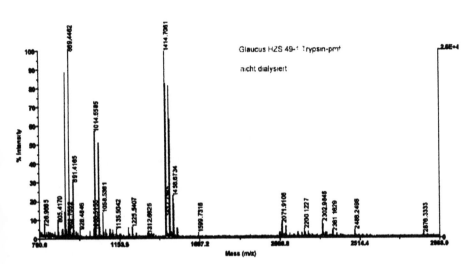

**Sample 49:**

MRGSHHHHHH KVPAQHDEAV DNKFNKEQQN AFYEILHLPN LNEEQRNAFI QSLKDDPSQS
ANLLAEAKKL NDAQAPKVDN KFNKEQQNAF YEILHLPNLN EEQRNAFIQS LKDDPSQSAN
LLAEAKKLND AQAPKVDASL ESTSLYKKAG LNKNFQRDLQ FFFNFCDFRS RDDDYETIAM
STMHTDVSKT SLKQASPVAF KKINNNDDNE KIYPAFLYKV VSAWSHPQFE K

**Figure 33**   Confirmation of protein identity by peptide mass fingerprinting (PMF). High
purity and high protein concentration allowed all but three amino acids to be represented
within this particular PMF.

to clone or problematic when being expressed as recombinant proteins. The latter
can currently be defined as extremes within the following ORF populations:

- Bias in AT/GC content
- Variation in melting temperature $T_m$ for primers and PCR products across
  the whole genome
- Codon Adaptation Index (of ORF and with respect to vector host)
- Low-complexity proteins and regions within proteins
- Use of rare codons in expression vector host
- Hydrophobic and insoluble proteins

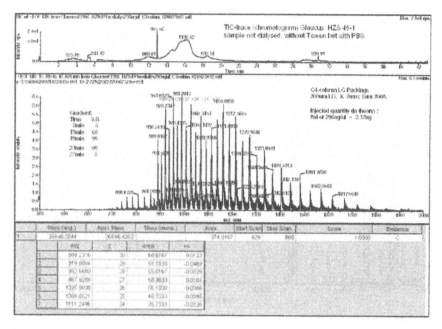

**Figure 34** An ESI-MS-MS spectrograph obtained from the same protein sample presented in Fig. 33.

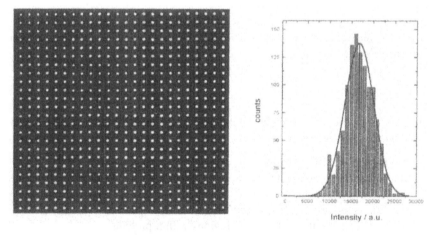

**Figure 35** Biochip quality assessment based on the distribution plot of signal intensities observed across an array.

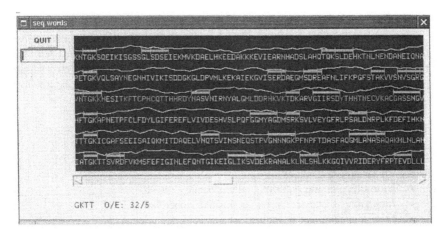

**Figure 36** A graphic user interface designed to visualize any protein within a fully sequenced genome so as to highlight cotranslational and posttranslational modifications as defined by ProSite (horizontal bars), the probability of surface exposure as predicted by Kyte–Doolittle hydrophobicity plots, color-coded protein sequence indicating commonness within a given genome of the amino acid string (expected frequencies calculated from randomized protein sequence within coding regions of the entire genome), and a hyperactive cursor to assess the phylogenetic similarity of a selected peptide string detected as common by database mining.

- Toxic genes
- Unstable clones
- Exon-rich ORFs
- Whole or partial gene-duplication events
- Organism-specific excretion pathways
- Previously undetected small ORFs
- Cleaved protein products
- Variability in induction times
- mRNA stability
- Tendency to form inclusion bodies

Efforts are ongoing to establish predictive tools based on sufficiently large learning sets so as to better anticipate genes for which closure strategies will be required in advance of problems being encountered in the molecular laboratory.

## IX. ADVANTAGES OF PROTEIN ARRAYS

Recombinant proteins once produced and purified are then immobilized onto a standard microscope slide. Contact printing procedures allow for up to 5000 to

6000 different elements to be placed on individual chips, be they different proteins or more replicates of less proteins. The latter is most desirable if one is intending to reliably rank experimental outcomes. Nanotechnologies and noncontact printing methodologies can further increase the number of elements included on a single protein biochip. The virtues of array-based assays with respect to many competing technologies include the following:

- Protein purity ensured (critical to data interpretation)
- Standardized protein abundance and accessibility
- On-array replicates of target (reproducibility of assay)
- Interarray reproducibility
- Biomolecular interactions mathematically ranked
- Inclusion of known targets as positive controls
- Inclusion of target homologs to assess target selectivity

In addition, an important advantage is the ability to titrate the concentration and time of potential ligands across the array, as opposed to the more simplistic, biologically naïve "Yes/No" responses obtained from techniques such as affinity capture [215–217] or the yeast two-hybrid approach [209,218–224]. Furthermore, protein arrays have some conspicuous advantages over methods employing cell-based bioassays [174–177]. These include the consistency and reproducibility of the assay with respect to the following:

- Temporal expression (i.e., variation in heterologous DNA sequence and the intracellular physiological effect engendered means that maximal expression of recombinant proteins is rarely synchronous)
- Location of protein gene product
- Target accessibility
- Multiple batches of arrays constructed from the same proteins (not so for recombinant proteins reinduced on several occasions as in cell-based assays)
- Guaranteed absence of 5′ and 3′ UTRs being incorporated in recombinant protein sequence

Tissue arrays, microdissected tissue slices, cell lysates, serum and Western blots of 2D gels each suffer from similar shortcomings. These include a significant diversity in protein abundance and accessibility, cell and tissue heterogeneity, and the same high-abundance proteins being encountered in all cells and tissues (most evident on images of 2D electrophoresis gels). Variability in abundance of cellular constituents can translate into a higher signal being obtained from a low-affinity binder interacting nonspecifically with, for example, enolase, ribosomal proteins, or heat shock proteins found in high abundance in all living cells. This situation is contrasted with a signal going undetectable as a result of an interaction between a critically important, low-abundance "housekeeping" gene interacting with its high-affinity binder (Fig. 37). Using currently available technologies, the latter could go undetected during lead optimization studies and possibly even following toxicological testing and clinical trials with the result and obvious serious ramifications to patients and the pharmaceutical group involved. Noteworthy is the fact that serum is routinely employed as a means of gauging the occurrence of nonspecific target binding.

## X. PROTEIN BIOCHIPS FOR LEAD OPTIMIZATION: AN IMMEDIATE VALUE-ADD FROM THE HUMAN GENOME PROJECT

Much of the current activity in proteomics has concentrated on the discovery of new drug targets or novel diagnostics markers for a particular disease entity, whereas genomics activity has been criticized for having elucidated a plethora of potentially interesting drug targets, each awaiting further validation. Neither has yet to revolutionize the pharmaceutical industry or replace traditional drug discovery pipelines. Arrays of purified recombinant proteins offer a means of rapidly verifying the extent of target selectivity exhibited by known therapeutic

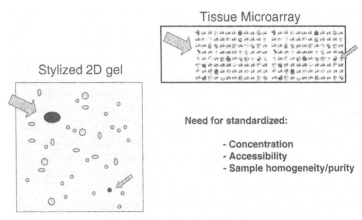

**Figure 37** The dilemma of a high-abundance protein producing a stronger signal than a low-abundance protein and thus the need for protein concentration to be standardized across an array in order to produce interpretable results.

molecules to known targets by employing technologies not previously accessible to the pharmaceutical industry (i.e., appropriately designed protein biochips based on a knowledge of the human genome). Protein microarrays can be used to screen lead molecules (antibodies, protein biomolecules, and small-molecule drugs) and iterations thereof for the most specific target binders prior to toxicological, pre-clinical, and clinical testing. Improvements in target selectivity linked to optimized recognition profiles following exposure to millions of potential binding sites derived from the human genome should translate into reduced adverse drug effects.

If these same protein microarray technologies can be applied to lead optimization, then the impact on improved drug development can be brought substantially nearer term than that seen to date in the Age of Genomics. More important still is the ability to impact at the higher-value end of the drug discovery chain (i.e., improved target selectivity during lead optimization). Such deliverables are nonetheless a by-product of a detailed knowledge of the human genome. Therapeutic molecules most often have their mode of action directed against the protein products of genes and not the nucleic acid code. The feasibility of transforming ORFs detected within the human genome, either directly or through amplification of cloned complementary DNA, into recombinant proteins will be demonstrated here with a view to better detailing target recognition in the presence of an increasingly significant number of human recombinant proteins present on-array. Indeed, more than an estimated 665,000,000 different 5-mer epitopes or drug-binding sites could be contained on a single protein microarray containing 5000 different recombinant proteins or domains of 300 amino residues each (Gestel and Humphery-Smith, unpublisehd data). A peptide array designed to display such diversity is not yet practicable with respect to the size of array required, time, and cost. To afford a good representation of these binding motifs, a population of recombinant proteins is randomly and covalently immobilized in a 3D hydrogel matrix atop of a glass substrate (Fig. 38). For maximal utility to lead optimization, the choice of proteins present on such arrays should include noncandidate proteins, known positive controls to allow ranking of results, and an expanded NTS, as shown in Fig. 39. The objective must then be directed toward enhanced specificity of binders as part of lead optimization (Fig. 40).

The likelihood of unforeseen side effects becoming apparent following the clinical release of new therapeutic molecules should be reduced as a result of improved techniques for target selectivity optimization (Fig. 41). With the availability of such tools, chemical iterations of lead molecules and/or members of particular family of molecules derived from screening chemical libraries should first be subjected to such screening. Examples relevant to the screening of therapeutic antibodies, protein biomolecules, and small-molecule drugs are presented here. However, for the latter to become feasible for large-scale screening of small molecules, these approaches must first be linked to nonlabeled parallel screening

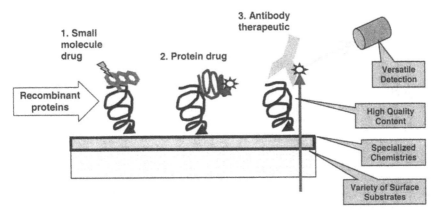

**Figure 38**  Schematic representation of small-molecule drug, protein biomolecules, and therapeutic antibody exposure to a protein microarray.

# Target Selectivity Screening:

**Lead Molecule + Chemical Iterations**          **Prescreened Drug Class**

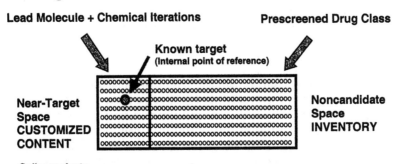

- Splice variants
- Domain homologies (distant & close, i.e. sequence & structural & threaded homology)
- Expression in: <u>E.coli</u>; Yeast; Insect; & Mammalian cells

**CRITICALITY OF:**

1) Speed and quality of synthesis for the near-target space
2) Purity of recombinant proteins for assessment of cross-reactivity
3) Iterative screening based upon rapid synthesis of new NTS
   based on positive hits

**Figure 39**  Schematic representation of the potential utility of protein arrays during lead optimization during drug development.

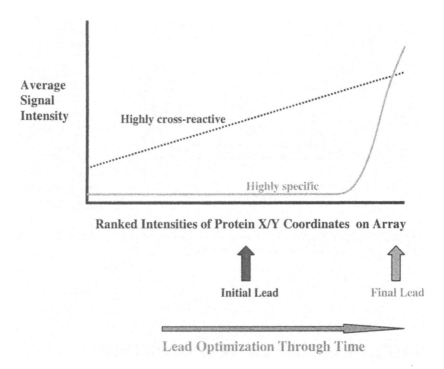

**Figure 40**  Overview of the lead optimization process and its desired outcome (i.e., a therapeutic agent exhibiting a high degree of target specificity).

technologies. Nonetheless, this is currently most easily practiced using radiolabeled small molecules, yet probably not for large numbers of lead molecules and their chemical iterations due the time and cost associated with radiolabeling procedures. The immobilization of domain structures only, as opposed to whole proteins, is likely to increase the chances of success in what is likely to be a most technologically demanding sector. Examples shown in Figs. 42–44 clearly demonstrate reliable mathematical ranking (i.e. on-array replicates) of an individual binder with respect to large numbers of potential targets. These reduction-to-practice experiments were conducted in parallel and combined with up to 12 on-array replicates to provide healthy levels of statistical confidence in the rankings obtained (Fig. 42).

If array-based proteomics can impact lead validation and lead optimization with respect to large numbers of recombinant proteins, there is real hope that improved target selectivity assessment of lead compounds prior to toxicological and certainly prior to clinical phase testing should result in respectively, time saving, safer drugs, and more rapid drug registration as a result of diminished

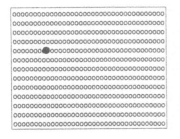
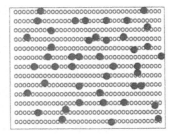

'*Good Drug*' candidate          *Probable* '*Bad Drug*'

- **Need to maintain target**
  **specificity in presence of**
  **numerous other Human proteins**

- **Increased likelihood**
  **of toxic side-effects**

**Figure 41** Exposure of novel therapeutic agents during lead optimization to human proteome ''look-alike'' biochips is likely to result in improvements in target selectivity and safer drugs and more rapid registration.

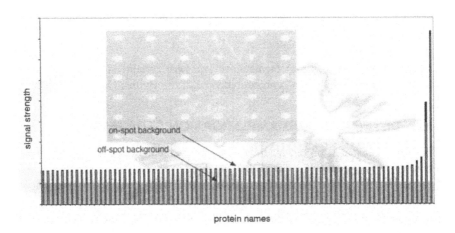

**Figure 42** An example of the fluorescent signal obtained for a highly specific ligand recognizing its target in the absence of cross-reactivity among 130 other recombinant proteins immobilized on the same array. The mean signal intensity obtained from six replicates is ranked lowest to the highest (left to right). The standard error appears as dark vertical bars. This is a graphic representation of the result shown in Fig. 12a.

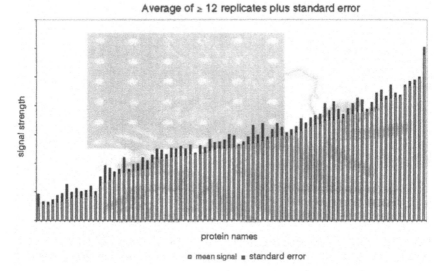

Average of ≥ 12 replicates plus standard error

□ mean signal ■ standard error

**Figure 43** An example of the fluorescent signal obtained for a highly cross-reactive ligand recognizing each of 131 recombinant proteins immobilized on array. The mean signal intensity obtained from up to 12 replicates is ranked lowest to the highest (left to right). The standard error appears as dark vertical bars. This is a graphic representation of the result shown in Fig. 19.

adverse side effects due to medication. This will certainly represent a major benefit to humankind, as a concrete outcome of the Human Genome Project.

## XI. DETECTION AND AMPLIFICATION STRATEGIES

For several decades, protein scientists have been able to detect antibody–antigen complexes and amplify the signal obtained through a variety of approaches linked to the generation of color and light passing from sandwich ELISAs to enhanced chemiluminescence [215,216]. These procedures will not be reviewed here. However, as the field of array-based proteomics evolves, the search for improved detection strategies will strive for enhanced sensitivity. Techniques such as rolling circle amplification [143] and resonance light scattering [217,218] are proving most exciting in this arena, but most attractive will be nonlabeled detection strategies. Unfortunately, there is a likely trade-off between sensitivity and unencumbered binding; yet, electronic detection of molecular binding events may produce surprising and spectacular results. Such an example is the nonlabeled detection by thermal lens microscopy of single-molecule interactions as a result of the heat

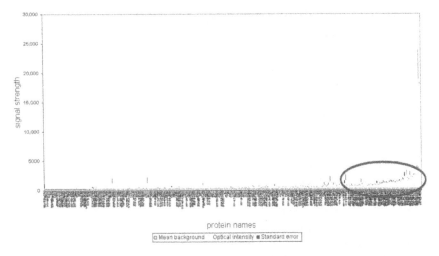

**Figure 44** An example of the fluorescent signal obtained following exposure of a therapeutic small molecule of 400 Da along with a 390 Da linker molecule attached on its nonactive side and subsequently linked to a fluorescent label following exposure to 6 replicates of 361 different recombinant human proteins on a single array. The encircled area indicates those binders showing above-background target interactions, but with less signal intensity than the biotin–streptavidin positive control. It is noteworthy that small molecule–protein interactions will possess far lower affinities than the latter control positive.

of reaction produced [219]. If this or other nonlabeled technologies can also accommodate extended dynamic range, then the future is indeed bright for array-based proteomics.

Any label attached to a binding partner (target or ligand) is likely to interfere with the molecular interactions being monitored. The need for nonlabeled detection becomes increasingly important as the molecular mass of the ligands being monitored decreases, as in the case of candidate small-molecule therapeutics exposed to protein arrays during lead validation and lead optimization [48]. For many years, the nonlabeled detection of biomolecules has been possible by a variety of techniques, but they are increasingly being adapted for of multiplexed assays [220]. However, a major challenge will be to miniaturize and parallelize these systems. Here again, surface chemistry is likely to play an important role in enhancing site occupancy and the signal-to-noise ratio [221,222]. Each increases in importance as the molecular mass and the concentration of the ligand in question decreases. Technologies thought to offer promise for parallel, on-array, nonlabeled detection include those detailed below. It will be interesting to see over time which of the following technological platforms performs best in an array-based setting due to their respective merits and shortcomings.

- Surface plasmon resonance [223,224]
- Grating coupled surface plasmon resonance [225–227]
- Colorimetric resonant reflection [228,229]
- Colorimetric gold nanoparticle sensors [230]
- Reflectometric interference spectroscopy [231]
- Quartz crystal microbalance [232]
- Magnetic tweezers [233]
- Optical tweezers [234]
- Atomic force microscopy [128,130,235,236]
- Nanocantilevers [237–239]
- Mach–Zehnder [240]
- Ellipsometry [241]
- Resonant mirrors [242]
- Fiber optics [243]
- Surface acoustic waves [244]
- Microcalorimetry [245]
- Electrochemical detection [246]
- Thermal lens microscopy [219,247]
- Hartman interferometry [221,222]
- Mass spectrometry [248,249]

## XII.  PATIENT COHORTING AND THE STRENGTH OF NUMEROUS DESCRIPTORS

Belov et al. [139], Petricoin et al. [185,186], and Zhang et al. [187] have shown the ability to provide numerical descriptors capable of patient cohorting with respect to disease outcome and precocious diagnosis. In all of these systems, the number of attributes being examined has been quite low; yet in the case of Petricoin et al. [185,186], the specificity of diagnosis has been remarkably high. Protein arrays offer the likelihood of expanding these patient groupings to a point whereby the mathematics should always be capable of discerning descriptors from within training sets. As the number of patients increases, concomitantly the number of useful numerical descriptors is likely to descend, however, with a 1000-element protein or antibody chip, it is probable that most clinical sample sets could be discerned. The reason for this is the inordinately large number of potentially distinct combinations able to be deciphered, namely $1.07 \times e^{301}$ possible combinations of 1's, 2's, 3's, 4's, 5's, and so forth out to one group of 1000 (calculation undertaken by Guilhuis-Pedersen). By default, nature will not adopt all of these possibilities, and not all antibodies or proteins could provide useful information when exposed to a tissue or cell lysate or body fluid. Nonetheless, even antibodies with unknown specificities are likely to be capable of discerning patient groups. This, in turn, could allow for better use of existing health care budgets. Although clinicians see patients suffering in their clinical ward today, new therapeutic targets or drugs for a particular disease state are always many years away and offer little hope for today's patient group. However, if protein and antibody arrays, along with other differentiating systems, can be applied to better discern those patient sets most likely to respond well to high-cost treatments from those best sent home for palliative care, then health care delivery may well undergo a dramatic transformation. Unlike DNA, the proteins themselves or affinity ligands recognizing the protein targets can provide invaluable information to the clinician. Similarly, protein and antibody arrays are likely to find applications during clinical phase testing of new drugs, preclinical animal-based research, and response and disease-progression monitoring. Of course, the major advantage of chip-based technologies here is their ability to be implemented inexpensively and on a large scale. Targets linked to disease genesis of multigenic/multifactorial diseases may become accessible through these approaches provided a sufficiently high number of parameters are employed to establish the learning set. Although it remains to be implemented, the hope of "individualized medicines" is often held up as a goal of the pharmaceutical industry in the future. Such medicines will depend on cost-effective solutions for the delivery of sufficiently robust numerical attributes to describe patient sets most likely to react well to a given therapeutic or for whom a particular drug should not be prescribed. This, too, will depend on the availability of low-cost, disposable diagnostics for home use or for use in the clinic to supplement drug administration.

## XIII.  CAVEATS OF PROTEIN AND ANTIBODY ARRAYS

No one technology is likely to supply the pharmaceutical industry with the knowledge required to confirm target selectivity with respect to all possible potential targets presented within the human proteome. Thus, one must insist that, at all times, results obtained on-array are confirmed by orthogonal approaches both in vitro and in vivo. In any case, this need to confirm experimental findings is likely to represent the status quo within the pharmaceutical industry. If one is employing recombinant proteins alone or in parallel, there will be a number of caveats needing to be considered, be they employed on an array or in solution. Recombinant proteins studied structurally one at a time by NMR or x-ray crystallography each suffer from similar caveats; that is, these problems are not unique to array-based proteomics. Highly insoluble and/or membrane-associated proteins remain a major challenge at every turn within the protein sciences. However, Fang, Frutos, and Lahiri [150] have suggested a path forward through the use of lipid arrays. Cellular compartmentalization can mean that interactions due to improved accessibility of targets are never encountered within living cells and can thus give rise to false positives on arrays. Protein complexes are thought to be important in driving much of biology; yet, these complexes cannot be easily synthesized and/or immobilized. A saving grace with respect to the latter is that one can expect differential assays dependent on interaction partners (total or partial) to produce a higher signal during differential screening than molecules not involved in interactions (i.e., between molecules associated as a complex or between motifs found on individual members of a protein complex and on-array targets). Whenever interaction partners are immobilized, there exist caveats with respect to in-solution assay. These can, however, be minimized through the use of random immobilization (as opposed to strategies which present only one side of a molecules for interaction assay) and the immobilization of targets in a 3D, highly hydrophilic hydrogel environment. These hydrogel substrates are thought to best emulate solutionlike properties. Cotranslational and posttranslational modifications of proteins need to be addressed during the synthesis of recombinant proteins. This is best achieved through the use of different expression vector hosts such as bacterial, yeast, insect, and mammalian cells for each ORF. Thereafter, the challenge for all recombinant techniques is to synthesize appropriately folded and conformationally correct recombinant proteins (i.e., to emulate the structural/binding integrity of the native protein). (*Note*: Emulation of, for example, enzymatic functional integrity may not be so easily emulated for numerous on-array analytes, whereby each possesses specific and distinct physiological requirements with respect to optimal pH, substrate, cleavage and activation of precursors.) Production procedures for recombinant proteins should be designed to minimize each of the above-mentioned caveats. In so doing, a powerful new parallel technology can be applied to lead optimization. Previously, such tools were simply not available to the pharmaceutical industry and, thus, information gathering on a similar scale would have been painstakingly slow.

## XIV. BIOCOMPUTING AND HIGH-DIMENSIONAL SPACE

Efforts to sequence the DNA of living organisms over the last decade have taught us that advances in the genomic sciences must go hand-in-hand with access to high-end computing infrastructure and the development of appropriate software tools to facilitate the storage and analysis of these ever-expanding datasets. The same is true for proteomics; biocomputing can neither be trivialized or ignored, particularly now that analysis of the human proteome has been clearly placed on the scientific horizon as the next major objective of better understanding the workings of whole human beings in health and disease. The Human Proteome Organization was formed in 2001 with the task of promoting this endeavor internationally.

Both cDNA and protein biochips have the potential to expand the world of biological research into the realm of high-dimensional space. Such is afforded by the acquisition of large numbers of high-quality and reproducible results. Biochips offer an exciting vista whereby the simplistic manner in which biological experiments have been designed and results examined in the past might be significantly transformed. The only exception here might be megaepidemiological studies, wherein data pertaining to hundreds of thousands of individuals have been examined for underlying trends with respect to numerous data points for each individual. As genomic and proteomic analyses are combined to interrogate the molecular mechanisms underlying multigenic phenomena, we can no longer be satisfied with low numbers of experimental replicates to examine large numbers of independent variables (cf. discussion in Ref. 50). These latter authors discovered some 23 proteins that behaved significantly ($p < 0.05$) differently between an asthma mouse model and controls from among 2115 protein spots followed on 24 different 2D electrophoresis gels. In other words, among any 20 such observations, one apparently significant observation could be expected to have occurred by chance alone. Thus, the questions arises as to whether or not, at this level of significance, one should not expect 106 occurrences (1/20th) from within 2115 to have behaved differently by chance alone. Whenever a large number of variables is being examined, the number of replicates needs to be increased accordingly to afford better statistical confidence so as to avoid this dilemma. In so doing, the number of observations about which meaningful conclusions can be reached also increases. In the experimental setup reported by Houtman et al. [50], only 830 of the 2115 spots included in the reference dataset were seen often enough (6 or more times) (i.e., at the limit of a possible statistically valid conclusion for the statistical test employed). Although further experimental validation would increase the statistical confidence in the results obtained, a reassuring factor was that most of the upregulated or downregulated proteins identified (18 out of 20) could be linked to known asthma symptomology, the likelihood of which is infinitesimally low to occur by chance alone.

Genomic- and proteomic-scale experiments will increasingly demand control datasets and test observations linked to multiple variables across exponentially expanding contingency tables in high-dimensional space (Fig. 45). The dilemma then becomes how best to interrogate the resulting datasets in high-dimensional space. It literally becomes impossible to examine each and every possible combination of data for significance with respect to each and every other piece of data contained within the master dataset or corresponding control dataset. This is because the problems at hand are nonpolynomially complete and will remain beyond the computation capacity of current and future computing well into the future. Thus, one must exploit analysis tools designed specifically for looking into complex data systems, whereby the available computing capacity is exploited maximally (i.e., not in a wasteful manner). For this reason, algorithm development and, more particularly, the development of parallelized computer algorithms are is most likely to bear fruits in the sector. Statistical analysis is a mature science. Thus, in order to interrogate biochip data, there is little need to reinvent methods, but, rather, focus should be placed on appropriate experimental design and the parallelization of existing test procedures. These parallelized computer algorithms

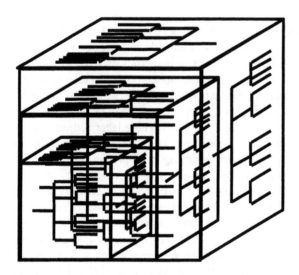

**Figure 45** Ever-expanding, high-dimensional space able to be analyzed by large number of reproducible protein and antibody biochips sampling matched controls for numerous parameters, such as days of the week, therapeutic agent employed, sex, age class, tissue type, and so forth The nonpolynomial complete nature of such experiments necessitates the use of data-mining algorithms as opposed to comparisons of all possible pairwise combinations across the complete dataset.

are best placed to efficiently extract trends from the data produced by biochip experiments.

For many centuries and still recently, the success and evolution of modern medicine can be attributed often to the intuitive powers of our healers and the inherent self-healing capacity of the human body, more so than directly to a clear understanding of the disease processes involved in ill-health. Without doubt, the complexity of human aliments will most often extend well beyond the powers of intuitive reasoning. Therefore, biochip technology and its ability to measure large numbers of independent variables may provide the means to transform modern medicine into a less intuitive and more mathematically driven discipline. The latter may also help provide the necessary economic stimulus to further evolve the mathematics of complexity analysis. These exciting potential changes in health care and health care delivery are likely to be witnessed in areas of disease diagnosis both at home and in the clinic, the monitoring of disease progression, the administration of therapeutics best suited to a particular individual, and analysis of disease predisposition. Access to clinicians is increasing, likely to be comple-mented by complex datasets acquired initially by specialised laboratories and later by user-friendly medical devices in a homecare environment (e.g. pregnancy testing today). Here again, low-cost, miniaturized, and highly parallelized analyses are likely to dominate. The need is to move beyond one-at-a-time testing to the analysis of multiple health-related variables in parallel. The primary detection of patterns of relevance to a particular disease state may demand supercomputing; however, once these critical number sets have been identified, subsequent recognition of tell tale parameters is already well within the computational capacity of extant pocket calculator microprocessors.

As one looks toward the future, the reality of contemporary biocomputing solutions demands a significant commitment to data management, data storage and data interrogation. Biochips are already capable of significant data production; however, increasingly large-scale analyses of hundreds and thousands of patients will demand that increasingly elaborate solutions are implemented. Current-generation biochip readers (e.g., the Tecan LS Series scanner) can be linked to slide-handling robotics, whereby it can be configured to automatically acquire image files from 320 chips in a little less than 24 h. Such files are routinely associated with approximately 25-Mbyte files. Unfortunately, one can now only dream of batteries of such engines in the protein biochip laboratory, but once in existence, the immediate priority must be data storage. Infrastructure established at Glaucus Proteomics BV in The Netherlands and rendered operational in late 2001 linked low-end disk storage robots to off-site fire and waterproof secure storage facilities via a dedicated experimental broad-band width communication link to the SARA supercomputing facility in Amsterdam (Figs. 46–48). Automated protocols were required for writing the information to disk in-house and thereafter transmitting large data files under a secure (firewall protected) environment. The task involved the laying of two dedicated optical dark fibers some 10 km long so as to establish links with this national network. The measures taken afforded highly cost-effective access to high-end supercomputing solutions, as well as off-site data backup.

Once information has been securely stored and backed up, the priority immediately passes to effective data processing and interrogation. This is far from trivial in the world of biochip data for information derived from either cDNA or proteins. Indeed, the experimental design and methodologies employed should be little different. Several authors [250–255] have reviewed many of the sources of variation associated with biochip data and the numerous challenges confronting good data acquisition and interrogation. Variation linked to data acquisition and the data itself leaves one with the opinion that the biological nature of gene regulation and the potentially significant variations in upregulation and downregulation of biological endpoints may actually be producing highly skewed datasets (e.g., Fig. 49). Nonetheless, the mathematical reality is that the different underlying sources of variance have not always been subtracted from the raw data (i.e., optical intensities of gene and protein abundance). Fig. 50 shows a previous attempt to implement analysis of variance linked to five different chip-related parameters using parallelized code in a supercomputing environment.

# Glaucus Computer Architecture

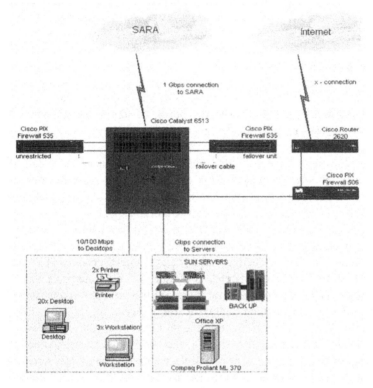

**Figure 46**   Schematic representation of the computing infrastructure installed by Glaucus Proteomics as a cost-effective solution to supercomputing requirements (i.e., the establishment of high-speed dedicated links to an existing high-performance computing infrastructure).

The same data employed to generate Figs. 49 and 50, Perou et al. [256], were examined as a test dataset to better asses the effectiveness of statistical methodologies. The study of Perou et al. [256] was based on data pertaining to pretreatment and posttreatment analysis of 20 breast cancer patients using cDNA biochips containing 9216 spots. Our analyses demonstrated 2790 spots behaved differently when analyzed by the Student's $t$-test and just 988 by the nonparametric Wilcoxon sign-ranked test, wherein 205 of the latter group were not included among the 2790 detected by the Student's $t$-test. The nonparametric calculation was more intolerant of the high variance inherent in both the pretreatment and posttreatment datasets, particularly one or two highly upregulated or downregulated observations (real or artifactual) occurring within the 20 observations. To

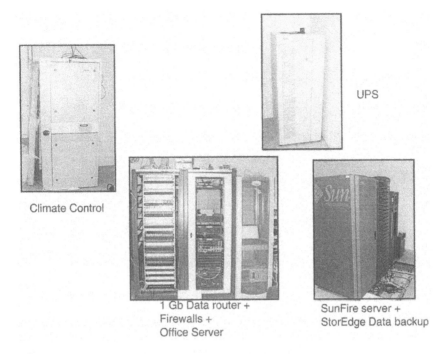

UPS

Climate Control

1 Gb Data router +
Firewalls +
Office Server

SunFire server +
StorEdge Data backup

**Figure 47**  In-house computing infrastructure including constant temperature, uninterrupted power supply (UPS), high-speed routing and firewalls for both Internet and broadband communications, and on-site disk and tape storage linked to off-site data backup.

examine these differences still further, all of the possible combinations of six paired observations before and after treatment were examined for statistical difference at a minimal threshold of $p < 0.05$ when using the Wilcoxon sign-ranked test. Of the 357,212,160 tests conducted on this dataset, an average of 335 pairs were seen as significant per paired group of 6 observations measured in parallel across the chip (as compared with 461 expected by chance alone at a similar level of statistical significance (i.e., 1/20th of the 9216 observations being examined on-chip). When the totality of the tests conducted (357,212,160) were summarized, 9084 of the spots represented on the arrays had been observed as statistically significant on at least one occasion. The latter result highlights the danger of statistical significance not measured below $p < 0.002$ or $p < 0.001$ when examining populations containing thousands of independent variables. Large numbers of independent variables dictate the need for large numbers of replicates so as to afford reliable outcomes. The same phenomenon has been commonplace knowledge for 2D electrophoresis data for more than a decade. Indeed, until the number of replicates is substantially increased, conclusions are not possible on the bulk of biological end points being examined (i.e., protein or cDNA abundance).

**Figure 48**  The other end of a cost-effective, high-performance computing infrastructure, namely a pre-existing infrastructure at national supercomputing facility, SARA, located in Amsterdam and connected to internal computing infrastructure (Fig. 47) by an experimental broad-band, optical-fiber link with a dedicated extension over 10 km. Pictured is the TERAS supercomputer consisting of two SGI Origin 3800 512 CPU systems with a maximum performance of 1Tflop per second. This is the latest upgrade among numerous other computing and storage solutions offered (www.sara.nl).

Elsewhere in proteomics, technologies are confronted by similar shortcomings. Although mass accuracy affords great confidence in the protein or peptide identification obtained (i.e., "mass never lies"); biology, on the other hand, has the habit of turning on or off or upregulating and downregulating many hundreds and even thousands of genes over a short time frame. This variance in gene or protein expression detected once at a given time point is unlikely to be the molecular mechanism underlying, for example, cancer pathogenesis when encountered in a cancer patient and not in a control individual [i.e., "but biology often does (lie)"]. This is particularly relevant before one sets about attributing still greater resources for further detailed molecular and therapeutic endeavors surrounding a given drug or its biological target. This problem is often encountered when examining whole biological samples by hybrid LC/CE/MS/MS-style analyses [54,257], multiple enzymatic digestion followed by MS [63–65], or in association with differential labeling with stable isotopes [258–266]. To date, the effort to obtain results once with stable isotopes on large numbers of differentially labeled

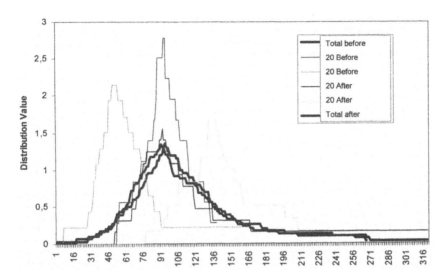

**Figure 49** Probability density functions (PDFs) produced for spot PRO1847 from data published by Perou et al. [256] on the analysis by cDNA biochips. The PDFs presented here are of subsets of 20 observations producing conflicting conclusions (either significantly upregulated or downregulated) when compared to the lack of difference between the grouped data. These results highlight the need for large numbers of replicates when examining datasets comprised of large numbers of independent variables.

peptides (i.e., one test and one control group) derived from complex protein mixtures has precluded any attempt to replicate the observations on numerous occasions so as to achieve statistically valid results (cf. Ref. 267).

The importance of experimental design is further highlighted when one considers clustering of "synexpression" (e.g., protein or cDNA expression levels exhibiting a similar pattern of variation in abundance across a given time line, experimental protocol, or patient set). One must inquire also as to the likelihood of a few genes or proteins behaving similarly within 10,000 distinct variables—in principle, quite high by chance alone for any pattern of interest. A similar scenario is encountered when the data obtained from plotting, for example, Cy3 fluorescent signals against the signal derived from Cy5 fluorescence or an alternate experimental procedure examined on the same biochip, be that for proteins or cDNA. Population outliers may be highly statistically significant from the other 10,000 paired observations on a particular chip (e.g., the test and control groups). However, the question remaining to be asked is whether or not these distinct data points are capable of maintaining their distinctiveness over multiple replicate chip experiments (i.e., the need for large number of replicates when employing biochips

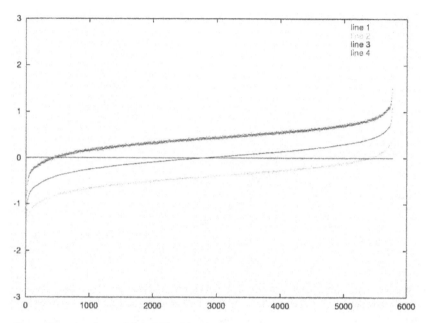

**Figure 50** A subset of the cDNA biochip data published by Perou et al. [256] wherein each of just less than 6000 spots used to examine 40 breast cancer samples both before and after treatment were subjected to analysis of variance (ANOVA). Here, linear code for classical ANOVA statistics against five independent parameters were computed using high- performance computing and a parallelized computer algorithm designed to increase the speed and efficiency of calculation for all spots compared with themselves as independent experiments, rather than as grouped datasets. The spot number appears on the $x$ axis and significant data points are those that do not include zero within their confidence limits.

encompassing thousands of independent data points). Simply put, biologists must become more demanding of the statistical tools employed on their biochip datasets.

When the data relevant to individual spots on a biochip have been accurately acquired and the sources of background variance have been removed, the task then turns to one of effective data processing to discover the underlying significant datasets in high-dimensional space. Here is not the place to enter into a treatise on statistical methods and experimental design, but both remain of critical importance to the field. As more and more independent parameters are measured on biochips, statistical tools designed to detect significant patterns become obligatory as pairwise comparisons become increasingly impractical on all possible pairs (see the above discussion on nonpolynomially complete problems, this section). Tools for pattern recognition include, among others, heuristic and principle component analyses, Fourier wavelet decomposition tools, and a variety of both supervised and unsupervised self-learning algorithms. Recently, a number of authors [185,268,269] have exploited these latter tools to detect significant trends in protein and gene expression data dealing respectively with ovarian cancer and diffuse large B-cell lymphomas. Among these self-learning engines are entities known as genetic algorithms. By amusing coincidence, they are now being applied to genomic and proteomics datasets. Indeed, the jargon being employed in a mathematically context will be familiar to most biomedical researchers and includes terms such as chromosomes, evolution, fitness, reproduction, adaptive behaviour, gametes, mutation, and haploid.

Patterns detected within biochip data can, in themselves, be excessively complex. Tables containing many columns and rows can be ranked, simplified, and color coded to represent significant trends, but these, too, can easily defy intuitive reasoning if one is wishing to display more than the highest-order data associations. Numerous tables and long lists of significant genes or proteins and their respective biological significance can rapidly camouflage important underlying trends. Thus, without doubt, an area requiring significant improvement is the visualization of complex biological data and the innumerable associated interactions, relative affinities, and/or levels significance. Here, novel visualization tools are required to assist the poor biologist to better fathom the complexity of human health and disease, even for statistical nodes in a uniplanar environment.

## XV. CONCLUSION

Protein and antibody arrays are likely to find immediate application in many areas of drug development and biomedical research such as target discovery, validation of targets discovered by the genomic sciences, precocious diagnosis of disease, patient cohorting with respect to disease and treatment outcomes, and replacement of diagnostic assays not currently conducted in such a highly parallelized fashion (e.g., ELISAs in a clinical and research setting). However, the greatest immediate advantage to the development of novel therapeutic agents likely to be derived from an increased knowledge of the human genome will be through the use of protein chips emulating increasingly large portions of the human proteome for applications directed toward improved target selectivity during lead optimization of novel therapeutics. Drug registration authorities globally remain on the lookout for such improvements in target selectivity testing procedures (i.e., so as to help reduce the likelihood of adverse drug effects associated with novel therapeutic agents). Indeed, the use of protein arrays during lead optimization has the potential of offering up a reliable "early cull" technology, more reliable than their cDNA counterparts, and, most importantly, help ensure against potentially deleterious interactions going undetected prior to clinical testing and market release.

The Human Genome Project is nearing completion. An increased understanding of the workings of the human body is likely to be afforded by discovery proteomics. This will demand large numbers of experimental replicates conducted on large numbers of clinical samples. Such remains beyond current generation approaches employed in proteomics and dependent on the separation sciences. Indeed, lessons learned during the genomic revolution over the last decade must now be applied to proteomics, namely parallelized, miniaturized, and automated analyses in the hunt for therapeutic targets and/or improved intervention strategies. These factors are equally well afforded by protein biochips as by their cDNA counterparts. However, the former represents a value-add in that proteins are the molecular workhorses within cells and thus the critical targets for disease intervention strategies.

Nonetheless, significant hurdles need to be overcome first. Apart from those linked to the detection technologies themselves, most dominant is the need for high-quality chip content (i.e., the reagents to be immobilized on chip). It will be content provision that is most likely to dominate advances in array-based proteomics over the next decade with respect to both commercial and scientific endeavors. This content refers to recombinant proteins expressed at high purity and uniform on-chip concentrations and the affinity ligands manifesting the highest possible target specificity. In order to generate the latter, large numbers of recombinant proteins must first be generated for each ORF in the human genome and corresponding to domains common to splice variants and not cleaved in the mature state. In turn, affinity ligands must then be generated to each of these in a cost- and time-effective manner. The existence of such ligands will enhanced performance of proteomic technologies both on-chip and via traditional approaches based on the separation sciences. Equally important will be improved affinity enrichment of proteins during disease diagnosis and detailed characterization.

Lastly, protein biochip performance will be greatly influenced by the quality of surface chemistry solutions designed specifically for applications in protein science. Here, site occupancy and the associated mass action need to be combined with low nonspecific binding so as to afford maximal signal-to-noise detection of biomolecular interactions. As in all of the analytical sciences, the signal-to-noise ratio, reproducibility and molecular soundness of the approach employed are critical to success, yet these aspects must be coupled with scalability to achieve relevance to the Human Proteome and its inherent multitude of protein isoforms manifested both temporally and physically with respect to cell and tissue type. Nonlabeled detection of nonimmobilized molecules in an array format designed to provide parallelization of independent assays clearly represent the Holy Grail of this discipline. The elimination of substrate takes this process one step further and provides access to reaction kinetics unencumbered by the steric hindrance due to substrate. Current technological advances would suggest that such functionality is not far removed. These advances are likely to herald more accurate predictions concerning molecular function and interaction.

## ACKNOWLEDGMENTS

I wish to extend my sincerest thanks to those who have helped implement and evolve the ideas extolled in this chapter since my arrival in The Netherlands from Australia in August 1999. I am indebted to colleagues at Glaucus Proteomics B.V., whose untimely and unfortunate demise has caused us all much angst (cf. *Genomika*, Volume 3, Issue 23, 27 November 2002), but also served to demonstrate the utility of array-based proteomics as a scalable and reproducible technology with applications in discovery proteomics, diagnostics, and drug development. It was an exciting adventure into biotech space shared with the following colleagues who demonstrated fantastic enthusiasm and commitment to the task at hand: Michel Arotcarena, Erik Baas, Johnnie Batten, Jenny Birch, Judy Bos-de Ruijter, Peter van de Broek, Gert-Jan Caspers, Mustapha Chalabi, Marinus Dansen, Menno Deij, Jochem Eigenhuisen, David Englert, Roel van Eijk, Soesila Ganeshie, Odile Grulet, Mark Hijmann, Kevin Jack, Luigi Jonk, Timo Kreike, Bart van Leeuwen, Rutger Lens, Scott Marshall, Margaret Menkveld, Thomas Mikosch, Hykel Mughal, Mieke Roefs, Markwin Velders, Arnold Vos, Erik Wischerhoff, Peter Verweyan, Rutger Verweyan, and, last but not least, Thomas Zacher. I am equally indebted to my students and colleagues from the Department of Pharmaceutical Proteomics at the Universiteit Utrecht who have contributed much to this collaborative effort, namely Yair Benita, Daniella Fiechter, David Gestel, Lisa Gilhuis-Pederson, Ryuji Hashimoto, Martin Lok, Jose-Luis Lopez, Ronald Oosting, Rene Houtman, and Jasmijn Stegeman. Work subcontracted to WITA in Berlin and Toplab in Munich was also presented here in Figs. 1a and 1b, and Figs. 32–34, respectively. My heartfelt gratitude is also extended to colleagues from Glaucus Industries Australia who made all this endeavour possible, namely Steven Prassas, John Comino, Peter Coumbis, and Sam Miller; Fujisawa Pharmaceutical Corporation, the Universiteit Utrecht and Atlas Venture are thanked for their commitment and support. Above all, I am indebted to Aart Brouwer for his friendship, tutelage, and worldly advice on the business of science and for the gratuitous manner in which these were proffered. Input from the following commercial partners is also gratefully acknowledged: Applied Biosystems, Daiichi Pharmaceutical Corporation, GenMab, Medarex, Philips Electronics, Sun and SARA.

## REFERENCES

1.  Humphery-Smith, I.; Ward, M.A. Proteome Research: Methods for Protein Characterization. In *Functional Genomics* Hunt SP, Livesey R, eds; Oxford University Press: Oxford, 2000, 197–241.
2.  Wasinger, V.C.; Pollack, J.D.; Humphery-Smith, I. The proteome of *Mycoplasma genitalium*. Chaps-soluble component. Eur. J. Biochem. 2000, 267, 1571–1582.

3.  Humphery-Smith, I. Replication-induced protein synthesis and its importance to proteomics. Electrophoresis. 1999, 20, 653–659.
4.  McKusick, V.A. Mendelian Inheritance in Man. Catalogs of Human Genes and Genetic Disorders. 12th ed; 1998, Available from OMIM, http://www.ncbi.nlm.nih.gov/omim.
5.  Venter, J.C. The sequence of the human genome. Science. 2001, 291, 1304–1351.
6.  Lander, E.S., et al. Initial sequencing and analysis of the human genome. Nature. 2001, 409, 860–921.
7.  Bang, M.-.T., et al. The complete gene sequence of titin, expression of an unusual ≈700-kDa titin isoforms, and its interaction with obscuring identify a novel Z-line to I-band linking system. Circ. Res. 2001, 89, 1065–1072.
8.  Albala, J.S.; Humphery-Smith, I. Array-based proteomics: High-throughput expression and purification of the IMAGE consortium clones. Curr. Opin. Mol. Therap. 1999, 1, 680–684.
9.  Emili, A.Q.; Cagney, G. Large-scale functional analysis using peptide or protein arrays. Nature Biotechnol. 2000, 18, 393–397.
10. Holt, L.J., et al. The use of recombinant antibodies in proteomics. Curr. Opin. Biotechnol. 2000, 11, 445–449.
11. Mac Beath, G.; Schreiber, S.L. Printing protein microarrays for high-throughput function determination. Science. 2000, 289, 1760–1763.
12. Service, R.F. Protein arrays step out of DNA's shadow. Science. 2000, 289, 1673.
13. Borrebaeck, C.A.K. Antibodies in diagnostics—from immunoassays to protein chips. Immunol. Today. 2000, 21, 379–382.
14. Clewley, J.P. Recombinant protein arrays. Commun. Dis. Public Health. 2000, 3, 311–312.
15. Irving, R.A.; Hudson, P.J. Proteins emerge from disarray. Nature Biotechnol. 2000, 18, 932–933.
16. Weinberger, S.R.; Morris, T.S.; Pawlak, M. Recent trends in protein biochip technology. Pharmacogenomics. 2000, 1, 395–416.
17. Albala, J.S. Array-based proteomics: the latest chip challenge. Expert Rev. Mol. Diagn. 2001, 1, 145–152.
18. Zhu, H.; Snyder, M. Protein arrays and microarrays. Curr. Opin. Chem. Biol. 2001, 5, 40–45.
19. Haab, B.B.; Dunham, M.J.; Brown, P.O. Protein microarrays for highly parallel detection and quantitation of specific proteins and antibodies in complex solutions. Genome Biol. 2001, 2, Research 0004.1–0004.13.
20. Tomlinson, I.M.; Holt, L.J. Protein profiling comes of age. Genome Biol. 2001, 2, Research 000.4.1–0004.33.
21. Blagoev, B.; Pandey, A. Microarrays go live—new prospects for proteomics. Trends Biochem. Sci. 2001, 26, 639–641.
22. Service, R.F. Proteomics. Searching for recipes for protein chips. Science. 2001, 294, 2080–2082.
23. Cahill, D.J. Protein and antibody arrays and their medical applications. J. Immunol. Methods. 2001, 250, 81–91.
24. Liotta, L.A.; Kohn, E.C.; Petricoin, E.F. Clinical proteomics. Personalized molecular medicine. JAMA. 2001, 286, 2211–2214.
25. Kodadek, T. Protein microarrays: prospects and problems. Chem. Biol. 2001, 8, 105–115.

26. Hebestreit, H.F. Proteomics: an holistic analysis of nature's proteins. Curr. Opin. Pharmacol. 2001, 1, 513–520.
27. Phizicky, E. Protein analysis on a proteome scale. Nature. 2001, 422, 208–215.
28. Mirzabekov, A.; Kolchinsky, A. Emerging array-based technologies in proteomics. Curr. Opin. Chem. Biol. 2001, 6, 70–75.
29. Jenkins, R.E.; Pennington, S.R. Arrays for protein expression profiling: Towards a viable alternative to two-dimensional gel electrophoresis? Proteomics. 2001, 1, 13–29.
30. Wulfkuhle, J.D., et al. New approaches to proteomics analysis of breast cancer. Proteomics. 2001, 1, 1205–1215.
31. MacBeath, G. Proteomics comes to the surface. Nature Biotechnol. 2001, 19, 828–829.
32. Zhou, H.; Roy, S.; Schulman, H.; Natan, M.J. Solution and chip arrays in protein profiling. TIBTECH. 2001, 19, S34–S39.
33. Bussow, K., et al. Protein array technology. Potential use in medical diagnostics. Am. J. Pharmacogenomics. 2001, 1, 37–43.
34. Templin, M.F., et al. Protein microarray technology. Trends Biotechnol. 2002, 20, 160–166.
35. Burbaum, J.; Tobal, G.M. Proteomics in drug discovery. Curr. Opin. Chem. Biol. 2002, 6, 427–433.
36. MacBeath, G. Protein microarrays and proteomics. Nature Genet. 2002, 32, 526–532.
37. Mitchell, P. A perspective on protein microarrays. Nature Biotechnol. 2002, 20, 225–229.
38. Schweitzer, B.; Kingsmore, S.F. Measuring proteins on microarrays. Curr. Opin. Biotechnol. 2002, 13, 14–19.
39. Khandurina, J.; Guttman, A. Microchip-based high-throughput screening analysis of combinatorial libraries. Curr. Opin. Chem. Biol. 2002, 6, 359–366.
40. Schlessinger, J. A solid base for assaying protein kinase activity. Nature Biotechnol. 2002, 20, 232–233.
41. Figeys, D. Adapting arrays and lab-on-a-chip technology for proteomics. Proteomics. 2002, 2, 373–382.
42. Stoll, D., et al. Protein microarray technology. Front. Biosci. 2002, 7, 13–32.
43. Talapatra, A.; Rouse, R.; Hardiman, G. Protein microarrays: challenges and promises. Pharmacogenomics. 2002, 3, 527–536.
44. Joos, T.O.; Stoll, D.; Templin, M.F. Miniaturised multiplexed immunoassays. Curr. Opin. Chem. Biol. 2002, 6, 76–80.
45. Ng, J.H.; Ilag, L.L. Biomedical applications of protein chips. J. Cell. Mol. Med. 2002, 6, 329–340.
46. Walter, G.; Bussow, K.; Lueking, A.; Gloker, J. High-throughput protein arrays: Prospects for molecular diagnostics. Trends Mol. Med. 2002, 8, 250–253.
47. Eickhoff, H., et al. Protein array technology: The tool to bridge genomics and proteomics. Adv. Biochem. Eng. Biotechnol. 2002, 77, 103–112.
48. Humphery-Smith, I.; Wischerhoff, E.; Hashimoto, R. Protein arrays for assessment of target selectivity: Transforming knowledge of the human genome into a lead optimization tool. Drug Discovery World. 2003, 1–8.

49. Voss, T.; Haberl, P. Observations on the reproducibility and matching efficiency of two-dimensional electrophoresis gels: consequences for comprehensive data analysis. Electrophoresis. 2000, 21, 3345–3350.

50. Houtman, R., et al. Lung proteome alterations in a mouse model for non-allergic asthma. Electrophoresis. (In press).

51. Humphery-Smith, I.; Guyonnet, F.; Chastel, C. Polypeptide cartography of *Spiroplasma taiwanense*. Electrophoresis. 1994, 15, 1212–1217.

52. Miklos, G.L.; Maleszka, R. Protein functions and biological contexts. Proteomics. 2001, 1, 169–178.

53. Washburn, M.P.; Wolters, D.; Yates, J.R. Large-scale analysis of the yeast proteome by multidimensional protein identification technology. Nature Biotechnol. 2001, 19, 242–247.

54. Gygi, S.P.; Corthals, G.L.; Zhang, Y.; Rochon, Y.; Aebersold, R. Evaluation of two-dimensional gel electrophoresis-based proteome analysis technology. PNAS. 2000, 97, 9390–9395.

55. Buttner, K., et al. A Comprehensive two-dimensional map of the cystolic proteins of *Bacillus subtilis*. Electrophoresis. 2001, 22, 2908–2935.

56. Bumann, D.; Meyer, T.F.; Jungblut, P.R. Proteome analysis of the common human pathogen *Helicobacter pylori*. Proteomics. 2001, 1, 473–479.

57. Shaw, A.C., et al. Comparative proteome analysis of *Chlamydia trachomatis* serovar A, D, and L2. Proteomics. 2002, 2, 164–186.

58. Cho, M.J., et al. Identifying the major proteome components of *Helicobacter pylori* strain 26695. Electrophoresis. 2002, 23, 1161–1173.

59. Uebarle, B.; Frank, R.; Herrmann, R. The proteome of the bacterium *Mycoplasma pneumoniae*: Comparing predicted open reading frames to identified gene products. Proteomics. 2002, 2, 754–764.

60. Gevaert, K., et al. Chromatographic isolation of methionine-conating peptides for gel-free proteome annalysis. Mol. Cell. Proteomics. 2002, 1, 896–903.

61. Lipton, M.S., et al. Global analysis of the *Deinococcus radiodurans* proteome by using accurate mass tags. PNAS. 2002, 99, 11,049–11,054.

62. Karaoglu, H.; Humphery-Smith, I. Signature peptides. From analytical chemistry to functional genomics. Methods Mol. Biol. 2000, 146, 63–94.

63. Geng, M.; Ji, J.; Regnier, F.E. Signature-peptide approach to detecting proteins in complex mixtures. J. Chromatogr. A. 2000, 870, 295–313.

64. Ji, J., et al. Strategy for qualitative and quantitative analysis in proteomics based on signature peptides. J. Chromatogr. B: Biomed. Sci. Appl. 2000, 745, 197–210.

65. Geng, M., et al. Proteomics of glycoproteins based on affinity selection of glycopeptides from tryptic digests. J. Chromatogr. B: Biomed. Sci. Appl. 2001, 752, 293–306.

66. Atassi, M.Z.; Smith, J.A. A proposal for the nomenclature of antigenic sites in peptides and proteins. Immunochemistry. 1978, 15, 609–610.

67. Barlow, D.J.; Edwards, M.S.; Thornton, J.M. Continuous and discontinuous protein antigenic determinants. Nature. 1986, 322, 747–748.

68. Thornton, J.M.; Edwards, M.S.; Taylor, W.R.; Barlow, D.J. Location of continuous antigenic determinants in the protruding regions of proteins. EMBO J. 1986, 5, 409–413.

69. Meloen, R.H.; Puijk, W.C.; Sloostra, J.W. Mimotopes: realization of an unlikely concept. J. Mol. Recogn. 2000, 13, 352–359.

70. Hayhurst, A.; Georgiou, G. High-throughput antibody isolation. Curr. Opin. Chem. Biol. 2001, 5, 683–689.

71. Berger, M.; Shankar, V.; Vafai, A. Therapeutic applications of monoclonal antibodies. Am. J. Med. Sci. 2002, 324, 14–30.

72. Stone, M.J. Monoclonal antibodies in the prehybridoma era: A brief historical perspective and personal reminiscence. Clin. Lymphoma. 2001, 2, 148–154.

73. Goldman, R.D. Antibodies: indispensable tools for biomedical research. Trends Biochem. Sci. 2000, 25, 593–595.

74. Schier, R., et al. Isolation of high affinity monomeric human anti-c-erbB-2 single chain Fv using affinity driven selection. J. Mol. Biol. 1996, 255, 28–43.

75. Griffiths, A.D.; Duncan, A.R. Strategies for selection of antibodies by phage display. Curr. Opin. Biotechnol. 1998, 9, 102–108.

76. Chowdhury, P.S.; Pastan, I. Analysis of cloned Fvs from phage display library indicates that DNA immunization can mimic antibody response generated by cell immunizations. J. Immunol. Methods. 1999, 231, 83–91.

77. Maynard, J.; Georgiou, G. Antibody engineering. Annu. Rev. Biomed. Eng. 2000, 2, 339–376.

78. Pini, A.; Bracci, L. Phage display of antibody fragments. Curr. Protein Peptide Sci. 2000, 1, 155–169.

79. Hudson, P.J.; Souriau, C. Recombinant antibodies for cancer diagnosis and therapy. Expert Opin. Biol. Ther. 2001, 1, 845–855.

80. Gao, C., et al. A method for the generation of combinatorial antibody libraries using pIX phage display. PNAS. 2002, 99, 12,612–12,616.

81. O'Connel, D., et al. Phage versus phagemid libraries for generation of human monoclonal antibodies. J. Mol. Biol. 2002, 321, 49–56.

82. Kronvall, G.; Jonsson, K. Receptins: a novel term for an expanding spectrum of natural and engineered microbial proteins with binding properties for mammalian proteins. J. Mol. Recogn. 1999, 12, 38–44.

83. Wernerus, H.; Lehtio, J.; Samuelson, P.; Stahl, S. Engineering of staphylococcal surfaces for biotechnological applications. J. Biotechnol. 2002, 96, 67–78.

84. Ronnmark, J.; Gronlund, H.; Uhlen, M.; Nygren, P.A. Human immunoglobulin A (IgA)-specific ligands from combinatorial engineering of protein A. Eur. J. Biochem. 2002, 269, 2647–2655.

85. Eklund, M.; Axelsson, L.; Uhlen, M.; Nygren, P.A. Anti-idiotypic protein domains selected from protein A-based affibody libraries. Proteins. 2002, 48, 454–462.

86. Skerra, A. Lipocalins as a scaffold. Biochim. Biophys. Acta. 2000, 1482, 337–350.

87. Skerra, A. Engineered protein scaffolds for molecular recognition. J. Mol. Recogn. 2000, 13, 167–187.

88. Skerra, A. "Anticalins": A new class of engineered ligand-binding proteins with antibody-like properties. J. Biotechnol. 2001, 74, 257–275.

89. Schlehuber, S.; Skerra, A. Duocalins: Engineered ligand-binding proteins with dual specificity derived from lipocalin fold. Biol. Chem. 2001, 382, 1335–1342.

90. Mercader, J.V.; Skerra, A. Generation of anticalins with specificity for a nonsymmetric phthalic acid ester. Anal. Biochem. 2002, 308, 269–277.

91. Kreider, B.L. PROfusion: genetically tagged proteins for functional proteomics and beyond. Med. Res. Rev. 2000, 20, 212–215.

92.  Xu, L., et al. Directed evolution of high-affinity antibody mimics using mRNA display. Chem. Biol. 2002, 9, 933.

93.  Jayasena, S.D. Aptamers: an emerging class of molecules that rival antibodies in diagnostics. Clin. Chem. 1999, 45, 1628–1650.

94.  Brody, E.N., et al. The use of aptamers in large arrays for molecular diagnostics. Mol. Diagn. 1999, 4, 381–388.

95.  O'Sullivan, C.K. Aptamers—the future of biosensing? Anal. Bioanal. Chem. 2002, 372, 44–48.

96.  Liss, M.; Petersen, B.; Wolf, H.; Prohaska, E. An aptamer-based quartz crystal protein biosensor. Anal. Chem. 2002, 74, 4488–4495.

97.  Clark, S.L.; Remcho, V.T. Aptamers as analytical reagents. Electrophoresis. 2002, 23, 1335–1340.

98.  Towbin, H.; Staehelin, T.; Gordon, J. Electrophoretic transfer of proteins from polyacrylamide gels to nitrocellulose sheets: procedure and some applications. PNAS. 1979, 76, 4350–4354.

99.  Burnette, W.N. "Western blotting": Electrophoretic transfer of proteins from sodium dodecyl sulfate–polyacrylamide gels to unmodified nitrocellulose and radiographic detection with antibody and radioiodinated protein. A. Anal. Biochem. 1981, 112, 195–203.

100. Harper, D.R.; Kit, M.L.; Kangro, H.O. Protein blotting: Ten years on. J. Virol. Methods. 1990, 30, 25–39.

101. Chang, T.-.W. Binding of cells to matrixes of distinct antibodies coated on solid surface. J. Immunol. Methods. 1983, 65, 217–223.

102. Ekins, R.; Chu, F.; Micallef, J. High specific activity chemiluminescent and fluorescent markers: their potential application to high sensitivity and "multianalyte" immunoassays. J. Biolumin. Chemilumin. 1989, 4, 59–78.

103. Ekins, R.P. Multi-analyte immunoassay. J. Pharm. Biomed. Anal. 1989, 7, 155–168.

104. Ekins, R.; Chu, F. Multianalyte microspot immunoassay—Microanalytical "compact disk" of the future. Clin. Chem. 1991, 37, 1955–1967.

105. Ekins, R.; Chu, F.; Biggart, E. Fluorescence spectroscopy and its application to a new generation of high sensitivity, multi-microspot, multianalyte immunoassay. Clin. Chim. Acta. 1990, 194, 91–114.

106. Ekins, R.; Chu, F. Multianalyte testing. Clin. Chem. 1993, 39, 369–370.

107. Ekins, R.P.; Chu, F. Developing multianalyte assays. TIBTECH. 1994, 12, 89–94.

108. Ekins, R. Immunoassay: recent developments and future directions. Nucl. Med. Biol. 1994, 21, 495–521.

109. Ekins, R.P. Ligand assays: From electrophoresis to miniaturized microarrays. Clin. Chem. 1998, 44, 2015–2030.

110. Geysen, H.M.; Meloen, R.H.; Barteling, S.J. Use of peptide synthesis to probe viral antigens for epitopes to a resolution of a single amino acid. PNAS. 1984, 81, 3998–4002.

111. Geysen, H.M.; Rodda, S.J.; Mason, T.J. The delineation of peptides able to mimic assembled epitopes. Ciba Found. Symp. 1986, 119, 130–149.

112. Maeji, N.J.; Bray, A.M.; Geysen, H.M. Multi-pin peptide synthesis strategy for T-cell determinant analysis. J. Immunol. Methods. 1990, 134, 23–33.

113. Cwirla, S.E.; Peters, E.A.; Barrett, R.W.; Dower, W.J. Peptides on phage: A vast library of peptides for identifying ligands. PNAS. 1990, 87, 6378–6382.

114. Devlin, J.J.; Panganiban, L.C.; Devlin, P.E. Random peptide libraries: a source of specific protein binding molecules. Science. 1990, 249, 404–406.

115. Lam, K.S., et al. A new type of synthetic peptide library for identifying ligand-binding activity. Nature. 1991, 354, 82–84.

116. Fodor, S.P., et al. Light-directed, spatially addressable parallel chemical synthesis. Science. 1991, 251, 767–773.

117. Pease, A.C., et al. Light-generated oligonucleotide arrays for rapid DNA sequence analysis. PNAS. 1994, 91, 5022–5026.

118. Bussow, K., et al. A method for global protein expression and antibody screening on high-density filters of arrayed cDNA library. Nucleic Acids Res. 1998, 26, 5007–5008.

119. Lueking, A., et al. Protein microarrays for gene expression and antibody screening. Anal. Biochem. 1999, 270, 103–111.

120. Walter, G., et al. Protein arrays for gene expression and molecular interaction screening. Curr. Opin. Microbiol. 2000, 3, 298–302.

121. Walter, G.; Konthur, Z.; Lehrach, H. High-throughput screening of surface displayed gene products. Comb. Chem. High Throughput Screen. 2001, 4, 193–205.

122. De Wildt, R.M.T.; Mundy, C.R.; Gorick, B.D.; Tomlinson, I.M. Antibody arrays for high-throughput screening of antibody-antigen interactions. Nature Biotechnol. 2000, 18, 989–994.

123. Uetz, P., et al. A comprehensive analysis of protein-protein interactions in *Saccharomyces cerevisiae*. Nature. 2000, 403, 623–627.

124. Tong, A.H.Y., et al. Systematic genetic analysis with ordered arrays of yeast deletion mutants. Science. 2001, 294, 2364–2368.

125. Mahlknecht, U.; Ottmann, O.G.; Hoelzer, D. Far-Western based protein–protein interaction screening of high density protein filter arrays. J. Biotechnol. 2001, 88, 89–94.

126. Rowe, C.A., et al. Array biosensor for simultaneous identification of bacterial, viral, and protein analytes. Anal. Chem. 1999, 71, 3846–3852.

127. Rowe, C.A., et al. An array immunosensor for simultaneous detection of clinical analytes. Anal. Chem. 1999, 71, 433–439.

128. Silzel, J.W., et al. Mass-sensing, multianalyte microarray immunoassay with imaging detection. Clin. Chem. 1998, 44, 2036–2043.

129. Mendoza, L.G., et al. High-throughput microarray-based enzyme-linked immunosorbent assay (ELISA). Biotechniques. 1999, 27, 778–788.

130. Arenkov, P., et al. Protein microchips: use for immunoassay and enzymatic reactions. Anal. Biochem. 2000, 278, 123–131.

131. Brizzolara, R.A. Patterning multiple antibodies on polystyrene. Biosensor Bioelectronics. 2000, 15, 63–68.

132. Schweitzer, B., et al. Immunoassays with rolling circle DNA amplification: A versatile platform for ultrasensitive antigen detection. PNAS. 2000, 97, 10,113–10,119.

133. Sapsford, K.E.; Liron, Z.; Shubin, Y.S.; Ligler, F.S. Kinetics of antigen binding to arrays of antibodies in different sized spots. Anal. Chem. 2001, 73, 5518–5524.

134. Moody, M.D., et al. Array-based ELISAs for high-throughput analysis of human cytokines. Biotechniques. 2001, 31, 186–194.

135. Huang, R.P. Detection of multiple proteins in an antibody-based protein microarray system. J. Immunol. Methods. 2001, 255, 1–13.

136. Huang, R.P.; Huang, R.; Fan, Y.; Lin, Y. Simultaneous detection of multiple cytokines from conditioned media and patient's sera by an antibody-based protein array system. Anal. Biochem. 2001, 294, 55–62.

137. Knezevic, V. Proteomic profiling of the cancer microenvironment by antibody arrays. Proteomics. 2001, 1, 1271–1278.

138. Goodey, A., et al. Development of multianalyte sensor arrays composed of chemically derivatized polymeric microspheres localized in micromachined cavities. J. Am. Chem. Soc. 2001, 123, 2559–2570.

139. Belov, L., et al. Immunophenotyping of leukemias using a cluster differentiation antibody microarray. Cancer Res. 2001, 61, 4483–4489.

140. Huang, R.P. Simultaneous detection of multiple proteins with an array-based enzyme-linked immunosorbent assay (ELISA) and enhanced chemiluminescence (ECL). Clin. Chem. Lab. Med. 2001, 39, 209–214.

141. Sapsford, K.E.; Charles, P.T.; Patterson, C.H.; Ligler, F.S. Demonstration of four immunoassay formats using the array biosensor. Anal. Chem. 2002, 74, 1061–1068.

142. Pawlak, M., et al. Zeptosens' protein microarrays: A novel high performance microarray platform for low abundance protein analysis. Proteomics. 2002, 2, 383–393.

143. Schweitzer, B., et al. Multiplexed protein profiling on microarrays by rolling-circle amplification. Nature Biotechnol. 2002, 20, 359–365.

144. Ge, H. UPA, a universal protein array system for quantitative detection of protein–protein, protein–DNA, protein–RNA and protein–ligand interactions. Nucleic Acids Res. 2000, 28, E1–7.

145. Avseenko, N.V.; Morozova, T.Y.a.; Ataullakhanov, F.I.; Morozov, V.N. Immobilization of proteins in immunochemical microarrays fabricated by electrospray deposition. Anal. Chem. 2001, 73, 6047–6052.

146. Zhu, H., et al. Global analysis of protein activities using proteome chips. Science. 2001, 293, 2101–2105.

147. Avseenko, N.V.; Morozova, T.Y.; Ataullakhanov, F.I.; Morozov, V.N. Immunoassay with multicomponent protein microarrays fabricated by electrospray deposition. Anal. Chem. 2002, 74, 927–933.

148. Joos, T.E., et al. A microarray enzyme-linked immunosorbent assay for autoimmune diagnostics. Electrophoresis. 2000, 21, 2641–2650.

149. Robinson, W.H., et al. Autoantigen microarrays for multiplexed characterization of autoantibody responses. Nature Med. 2002, 8, 295–301.

150. Fang, Y.; Frutos, A.G.; Lahiri, J. Membrane protein arrays. J. Amer. Chem. Soc. 2002, 124, 2394–2395.

151. Espejo, A., et al. A protein-domain microarray identifies novel protein–protein interactions. Biochem. J. 2002, 367, 697–702.

152. Toepert, F., et al. Synthesis of an array comprising 837 variants of the hYAP WWW protein domain. Angew. Chem. Int. Ed. 2001, 40, 897–900.

153. Zhu, H., et al. Analysis of yeast protein kinases using protein chips. Nature Genet. 2000, 26, 283–289.

154. Curey, T.E., et al. Characterization of multicomponent monosaccharide solutions using an enzyme-based sensor array. Anal. Biochem, 293, 178–184.
155. Houseman, B.T.; Mrksich, M. Towards quantitative assays with peptide chips: A surface engineering approach. TIBTECH. 2002, 20, 279–281.
156. Houseman, B.T.; Huh, J.H.; Kron, S.J.; Mrksich, M. Peptide chips for the quantitative evaluation of protein kinase activity. Nature Biotechnol. 2002, 20, 232–233.
157. Houseman, B.T.; Mrksich, M. Carbohydrate arrays for the evaluation of protein binding and enzymatic modification. Chem. Biol. 2002, 9, 443–454.
158. Lee, J.; Bedford, M.T. PABP1 identified as an arginine methyltransferase substrate using high-density protein arrays. EMBO Rep. 2002, 3, 268–273.
159. Kononen, J., et al. Tissue microarrays for high-throughput molecular profiling of tumor specimens. Nature Med. 1998, 4, 844–847.
160. Rimm, D.L., et al. Tissue microarray: a new technology for amplification of tissue resources. Cancer J. 2001, 7, 24–31.
161. Hoos, A.- Tissue microarray profiling of cancer specimens and cell lines: opportunities and limitations. Lab. Invest. 2001, 81, 1331–1338.
162. Rimm, D.L., et al. Amplification of tissue by construction of tissue microarrays. Exp. Mol. Pathol. 2001, 70, 255–264.
163. Fejzo, M.S.; Slamon, D.J. Frozen tumour tissue microarray technology for analysis of tumour RNA, DNA and proteins. Am. J. Pathol. 2001, 159, 1645–1650.
164. Horvath, L.; Henshall, S. The application of tissue microarrays to cancer research. Pathology. 2001, 33, 125–129.
165. Wenschuh, H., et al. Coherent membrane supports for parallel microsynthesis and screening of bioactive peptides. Biopolymers. 2000, 55, 188–206.
166. Reineke, U.; Volkmer-Engert, R.; Schneider-Mergener, J. Applications of peptide arrays prepared by SPOT-technology. Curr. Opin. Biotechnol. 2001, 12, 59–64.
167. Reineke, U., et al. Identification of distinct epitopes and mimotopes from a peptide array of 5520 randomly generated sequences. J. Immunol. Methods. 2002, 267, 37–51.
168. Melnyk, O., et al. Peptide arrays for highly sensitive and specific antibody-binding fluorescene assays. Bioconjug. Chem. 2002, 13, 713–720.
169. Paweletz, C.P., et al. Reverse phase protein microarrays which capture disease progression show activation of pro-survival pathways at the cancer invasion front. Oncogene. 2001, 20, 1981–1989.
170. Srinivas, P.R.; Srivastava, S.; Hanash, S.; Wright, G.L. Proteomics in early detection of cancer. Clin. Chem. 2001, 47, 1901–1911.
171. Madoz-Gurpide, J., et al. Protein based microarrays: A tool for probing the proteome of cancer cells and tissues. Proteomics. 2001, 1, 1279–1287.
172. Ericsson, D., et al. Downsizing proteolytic digestion and analysis using dispenser-aided sample handling and nanovial matrix-assisted laser/desorption ionization-target arrays. Proteomics. 2001, 1, 1072–1081.
173. Ekstrom, S., et al. Signal amplification using "spot-on-a-chip" technology for the identification of proteins via MALDI-TOF MS. Anal. Chem. 2001, 73, 214–219.
174. Ziauddin, J.; Sabatini, D.M. Microarrays of cells expressing defined cDNAs. Nature. 2001, 411, 107–110.

175. Link, D., et al. A model system for studying postnatal myogenesis with tetracycline-responsive, genetically engineered clonal myoblasts in vitro and in vivo. Exp. Cell Res. 2001, 270, 138–150.
176. Bochner, B.R.; Gadzinski, P.; Panomitros, E. Phenotype microarrays for high-throughput phenotypic testing and assay of gene function. Genome Res. 2001, 11, 1246–1255.
177. Wu, R.Z.; Bailey, S.N.; Sabatini, D.M. Cell-biological applications of transfected-cell microarrays. Trends Cell Biol. 2002, 12, 485–488.
178. Austen, B.M.; Frears, E.R.; Davies, H. The use of SELDI proteinchip arrays to monitor production of Alzheimer's beta-amyloid in transfected cells. J. Peptide Sci. 2000, 9, 459–469.
179. Paweletz, C.P., et al. Proteomic patterns of nipple aspirate fluids obtained by SELDI-TOF: potential for new biomarkers to aid in the diagnosis of breast cancer. Dis. Markers. 2001, 17, 301–307.
180. Fung, E.T.; Thulasiraman, V.; Weinberger, S.R.; Dalmasso, E.A. Protein biochips for differential profiling. Curr. Opin. Biotechnol. 2001, 12, 65–69.
181. Adam, B.-.L.; Vlahou, A.; Semmes, O.J.; Wright, G.L. Proteomic approaches to biomarker discovery in prostate and bladder cancers. Proteomics. 2001, 1, 1264–1270.
182. Rosty, C., et al. Identification of hepatocarcinoma-intestine-pancreas/pancreatitis-associated protein I as a biomarker for pancreatic ductal adenocarcinoma by protein biochip technology. Cancer Res. 2002, 62, 1868–1875.
183. Li, J., et al. Proteomics and bioinformatics approaches for identification of serum biomarkers to detect breast cancer. Clin. Chem. 2002, 48, 1296–1304.
184. Wellmann, A., et al. Analysis of microdissected prostate tissue with ProteinChip arrays—A way to new insights into carcinogenesis and to diagnostic tools. Int. J. Mol. Med. 2002, 9, 341–347.
185. Petricoin, E.F., et al. Serum proteomic patterns for detection of prostate cancer. J. Natl. Cancer Inst. 2002, 94, 1576–1578.
186. Petricoin, E.F., et al. Use of proteomic patterns in serum to identify ovarian cancer. Lancet. 2002, 359, 572–577.
187. Zhang, L., et al. Contribution of human alpha-defensin 1, 2, and 3 to the anti-HIV-1 activity of CD8 antiviral factor. Science. 2002, 298, 995–1000.
188. Rai, A.J., et al. Proteomic approaches to tumor marker discovery. Arch. Pathol. Lab. Med. 2002, 126, 1518–1526.
189. Weinberger, S.R.; Dalmasso, E.A.; Fung, E.T. Current achievements using ProteinChip Array technology. Curr. Opin. Chem. Biol. 2002, 6, 86–91.
190. Winssinger, N.; Ficarro, S.; Schultz, P.G.; Harris, J.L. Profiling protein function with small molecule arrays. PNAS. 2002, 99, 11,139–11,144.
191. O'Brien, C. Protein fingerprints, proteome projects and implications for drug discovery. Mol. Med. Today. 1996, 2, 316.
192. Zacher, T.; Wischerhoff, E. Real-time two-wavelength surface plasmon resonance as a tool for the vertical resolution of binding processes in biosensing hydrogels. Langmuir. 2002, 18, 1748–1759.
193. Johnsson, B.; Lofas, S.; Lindquist, G. Immobilization of proteins to a carboxymethyldextran-modified gold surface for biospecific interaction analysis in surface plasmon resonance sensors. Anal. Biochem. 1991, 198, 268–277.

194. Yee, A., et al. An NMR approach to structural proteomics. PNAS. 2002, 99, 1825–1830.
195. Edwards, A.M., et al. Protein production: feeding the crystallographers and NMR spectroscopists. Nature Struct. Biol. 2000, 7, 970–972.
196. Davis, G.D.; Elisee, C.; Newham, D.M.; Harrison, R.G. New fusion protein systems designed to give soluble expression in Escherichia coli. Biotechnol. Bioeng. 1999, 65, 382–388.
197. Kapust, R.B.; Waugh, D.S. *Escherichia coli* maltose-binding protein is uncommonly effective at promoting the solubility of polypeptides to which it is fused. Protein Sci. 1999, 8, 1668–1674.
198. Larsson, M., et al. High-throughput protein expression of cDNA products as a tool in functional genomics. J. Biotechnol. 2000, 80, 143–157.
199. Hartley, J.L.; Temple, G.F.; Brasch, M.A. DNA cloning using *in vitro* site-specific recombination. Genome Res. 2000, 10, 1788–1795.
200. Walhout, A.J., et al. GATEWAY recombinational cloning: Application to the cloning of large numbers of open reading frames or ORFeomes. Methods Enzymol. 2000, 328, 575–592.
201. Albala, J.S., et al. From genes to proteins: High-throughput expression and purification of the human proteome. J. Cell. Biochem. 2000, 80, 187–191.
202. Harrington, J.J., et al. Creation of genome-wide protein expression libraries using random activation of gene exporession. Nature Biotechnol. 2001, 19, 440–445.
203. Gilbert, M.; Albala, J.S. Accelerating code to function: sizing up the protein production line. Curr. Opin. Chem. Biol. 2001, 6, 102–105.
204. Hammarstrom, M., et al. Rapid screening for improved solubility of small human proteins produced as fusion proteins in *Escherichia coli*. Protein Sci. 2002, 11, 313–321.
205. Braun, P., et al. Proteome-scale purification of human proteins from bacteria. PNAS. 2002, 99, 2654–2659.
206. Glaser, V. Strategies for target validation streamline evaluation of leads. Genet. Eng. News. 1997, 17(September 15), 1.
207. Marten, M.R., et al. A biochemical genomics approach for identifying genes by the activity of their products. Science. 1999, 286, 1153–1155.
208. Grayhack, E.J.; Phizicky, E.M. Genomic analysis of biochemical function. Curr. Opin. Chem. Biol, 5, 34–39.
209. Fields, S.; Song, O. A novel genetic system to detect protein–protein interactions. Nature. 1989, 340, 245–246.
210. Baneyx, F. Recombinant protein expression in *Escherichia coli*. Curr. Opin. Biotechnol. 1999, 10, 411–421.
211. Harrison, R.G. Expression of soluble heterologous proteins via fusion with NusA protein. Innovations. 1999, 11, 4–7.
212. Brazma, A., et al. Minimum information about a microarray experiment (MIAME)-toward standards for microarray data. Nature Genet. 2001, 29, 365–371.
213. Spellman, P.T., et al. Design and implementation of microarray gene expression markup language (MAGE-ML). Genome Biol. 2000, 3, Research 0046.1–0046.9.
214. Natsume, T., et al. Rapid analysis of protein interactions: On-chip micropurification of recombinant protein expressed in *Escherichia coli*. Proteomics. 2002, 2, 1247–1253.

215. Porstmann, T.; Kiessig, S.T. Enzyme immunoassay techniques. An overview. J. Immunol. Methods. 1992, 150, 5–21.
216. Roda, A., et al. Bio- and chemiluminescence in bioanalysis. Fresenius J. Anal. Chem. 2000, 366, 752–759.
217. Yguerabide, J.; Yguerabide, E.E. Resonance light scattering particles as ultrasensitive labels for detection of analytes in a wide range of applications. J. Cell. Biochem. 2001, 37(Suppl.), 71–81.
218. Bao, P., et al. High-sensitivity detection of DNA hybridization on microarrays using resonance light scattering. Anal. Chem. 2002, 74, 1792–1797.
219. Tokeshi, M., et al. Determination of suboctomole amounts of nonfluorescent molecules using a thermal lens microscope: subsingle-molecule determination. Anal. Chem. 2001, 73, 2112–2116.
220. Cooper, M.A. Optical biosensors in drug discovery. Nature Rev. Drug Discovery. 2002, 1, 515–528.
221. Schneider, B.H., et al. Highly sensitive optical chip immunoassays in human serum. Biosensor Bioelectronics. 2000, 15, 13–22.
222. Schneider, B.H., et al. Optical chip immunoassay for hCG in human whole blood. Biosensor Bioelectronics. 2000, 15, 597–604.
223. Frostell-Karlsson, A., et al. Biosensor analysis of the interaction between immobilized human serum albumin and drug compounds for prediction of human serum albumin binding levels. J. Med. Chem. 2000, 43, 1986–1992.
224. Karlsson, R., et al. Biosensor analysis of drug–target interactions: Direct and competitive binding assays for investigation of interactions between thrombin and thrombin inhibitors. Anal. Biochem. 2001, 278, 1–13.
225. Lawrence, C.R.; Geddes, N.J.; Furlong, D.N. Surface plasmon resonance studies of immunoreactions utilizing disposable diffraction gratings. Biosensor Bioelectronics. 1996, 11, 389–400.
226. Bernard, A.; Bosshard, H.R. Real-time monitoring of antigen–antibody recognition on a metal oxide surface by an optical grating coupler sensor. Eur. J. Biochem. 1995, 230, 416–423.
227. Nellen, P.M.; Tiefenthaler, K.; Lukosz, W. Integrated optical input grating couplers as biochemical sensors. Sensors Actuators. 1988, 15, 285–295.
228. Cunningham, B.; Li, P.; Pepper, J. Colorimetric resonant reflection as a direct biochemical assay technique. Sensors Actuators B. 2002, 81, 316–328.
229. Lin, B., et al. A label-free optical technique for detecting small molecule interactions. Biosensor Bioelectronics. 2002, 17, 827–834.
230. Nath, N.; Chilkoti, A. A Colorimetric Gold nanoparticle sensor to interrogate biomolecular interactions in real time on a surface. Anal. Chem. 2002, 74, 504–509.
231. Piehler, J.; Brecht, A.; Geckeler, K.E.; Gauglitz, G. Surface modification for direct immunoprobes. Biosensor Bioelectronics. 1996, 11, 579–590.
232. Alberl, F.; Köβlinger, C.; Drost, S.; Wolf, H. Quartz crystal microbalance for immunosensing. Fresenius J. Anal. Chem. 1994, 349, 340–345.
233. Schmidt, F.G.; Ziemann, F.; Sackmann, E. Shear field mapping in actin networks by using magnetic tweezers. Eur. Biophys. Biophysics Lett. 1996, 24, 348–353.
234. Helmerson, K.; Kishore, R.; Philips, W.D.; Weetall, H.H. Optical tweezers-based immunosensor detects femtomolar concentrations of antigens. Clin. Chem. 1997, 43, 379–383.

235.  K, Lee, et al. Protein nanoarrays generated by dip-pen nanolithography. Science. 2002, 295, 1702–1705.
236.  Ruan, C.; Yang, L.; Li, Y. Immunobiosensor chips for detection of *Escherichia coil* O157:H7 using electrochemical impedance spectroscopy. Anal. Chem. 2002, 74, 4818–4820.
237.  Fritz, J., et al. Translating biomolecular recognition into nanomechanics. Science. 2000, 288, 316–318.
238.  Grogan, C., et al. Characterisation of an antibody coated microcantilever as a potential immuno-based biosensor. Biosensor Bioelectronics. 2002, 17, 201–207.
239.  McKendry, R., et al. Multiple label-free biotection and quantitative DNA-binding assays on a nanomechanical cantilever array. PNAS. 2002, 99, 9783–9788.
240.  Göpel, W.; Heiduschka, P. Interface analysis in biosensor design. Biosensor Bioelectronics. 1995, 10, 853–883.
241.  Striebel, C.; Brecht, A.; Gauglitz, G. Characterization of biomembranes by spectral ellipsometry, surface plasmon renosance and interferomtery with regard to biosensor application. Biosensor Bioelectronics. 1994, 9, 139–146.
242.  Cush, R., et al. The resonant mirror; a novel optical biosensor for direct sensing of biomolecular interactions. Part I: Principle of operation and associated instrumentation. Biosensor Bioelectronics. 1993, 8, 347–353.
243.  Bender, W.J.H.; Dessy, R.E.; Miller, M.S.; Claus, R.O. Feasibility of a chemical microsensor based on surface plasmon resonance on fiber optics modified by multilayer vapor deposition. Anal. Chem. 1994, 66, 963–970.
244.  Gizeli, E.; Lowe, C.R.; Liley, M.; Vogel, H. Detection of supported lipid layers with the acoustic Love waveguide device: application to biosensors. Sensors Actuators B. 1996, 34, 295–300.
245.  Doyle, M. Characterization of binding interactions by isothermal titration microcalorimetry. Curr. Opin. Biotechnol. 1997, 8, 31–35.
246.  Dijksma, M.; Kamp, B.; Hoogvliet, J.C.; van Bennekom, W.P. Development of an electrochemical immunosensor for direct detection of interferon-$\gamma$ at the attomolar level. Anal. Chem. 2001, 73, 901–907.
247.  Tamaki, E., et al. Single-cell analysis by a scanning thermal lens microscope with a microchip: Direct monitoring of cytochrome c distribution during apoptosis process. Anal. Chem. 2002, 74, 1560–1564.
248.  Borrebaeck, C.A., et al. Protein chips based on recombinant antibody fragments: a highly sensitive approach as detected by mass spectrometry. Biotechniques. 2001, 30, 1126–1132.
249.  Sonksen, C.P., et al. Combining MALDI mass spectrometry and biomolecular interaction analysis using a biomolecular interaction analysis instrument. Anal. Chem. 1998, 70, 2731–2736.
250.  Kerr, M.K.; Churchill, G.A. Bootstrapping cluster analysis: assessing the reliability of conclusions from microarray experiments. PNAS. 2001, 98, 8961–8965.
251.  Kerr, M.K.; Churchill, G.A. Statistical design and the analysis of gene expression microarray data. Genet. Res. 2001, 77, 123–128.
252.  Jin, W., et al. The contributions of sex, genotype and age to transcriptional variance in *Drosophila melanogaster*. Nature Genet. 2001, 29, 389–395.

253. Krajewski, P.; Bocianowski, J. Statistical methods for microarray assays. J. Appl. Genet. 2002, 43, 269–278.

254. Eisen, M.B.; Brown, P.O. DNA arrays for analysis of gene expression. Methods Enzymol. 1999, 303, 179–205.

255. Hastie, T., et al. "Gene shaving" as a method for identifying distinct sets of genes with similar expression patterns. Genome Biol. 2000, 1, Research 0003.1–0003.21.

256. Perou, C.M., et al. Molecular portraits of human breast tumours. Nature. 2000, 406, 747–752.

257. Opiteck, G.J., et al. Comprehensive two-dimensional high-performance liquid chromatography for the isolation of overexpressed proteins and proteome mapping. Anal. Biochem. 1998, 258, 349–361.

258. Martinovic, S., et al. Selective incorporation of isotopically labeled amino acids for identification of intact proteins on a proteome-wide level. J. Mass Spectrom. 2002, 37, 99–107.

259. Aebersold, R.; Mann, M. Mass spectrometry-based proteomics. Nature. 2002, 422, 198–207.

260. Lee, H., et al. Development of a multiplexed microcapillary liquid chromatography system for high-throughput proteome analysis. Anal. Chem. 2002, 74, 4353–4360.

261. Zhou, H.; Ranish, J.A.; Watts, J.D.; Aebersold, R. Quantitative proteome analysis by solid-phase isotope tagging and mass spectrometry. Nature Biotechnol. 2002, 20, 512–515.

262. Smith, R.D., et al. Rapid quantitative measurements of proteomes by Fourier transform ion cyclotron resonance mass spectrometry. Electrophoresis. 2001, 22, 1652–1668.

263. Munchbach, M., et al. Quantitation and facilitated *de novo* sequencing of proteins by isotopic N-terminal labeling of peptides with a fragmentation-directing moiety. Anal. Chem. 2000, 72, 4047–4057.

264. Mirgorodskaya, O.A., et al. Quantitation of peptides and proteins by matrix-assisted laser desorption/ionization mass spectrometry using ($^{18}$)O-labeled internal standards. Rapid Commun. Mass Spectrom. 2000, 14, 1226–1232.

265. Conrads, T.P.; Issaq, H.J.; Veenstra, T.D. New tools for quantitative phosphoproteome analysis. Biochem. Biophys. Res. Commun. 2002, 290, 885–890.

266. Smith, R.D., et al. An accurate mass tag strategy for quantitative and high-throughput proteome measurements. Proteomics. 2002, 2, 513–523.

267. Ideker, T., et al. Integrated genomic and proteomic analyses of a systematically perturbed metabolic network. Science. 2001, 292, 929–934.

268. Shipp, M.A., et al. Diffuse large B-cell lymphoma outcome prediction by gene-expression profiling and supervised machine learning. Nature Med. 2002, 8, 68–74.

269. Ball, G., et al. An integrated approach utilizing artificial neural networks and SELDI mass spectrometry for the classification of human tumours and rapid identification of potential biomarkers. Bioinformatics. 2002, 18, 395–404.

# 2

# Ultrasensitive Microarray-Based Ligand Assay Technology

ROGER EKINS and FREDERICK CHU
*University College London Medical School*
*London, England*

## I. INTRODUCTION

Albeit ill-defined, the terms "microarray" and "biochip" are now widely used in biomedical science in connection with a ubiquitous miniaturized assay technology that permits, in principle, the simultaneous, ultrasensitive, assay of tens, hundreds, or even thousands of substances of biological interest (e.g., DNA fragments, hormones, drugs, etc.) in a small sample (e.g., a drop of blood). Such a "microarray" comprises an array of "microspots" located—in known or identifiable positions—on a solid support,[*] each microspot comprising a minute area of a specific binding agent (typically an oligonucleotide probe or antibody) that recognizes and binds molecules of an individual target analyte. Although normally present in solution in the sample to which the microarray is exposed, target analytes may also constitute surface components of, for example, cell membranes or other small insoluble cellular structures. Microarray technology is clearly of major importance in medical research and diagnosis, but it is also potentially of value in many other areas, including, for example, the food industry, environmental monitoring, agriculture, forensic investigation, military defense, and so forth.

---

[*] In a field providing such scope for litigation and rich feeding grounds for patent lawyers, great caution is required in the use of words. For example, the term "solid" in this context is intended to describe a material which serves to support a microspot in a fixed and identifiable position such that the binding agent within the spot is readily accessible to a fluid sample and target molecules in the liquid or gaseous phase. Such a material may be smooth and impermeable or porous (e.g., sintered glass or nylon film), albeit the use of porous materials—by imposing diffusion constraints on reaction velocities and obviously inappropriate if spot sizes are extremely small—may be disadvantageous.

Conceived of, and initially developed, by the authors in the early to mid-1980s, the technology has subsequently aroused explosive interest throughout the world. This interest was stimulated by initiatives taken by U.S. government agencies that recognized the technology's potential application and importance to genetic analysis, leading to the establishment, in 1992, of the U.S. Genosensor Project* and Consortium†. Reputedly one of the most heavily government-funded biomedical projects in U.S. history, the Genosensor Project's creation followed from, and was essentially complementary to, the international Human Genome Project, constituting a means of exploiting, scientifically and industrially, the results this was anticipated to yield.

The early emphasis placed in the United States on the technology's application to nucleic acid analysis—together with U.S. government agencies' financial encouragement of industrial investment in this area—have subsequently led many later entrants to the field to assume that the technology was an invention of U.S. biotech companies. In particular, its genesis is commonly represented as stemming from, and crucially dependent on, the emergence in the early 1990s of a technique relying on "combinatorial chemistry" developed (for entirely different purposes) for the *in situ* synthesis (on arrays) of polypeptides and (subsequently) polynucleotides [1]. So entrenched are these perceptions of the technology's origins that its possible application to other classes of analyte (e.g., proteins) has not infrequently been portrayed by reviewers (e.g., Ref. 2) as a recently emerging prospect modeled on the technology's initial application to DNA/RNA analysis.

This notion not only misrepresents the technology's genesis and history [3], but also—of greater importance to this chapter's objectives—obscures the true nature of the scientific concepts (often described as counterintuitive) that led to the technology's original development. In particular, the recognition that miniaturized microspot-based "ligand assays" could, in contradiction to universally accepted ideas in the field, be of greater sensitivity and require shorter incubation times than those conforming to conventional formats constituted the crucial findings that initiated the authors' establishment of a microspot array development project in 1986.‡ Moreover, it was apparent that the development of ultrasensitive ligand assays relying on the use of a "vanishingly small" amount of the binding agent located on a spot of such minute area as to be scarcely visible to the naked eye

---

* Established as part of the U.S. National Institute of Standards and Technology's Advanced Technology Program and providing funding in excess of $50 million from grants from the National Institute of Standards and Technology, the National Center for Human Genome Research, and the Department of the (U.S.) Air Force.

† Members of the consortium included Beckman Instruments, Genometrix Inc, Genosys Biotechnologies, MicroFab Technologies, Laboratories for Genetic Services, Triplex Pharmaceuticals, Houston Advanced Research Center, Baylor College of Medicine, and MIT.

‡ Generously funded by a grant from the Wolfson Foundation received in that year.

opened up the prospect of high-density arrays comprising, in principle, thousands or even millions, of microspots per square centimeter.

The technology's use for "massive parallel testing" has since captured popular imagination, largely because of its obvious relevance to DNA analysis. In consequence, the potentially greater sensitivity of microspot-based assay methods (as compared with conventional methods) has thus been largely disregarded. However, this feature is of crucial importance in the context of array-based protein assays (for use in the immunodiagnostics and proteomics fields) and of the assay of other analytes where "target analyte amplification" techniques [such as the polymerase chain reaction (PCR)] are inapplicable. In particular, the recognition that protein expression is poorly correlated with mRNA levels has focused attention on the need for an array-based protein assay technology comparable to—but of far greater sensitivity than—current nucleic acid array-based analysis methods [4]. This chapter is, for this reason, primarily intended to clarify the fundamental principles underlying the attainment of this objective and to identify the events and "counterintuitive" ideas that underlay the technology's emergence rather than to catalog recent technical advances in this fast-moving area.

It is therefore appropriate first to briefly summarize the history of ligand assay methodology and to examine the concepts (some erroneous) that have governed its past evolution.

## II.  A BRIEF HISTORY OF LIGAND ASSAY AND REVIEW OF ITS GOVERNING PRINCIPLES

### A.  Ligand Assay

Microarray methods constitute the most recent major development in the field of ligand assay, the microanalytical technique—originally developed in the late 1950s and early 1960s—primarily used to determine the concentrations of hormones and other substances of biological interest present at very low levels in body fluids [5,6]. Ligand assays comprise the class of microanalytical methods relying on observation of the binding reaction between the target substance and a specific "molecular-recognition" reagent, such observation being commonly facilitated by the use of a high-specific-activity label* (attached to the binding agent or analyte), thereby increasing sensitivity.

---

* That is, a molecular label (e.g., a radioisotope, enzyme, fluorophor, or chemiluminescent marker) such that labeled molecules yield large numbers of detectable signals (e.g., photons) either spontaneously or when appropriately stimulated. The term "specific activity"—generally used to describe the number of disintegrations per unit time per unit mass (or number of atoms or molecules) of a radioisotope or radiolabeled substance—is employed here in a wider sense to represent the number of observable events per unit time per unit mass or molecular number yielded by the label under the experimental conditions used. For example the specific activity of a fluorescent label (i.e., the photons emitted/unit time/unit amount) depends, *inter alia*, on the intensity and wavelength of the light to which it is exposed.

For many years, antibodies to target antigens raised in laboratory or farm animals constituted the most widely used binding agents (in assays generally known as "immunoassays"); however, other binding agents (e.g., specific binding proteins, cell receptors, enzymes, etc.) were also well known and occasionally employed in closely analogous methods (e.g., "protein-binding assays"). Traditional methods of antibody synthesis were supplemented in the mid–late 1970s by the techniques of monoclonal antibody production developed by Köhler and Milstein [7] and subsequently by the introduction of phage display methods (e.g., by Winter et al. [8], Gao et al. [9], and others). Although antibodies synthesized by these largely in vitro methods generally possess lower binding affinities than those produced by traditional immunization techniques, their relative purity render them especially useful in so-called "noncompetitive" labeled antibody methods (see Sec. II.C), in which the deleterious effects of low binding affinity on assay sensitivity are often, in practice, of lesser importance. However, in attempts to circumvent the labor-intensive and time-consuming procedures involved in large-scale antibody production, other specific protein-binding agents have been proposed, such as protein-binding oligonucleotides or "aptamers" [10–12] and even plastics in which the molecular shapes of proteins are cast [13]. Some of these are presently under active industrial development [by companies such as SomaLogic (www.somalogic.com) [14]] albeit their ultimate place in the ligand assay methodological armamentarium is still unclear.

Meanwhile, assays involving binding reactions between complementary polynucleotides (also, by definition, ligand assays) have likewise been recognized for more than 20 years as closely comparable in principle and practice to immunoassays (see, e.g., Ref. 14), such methods attracting increasing attention in the past decade with the nearing completion of the Human Genome Project. Not unexpectedly, the molecular structures of the binding sites involved in the binding of complementary single-stranded polynucleotides (like those between other types of binding partners) differ in certain detailed respects from those of the sites involved in antibody–antigen binding [15]. Nevertheless the physicochemical laws governing the (reversible) binding reactions between complementary polynucleotides and between antibodies and antigens are essentially identical, as are many of the statistical and mathematical concepts that underlie the design and performance of assays relying on observation of the reactions between them. In short, many analytical concepts are common to both these particular examples of ligand assay, as well as to analogous assays relying on other types of binding pair.

Ligand assays have made a major impact on biomedical research and diagnostic medicine in the past 40 years, largely because of their simplicity and "exquisite sensitivity" [5]. The latter attribute enabled for the first time the assay of many substances of biological importance (e.g., hormones, vitamins, viruses, etc.) present in body fluids at concentrations much below the reach of previous analytical methods. Pioneers in the field thus focused particular attention on the attainment of high sensitivity, constructing, *inter alia*, theoretical models intended (generally with this specific objective primarily in mind) to provide guidance on the selection of binding reagents possessing appropriate physicochemical attributes and the optimal concentrations at which they should be used (see Refs. 16 and 17). Moreover, the quest for ever-higher sensitivity has long constituted a major factor driving these methods' ongoing development.

Paradoxically, these efforts have been seriously impeded by a still unresolved lack of agreement (in this and many other areas of science) regarding the definition of "sensitivity" and the concept this term represents. The effects of this phenomenon have been especially evident in the ligand assay field, generating uncertainty and heated controversy regarding optimal assay design [18,19]. In particular, it contributed to the incredulity often expressed by experienced practitioners in the field when the possibility of microarray technology was first proposed and demonstrated. Moreover, it is evident from recent publications that similar conflicts may again arise in regard to microarray design. It is therefore imperative, in an examination of the technology's underlying concepts, that this contentious issue should be clarified, notwithstanding the fact that it has been discussed at some length in recent publications [20–23], and therefore needs only relatively brief examination here.

## B. The Concept of "Sensitivity"

Notwithstanding the long-established meaning of "sensitivity" in the English language as indicative of the ability of an organism or instrument to determine, detect, or sense a small stimulus or quantity, differences among scientists regarding the determinants of a measuring instrument's sensitivity have long existed. Many prominent international and national organizations, including, for example, the International Union of Pure and Applied Chemistry (IUPAC) [24,25] and the American Chemical Society [26], have formally defined sensitivity in terms of the response–stimulus ratio (or, equivalently, the slope of the dose–response curve) yielded by a measuring instrument or assay. More importantly, this concept has, in practice, determined the approach to ligand assay design adopted by the majority of workers in this field.

For example, Berson and Yalow [in their many theoretical publications relating to immunoassay design (e.g., Refs. 16, 18, and 27)] repeatedly defined sensitivity as the slope of the curve relating the response variable [arbitrarily identified in their earlier publications as the ratio of antibody-bound to free labeled analyte ($B/F$); and later, as the fraction of labeled analyte bound ($b$)] to the unlabeled analyte concentration ($[H]$). Fundamental to Berson and Yalow's theoretical analyses was the assumption that the response-curve slope at zero dose [i.e., $(db/d[H])_0$ in their later publications] constitutes the *sole* determinant of a ligand assay's ability to determine low analyte concentrations, implying, according to these authors [27], that maximizing $(db/d[H])_0$ *necessarily* minimizes a ligand assay's lower limit of detection. Thus, in accordance with this view, the more "sensitive" of two ligand assays is that yielding the greater value of $(db/d[H])_0$.

In contrast, Ekins et al. (e.g., Refs. 17, 19, 20, 28, and 29) explicitly defined an assay's sensitivity as the (im)precision (i.e., standard deviation) of measurement of an analyte concentration of zero; that is, as $\sigma_{H_0} = \sigma_{R_0}/(dR/d[H])_0$, where $R$ is the response variable (however expressed), and $\sigma_{R_0}$ is the standard deviation of the response $R_0$ and $(dR/d[H])_0$ = the response-curve slope, both at zero dose. $\sigma_{H_0}$ constitutes the key determinant of the lower limit of detection of any measuring system and, thus, of an assay's ability to determine or detect low analyte amounts or concentrations. According to this definition, the more sensitive of two assays is that yielding the lower value of $\sigma_{H_0}$ {i.e., $\sigma_{R_0}/(dR/d[H])_0$}. In short, the more sensitive an assay, the smaller the amount it will detect. Thus, improvement of an assay's sensitivity as thus defined implies that steps are taken to decrease the quotient $\sigma_{R_0}/(dR/d[H])_0$.

The quantity $\sigma_{R_0}$ is often referred to as the "noise" generated within a measuring system, its magnitude in an assay determining the minimum response (i.e., "signal") that is, with reasonable confidence, attributable to the presence of the target analyte. Moreover, physicists have long been familiar with the proposition that the "signal–noise ratio" (which can be shown to be essentially equivalent to the detection limit[*]) is a measure of an instrument's ability to detect small amounts of that which it is designed to measure. As a corollary of this definition, sensitivity is expressed in units representing the measured quantity, enabling, *inter alia*, the sensitivities of two systems measuring the same quantity, but differing in their modes of operation, to be compared. In contrast, the slope definition expresses sensitivity in units possessing the physical dimensions of the "response/ dose" quotient, implying that the relative sensitivities of systems differing in the nature and/or dimensions of the dose and response variables cannot be assessed.

Clearly, the key distinction between these two concepts is that the latter is based on the proposition that an instrument's ability to determine a small amount of (or small difference in) the measured quantity depends on *two* factors: the response-curve slope and the random error incurred in the measurement of the response, both of which are likely to vary with the amount of the measured quantity. Moreover, it presupposes that any change in assay design is likely to alter either or both factors. For example, a reduction in the amount of binding

---

[*] The signal–noise ratio (i.e., $R/\sigma_{R_0}$, determines a measuring instrument's ability to detect a small amount of "that which it is designed to measure" ($H$), (where the signal generated by $H$ is $R$). The detection limit is given (approximately) by $H\sigma_{R_0}/R$. Hence, maximizing $R/\sigma_{R_0}$ (keeping $H$ constant) minimizes the detection limit.

agent in a ligand assay system reduces the amount of analyte bound and hence—when the fraction bound at zero dose falls below 33% (see next paragraph)—reduces the slope $\{(db/d[H])_0\}$ of the $b$ versus $H$ response curve. However, the standard deviation $(\sigma_{b_0})$ of the measurement of the response at zero dose $(b_0)$ is also likely to diminish in these circumstances, implying that the overall effect on the lower limit of detection (and hence the system's ability to determine small amounts) is unknown unless $\sigma_{b_0}$ and $(db/d[H])_0$ are both determined. In short, this definition of sensitivity requires statistical analysis of assay data [to estimate, either directly or indirectly, the magnitude of random errors in the measurement of the selected response variable at zero dose[*] $(\sigma_{R_0})$]. Moreover, an increase in the sensitivity of a system may be achieved either by increasing the response–dose ratio $(dR/d[H])_0$ without changing $\sigma_{R_0}$, or reducing $\sigma_{R_0}$ without alteration of $(dR/d[H])_0$ (or, of course, a combination of both).

The "response–stimulus ratio", or "response-curve slope," concept is nevertheless easier for nonmathematicians both to understand and to quantify (requiring only a simple calculation or visual inspection). For this and other reasons, most workers in the ligand assay field (encouraged by the IUPAC and other such bodies) have, in practice, been guided by observation solely of the response-curve slope when developing new ligand assay systems and selecting optimal reagent concentrations for use therein, disregarding concomitant effects on the magnitude of errors incurred in the measurement of the selected response variable. Moreover, theoretical conclusions relating to ligand assay design deriving from this concept have long been widely accepted—for example, the precept, originally enunciated by Berson and Yalow, that a conventional radioimmunoassay system is most sensitive when the amount of antibody used is such that 33% of a trace amount of the analyte is bound following incubation.[†] This (specious) proposition stems directly from the observation that when the results of such an assay are plotted graphically in terms of the fraction of labeled analyte bound against the target analyte concentration, the slope of the resulting dose–response curve at zero dose is, in such circumstances, maximal.

---

[*] It is becoming increasingly common for practitioners also to estimate the value of $\sigma R$ at all points along the response curve, the value at any point indicating the imprecision of the corresponding dose measurement.

[†] This proposition applies specifically to "competitive" assay systems (see below). This situation arises when the concentration of binding agent in the system is given by $0.5/K$, where $K$ is the (apparent, or "effective") affinity constant (L/mol) governing the reaction between binding agent and target analyte as measured (e.g., by Scatchard analysis;–see footnote on page 96) under the conditions used in the assay. In these circumstances, it is readily demonstrable that the slope (at zero dose) of the response curve relating the antibody-bound analyte fraction to analyte concentration $(db/d[H])$ is maximal. In the case of "noncompetitive" assay, it has been commonly believed that the amount of (capture) binding agent should be such as to bind all, or most, of the target analyte in the test sample (see, e.g., Ref. 30).

Nevertheless, notwithstanding its widespread acceptance and promulgation by national and international bodies, the ratio (or slope) definition is demonstrably untenable and, indeed, essentially meaningless. For example, which of two ligand assays is perceived as the more "sensitive" depends on the dose and response variables chosen for the plotting of response curves [20]. Thus, two assayists analyzing the same experimental data are as likely as not to reach opposite conclusions regarding the relative sensitivities of two systems, depending on how each plots the data (see Fig. 1). Such absurd contradictions (which, in the author's experience, are not uncommon) are obviated if a measuring system's sensitivity is explicitly defined as, and represented by, its lower limit of detection, the value of which is independent of the choice of assay dose and response variables and hence the coordinate frame in which the dose–response curve is plotted.

The general acceptance of the "slope" definition, despite its obvious contradictions, has, as a consequence, led to the adoption of, and widespread adherence to, principles governing ligand assay design that are largely specious and little more than myths. Of greatest importance in the context of this chapter is the delusion that an increase in the signal yielded by an assay system in response to a given amount of the target analyte *necessarily* increases its ability to determine smaller analyte amounts (i.e., to be of greater sensitivity).

This fallacy underlays the initial disbelief of the "counterintuitive" proposition that a microspot containing amounts of binding agent orders of magnitude

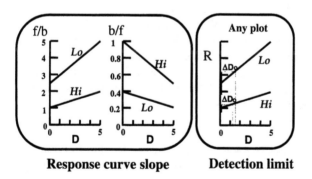

**Figure 1** Effect of the use of low and high antibody concentrations on the slopes of radioimmunoassay response curves plotted in terms of the free to bound (*f/b*) and bound to free (*b/f*) labeled antigen ratios (left). The antibody concentration yielding the assay judged as the more "sensitive" (as defined by IUPAC and others) depends on the choice of response variable. However, the calculated detection limits for the two assays are unaffected by this choice. In the circumstances hypothesized in this figure (right), the assay relying on the high antibody concentration detects lower analyte concentrations and is thus the more sensitive (authors' definition).

less than those traditionally regarded as necessary and binding only a very small proportion of the target analyte represented an assay strategy potentially yielding greater sensitivity* than that adopted in conventional ligand assay designs. The authors' demonstration in the mid-1980s of this revolutionary proposition's truth and of the corollary that miniaturized arrays comprising thousands of such microspots—each directed against a different analyte—could, in principle, be constructed represented the fundamental conceptual breakthrough that underlay their experimental studies at this time and that ultimately led to the emergence of microarray technology.

## C.  "Competitive" and "Noncompetitive" Ligand Assays

Binding of target analyte to a specific binding agent may, of course, be observed "directly" without the use of a label; for example, early immunoassays relied on simple visual observation of a precipitate of the antigen–antibody complex. However, the introduction in the late 1950s of the use in such assays of high-specific-activity reagent labels enabled observation of the binding reactions between far smaller numbers of molecules than had hitherto been possible, thus greatly increasing sensitivity. For many years, radioisotopes constituted the most popular label employed in this context, giving rise—in the case of immunoassays—to the terms "radioimmunoassay" (RIA) and "immunoradiometric assay" (IRMA).

The former refers to immunoassays involving the addition of exogenous radiolabeled analyte or analyte analog (usually in a known and standard amount), together with a small amount of specific antibody, to both test samples and calibrants prior to their incubation for a convenient (albeit generally long†) period, during which the binding reaction proceeds. Assuming the antibody's possession of appropriate physicochemical characteristics and its addition to the assay system in an appropriately "small" amount, the greater the target analyte concentration in the test sample, the less labeled analyte is bound to the antibody following a period of incubation. The distribution of labeled antigen between antibody-bound and free moieties (generally following their physical separation) thus provides a measure of the amount of unlabeled target analyte present in the sample. Immunoassays and other ligand assay systems conforming to this approach are therefore

---

* The term "sensitivity" of an assay system will henceforth here refer to the system's ability to detect small analyte amounts or concentrations, as represented by the assay detection limit.

† Commonly (in the past) in the order of 12–24 h. Incubation periods of this order are generally necessary in this class of assay because of the low concentrations of analyte and binding reagent, implying slow binding kinetics. For the highest sensitivity, experienced workers in the field have occasionally allowed reactions to proceed for several days.

often described as "competitive," unlabeled and labeled analyte molecules being perceived as "competing" for a limited number of analyte-binding sites. (Note, however, that objections to this terminology have occasionally been raised on the grounds that, in the case of certain assays of this genre, labeled and unlabeled analyte molecules are essentially chemically identical, labeled molecules being regarded as "tracers" of the unlabeled analyte's behavior in its reactions with the binding agent. In certain circumstances, this view is especially justified; see Sec. II.D.)

The term "immunoradiometric" assay (IRMA) was later coined to distinguish immunoassays—originally developed in the late 1960s [31–34]—also relying on radioisotopes as labels, the label being attached, in such methods, to the antibody. In this approach, labeled antibody binds to target analyte molecules following its addition to the test sample; the greater the amount of analyte present, the greater the amount of labeled antibody being analyte bound following incubation. For this reason, IRMAs were frequently described as "noncompetitive" (but see next paragraph). Considerable doubts and controversy nevertheless centered on the early assertion that labeled antibody methods would yield higher sensitivity [32]; this claim, which lacked any mathematical basis or persuasive experimental evidence in its support, was contested on theoretical grounds by Rodbard and Weiss [35] and others.

Debate on this issue was complicated by the prevailing confusion, referred to earlier, relating to the concept of sensitivity and its assessment. Moreover, certain workers in the field (see, e.g., Ref. 36) developed labeled antibody methods that they also described as "competitive", on the grounds that, in such methods, the signal generated by labeled antibody bound to an immunosorbent added to the reaction mixture is determined, the immunosorbent being regarded as "competing" with the target analyte for labeled antibody-binding sites. Nevertheless, these authors evidently failed to perceive (and hence did not clarify) the key distinction between so-called competitive and noncompetitive labeled antibody methods based on the use of a single labeled antibody (*all* such methods typically relying on an immunosorbent to separate analyte bound and unbound labeled antibody fractions).

It is therefore unsurprising that in the late 1970s and early 1980s, uncertainty and controversy centered on which of these two basic forms of ligand assay offered the prospect of improved analytical performance (i.e., greater sensitivity, precision, and specificity) and shorter incubation times. A key factor contributing to confusion on this issue was the comparison, by protagonists involved in the debate, of RIAs and IRMAs. However, the classification of ligand assays on the basis of which component of the system is labeled deflects attention from the fundamental feature that determines assay performance, particularly sensitivity. Nevertheless, for want of better terminology, the terms "competitive" and "noncompetitive" are retained here to describe two forms of assay whose performance characteristics basically differ, albeit—as indicated in Section II.D—these terms, when so used, require more rigorous definition than they have hitherto been accorded.

## D. The "Binding-Agent 'Fractional Occupancy'" Principle of Ligand Assay

The principles governing ligand assays may be portrayed in a variety of ways exemplified, as indicated in Section II.C, by the common representation of RIAs as based on competition between isotopically labeled analyte and unlabeled analyte for antibody-binding sites introduced into the assay system. However, an alternative and more useful portrayal is in terms of the "fractional occupancy" principle.

All ligand assays rely on observation of the fraction of binding sites (of a "structurally specific" binding agent) occupied by analyte following the binding agent's exposure to an analyte-containing medium. The value of this fraction is, in general, determined by several factors: the sample volume, the amount of binding agent, its effective binding affinity vis-à-vis the analyte, and the analyte concentration (including any labeled analyte added to the sample). If the first four of these parameters are held constant in both test samples and calibrants, binding-site fractional occupancy varies only with, and hence reflects, the analyte concentration present in individual samples.

Two approaches are available for the measurement of binding-site "fractional occupancy" by (unlabeled) analyte, these relying on the measurement of occupied or unoccupied sites, respectively. The classification of ligand assays in this manner conforms broadly to the descriptions noncompetitive and competitive (Fig. 2). However, in view of past confusion on this issue, it must be emphasized that not all ligand assays relying on the use of labeled binding agents (e.g., IRMAs) fall into the first category and can, for this reason, be classified as noncompetitive, nor likewise can all assays based on the use of exogenous labeled analyte or analyte analog (e.g., RIAs) be classified as competitive.

Figure 3 summarizes the principal traditional approaches employed in the immunoassay and protein-binding assay fields. As indicated earlier, single-site labeled antibody-based immunoassays may be described as either noncompetitive or competitive depending on whether occupied sites on the labeled antibody are determined directly or indirectly, respectively (i.e., by observation, in the latter case, of unoccupied sites). So-called two-site or "sandwich" labeled antibody assays (as generally performed) can be viewed as relying on the direct measurement of the occupancy of either a capture antibody located on a solid support or of the second, labeled antibody, reactive with the analyte used in the system. Thus, irrespective of how such a system is portrayed, it can be categorized as noncompetitive.

Competitive and noncompetitive
ligand assay designs

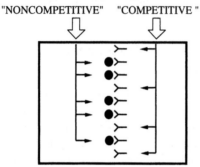

"NONCOMPETITIVE"      "COMPETITIVE "

*Measure occupied sites    Measure unoccupied sites*

**Figure 2**   Ligand assays rely on the determination of the fraction of binding sites of a specific capture binding agent (or "receptor") occupied by the target analyte (•). This may be effected by measurement of either the unoccupied ("competitive" assay) or occupied sites ("noncompetitive assay").

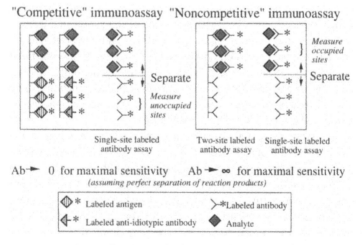

"Competitive" immunoassay "Noncompetitive" immunoassay

**Figure 3**   Principal competitive and noncompetitive immunoassay designs. Note that single-site labeled antibody assays can be classified as "competitive" or "noncompetitive" depending on the fraction of labeled antibody that is measured following their separation (using, e.g., an immunosorbent).

Note that all methods relying on the use of a labeled analyte (or labeled anti-idiotypic antibody) shown in Fig. 3 can be classified as competitive; the greater the occupancy of binding agent by analyte, the lower the number of unoccupied sites, and hence the lower the amount of labeled analyte ultimately bound (whether the labeled analyte is added following, simultaneously with, or even before, the addition of unlabeled analyte to the binding agent). Note also that Fig. 3 represents strategies that are equally applicable to the assay of polynucleotides, using, for example, oligonucleotides as specific binding agents. However a different strategy is permissible and generally employed in the latter context, [i.e., the inclusion in the system of a labeled polynucleotide produced by polymerase chain reaction (PCR) methods from the initially unlabeled target nucleic acid]. In this circumstance, the amount of labeled analyte present in the system is essentially proportional to the amount of unlabeled analyte initially present, labeled molecules acting as a true tracer (or "indicator") of the extent of binding of unlabeled material. In these circumstances, the greater the amount of analyte bound, the greater the labeled analyte bound; thus such a system can be classified as noncompetitive.

In summary, noncompetitive methods (as here defined) constitute the class of ligand assays in which the observed signal is generated from binding sites occupied by the target analyte. Conversely, competitive assays are those in which the signal is generated from sites *not* occupied by the target analyte, the signal emanating either directly from unoccupied sites or by a labeled material reactive with them.

This portrayal of the principles underlying the ligand assay enables the approach yielding the greater sensitivity to be readily identified without detailed theoretical analysis. In short, the controversy that centered on this issue in the 1970s and early 1980s could have been readily resolved simply by visualization of the assay sensitivity problem in terms of a simple analogy; that is, which of the two possible methods of measuring a length permits the accurate determination of the shortest length (see Fig. 4)? Clearly, the direct measurement of the distance AB (the "noncompetitive" approach to the measurement of length) is likely to be more precise than a determination of the difference of measurements of AC and BC (the "competitive" approach), both of which measurements are likely to be subject to greater random errors. Hence, the error in the direct measurement of AB (the "noncompetitive" approach to the measurement of length) is likely to be more precise than a determination of the difference of measurements of AC and BC (the "competitive" approach), both of which measurements are likely to be subject to greater random errors. In other words, the error in the direct measurement of AB ($\sigma_{AB}$) is likely to be less than the error in the indirect measurement ($\sqrt{[(\sigma_{AC})^2 + (\sigma_{BC})^2]}$). Thus, the noncompetitive approach permits shorter lengths to be determined and can, therefore, by analogy, be said to be the more "sensitive."

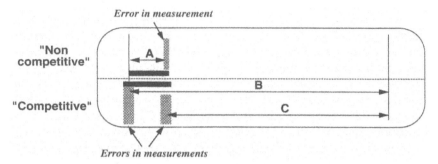

**Figure 4** "Competitive" and "noncompetitive" measurement strategies can be used to determine a length. The "noncompetitive" approach generally yields smaller errors in the determination of short lengths and is therefore the more "sensitive."

Moreover, formal theoretical analysis reveals that the noncompetitive approach can yield assay sensitivities orders of magnitude higher than competitive designs [37] provided (1) that nonspecific binding of the labeled binder is reduced to very low levels (ideally < 0.01%) and (2) that the label used is a nonisotopic label of much higher specific activity than that of the radioisotopes whose use had previously dominated the ligand assay field. These conclusions led in the late 1970s to one of the present authors' collaborative development with Wallac Oy, a Finnish instrument company, of a new class of fast "ultrasensitive" ligand assays relying on the use of fluorescent rare-earth chelate labels and their measurement using time-resolution techniques [38]. These became the model for a number of analogous ultrasensitive techniques subsequently developed by other manufacturers (based on similar principles, but using other high-specific-activity nonisotopic labels) that emerged in the mid–late 1980s and which subsequently totally transformed the immunodiagnostic field

In the light of these broad principles, we may now address the novel concepts on which microspot and microarray-based ligand assay methods depend.

## III. AMBIENT ANALYTE LIGAND ASSAY

### A. Basic Principle

The term "ambient analyte assay" was coined by one of the present authors [39] to represent a previously unrecognized physicochemical principle enabling the creation of volume-independent ligand assays that *directly* determine the analyte concentration to which a specific binding agent (coupled to a solid support) is exposed (Fig. 5) (i.e. not, as was previously customary, by measuring the *amount* of the target analyte *in a known volume of the test sample*). This principle (which,

"Ambient analyte" ligand assay

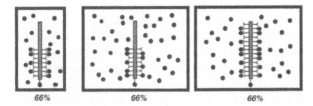

**Figure 5** The ambient analyte assay principle. The fractional occupancy of binding sites by analyte is given (approximately) by [An]/([An] + 1/$K$) [for all analyte concentrations ([An])] when the binding site concentration is < 0.1/$K$, preferably < 0.01/$K$. In this figure, [An] and F are assumed to be such that [An]/([An] + 1/$K$) = 0.66.

*inter alia*, also requires the use of a small amount of binding agent sequestering only an "insignificant" fraction of the target analyte in test samples) was originally exploited primarily to enable the determination of analyte concentrations in circumstances in which the measurement of sample volumes is impossible, difficult, or potentially hazardous (e.g., for the direct in vivo determination of analyte concentrations in body fluids such as saliva). However, although intended for a different purpose, it ultimately led to the authors' development of microspot and microarray-based assay methods when it emerged that ligand assays based on this principle could—if appropriately designed—yield higher sensitivities in shorter times than conventional methods.

Nevertheless, it must be emphasized that an ambient analyte assay system *per se* is not necessarily sensitive. To achieve sensitivities comparable to or higher than those of conventional ligand assay methodologies, the binding agent must be coated at high surface density (preferably as a molecular monolayer) within a minute spot situated on a solid support. Such a design distinguishes a "microspot" assay from assays of conventional design relying on the binding agent's attachment, in relatively large amount, to a correspondingly large area of a solid support (i.e., as a "macroarea" or "macrospot"), binding a relatively high fraction of the target analyte in accordance with previously accepted assay design principles.

Arrays of "macrospots" may be described as "macroarrays." Arrays conforming to this description (generally comprising a very limited number of assays) have long been known (see, e.g., Ref. 40), albeit they offer few, if any, advantages in terms of sensitivity and speed and have, therefore, evoked little interest in the past. For these reasons, the authors' view is that the term "microarray" should be restricted to arrays in which the individual assays in the array predominantly conform to ambient analyte assay principles.[*]

Certain investigators have recently resurrected the use of array-based assays of conventional design, referring to them—perhaps to capitalize on the interest generated by the "microarray revolution"—as "mass-sensing" microarrays [41]. The term "mass sensing" is presumably intended to convey the notion that individual binding areas within the array each capture a high proportion of the analyte against which they are directed (in accordance with conventional concepts), thereby distinguishing this approach from that underlying the "concentration-sensing" microarrays described in this chapter. Insofar as the readoption of this approach is based on a scientific objective (rather than on patent considerations [42]), it apparently reflects the traditional belief that such a strategy, by increasing the magnitude of the observed signal, increases sensitivity. Although the use of a large amount of binding agent within the spots in an array may be unavoidable when low-specific-activity labels are employed, the overall effect is a reduction in assay performance (i.e., sensitivity, specificity, and speed) as compared with that characterizing concentration-sensing arrays. Moreover, such an approach also effectively prohibits the construction of high-density microarrays.

As indicated earlier, all ligand assays depend on the determination of the fractional occupancy by the analyte of specific analyte-binding sites characterizing the binding agent. In general, the fractional occupancy ($F$) of such sites is dependent on the total analyte and binding-site concentrations present in the system, and the (effective[†]) affinity constant ($K$) governing the binding reaction, $F$ being given by the following equation (derived from the mass action laws):

$$F^2[S] - F(\tfrac{1}{K} + [An] + [S]) + [An] = 0 \tag{1}$$

or

$$F^2[S] - \frac{F}{K} - F[An] - F[S] + [An] = 0 \tag{2}$$

where [S] is the binding-site concentration (mol/L), [An] is the analyte concentration (mol/L), and $F < 1$. However, if the binding site concentration [S] $<<$ 1/$K$, it follows that, for all values of [An],

---

[*] Insofar as the detailed mode of operation of commercial microarray methods has been disclosed, this is invariably the case.

[†] It has been common in theoretical publications in the ligand assay field to express reagent concentrations in terms of the 1/$K$, where $K$ is the affinity constant as measured (e.g., by Scatchard analysis; see footnote on page 87) under the incubation conditions (and following the same incubation period) as those under which the assay is carried out. Note that the apparent affinity constant as determined in this manner increases with time, final equilibrium only being reached after infinite time. In practice, binding reactions are thus invariably terminated before the attainment of final equilibrium.

$$[An] - \frac{F}{K} - F[An] = 0 \tag{3}$$

or

$$F = \frac{[An]}{[An] + 1/K} \text{ for all values of } [An]. \tag{4}$$

In short, when the total binding-site concentration ([S]) approximates $0.1/K$ or (preferably) less, the fractional occupancy of binding-sites is essentially independent of the binding site concentration in the system, being solely dependent on the (original) analyte concentration ([An]) and the effective affinity constant governing the binding reaction. This phenomenon defines an ambient analyte assay. For example, if [An] $= 1/K$, then $F$ 0.5, (i.e., half the binding sites are occupied, irrespective of the number of sites present). In these circumstances, if the total number of sites in the system is $10^5$, $5 \times 10^4$ sites will be occupied; if the total number is $10^4$, $5 \times 10^3$ sites will be occupied; and if 10 sites are present, 5 will be occupied.

These conclusions are portrayed in Fig. 6 in which the value of $F$ corresponding to various analyte concentrations [calculated from Eq. (1)] is plotted against binding site concentration. Figure 6 illustrates the proposition that when a binding site concentration of less than $\sim 0.01/K$ –$0.1/K$ (the sites being preferably located on a solid support*) is exposed to an analyte-containing medium, the resulting (fractional) binding-site occupancy solely reflects the *initial* analyte concentration in the medium to which the sites are exposed and is independent both of the total number of binding sites and of sample volume. Analyte binding to binding sites inevitably causes some depletion of unbound analyte in the medium, but because the amount so bound is relatively small, the reduction in the ambient analyte concentration is insignificant. For example, if the binding-site concentration is less than $0.01/K$, the reduction in the ambient analyte concentration is invariably less than 1% (regardless of the analyte concentration), and the system is thus sample volume independent.

## B. Microspot Assay

As indicated earlier, the ambient analyte assay principle leads to a further important concept; that is, that the "vanishingly small" amount of binding agent used in an ambient analyte assay system may be confined at high surface density within a minute "microspot" located on a solid support, the total number of effective

---

* The binding site "concentration" (when the binding agent is located on a solid support) is given by the number of effective sites (i.e., those not impeded from reaction with analyte in the solution to which the binding agent is exposed) divided by the solution volume. Thus, if a system operates under ambient analyte assay conditions, a further increase in the volume of the test solution will diminish the binding-site concentration, but the fractional occupancy of these sites will remain essentially unchanged.

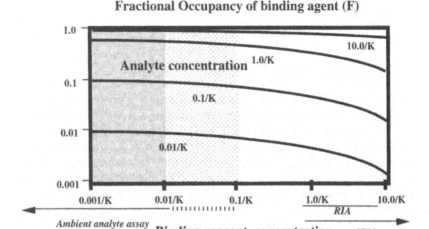

**Figure 6**  Curves showing $F$ as a function of the concentration of binding agent and target analyte. Binding reagent concentrations of $< 0.1/K$, ideally $< 0.01/K$ and less, alter ambient analyte concentrations to an insignificant extent.

binding sites within the spot being less than $v/K \times 10^{-5} \times N$ [where $v$ is the sample volume to which the microspot is exposed (mL) and $N$ is Avogadro's number ($6 \times 10^{23}$)]. (For example, if $v = 1$ and $K = 10^{11}$ L/mol, the maximum number of binding sites causing negligible disturbance ($<1\%$) to the ambient analyte concentration is $6 \times 10^7$, this number being greater if the binding agent is of lower effective binding affinity.) Assuming binding-site surface densities (when closely packed in the form of a monolayer) on the order of $10^4$–$10^5$ sites/$\mu m^2$, microspot areas on the order of 600–6000 $\mu m^2$ (i.e., spots on the order of 100 $\mu m$ in diameter or less) will accommodate numbers of sites that conform to the ambient analyte assay principle when exposed to sample volumes on the order of 1 mL[*]. Even if such microspots are exposed to sample volumes of 100 $\mu$L,

---

[*] These figures represent only an approximate guide to the construction of a microspot that operates under ambient analyte assay conditions. For example, a reduction in the "effective" equilibrium constant occurs if the binding reaction is terminated before equilibrium is reached (see footnote on page 96). Thus, the effective equilibrium constant depends, *inter alia*, on the shape of the reaction chamber and its effect on the diffusion of analyte molecules to the microspot, the sample size and extent of sample mixing, the incubation time, and so forth. Likewise, the number of effective binding sites depends on the extent to which potential binding sites located within the spot area are sterically prevented from binding to analyte molecules. In practice, whether or not a microspot operates in conformity with ambient analyte assay requirements can only be determined, in practice, by observing the change (if any) in ambient analyte concentrations at the termination of a typical assay.

the resulting reduction in the ambient analyte concentration is still likely to be insignificant.

A useful comparison can be made (see Fig. 7) between the introduction of a small amount of binding agent located on a microspot into an analyte-containing medium and that of a small, cold thermometer into a large heat-containing body (e.g., a volume of warm water). The thermometer, essentially a heat-measuring device, absorbs heat from its surroundings and ultimately reaches quasiequilibrium with, and closely reflects, the water's original temperature provided the thermometer is of relatively low thermal capacity and that the quantity of heat it absorbs is negligible compared with the water's total heat content. Such a thermometer constitutes an ambient temperature-measuring device, the principal *lower* limit on its size being that it should be large enough to be readable. Conversely, the introduction of a *large* thermometer into a relatively small amount of water implies the thermometer's absorption of a significant proportion of the water's heat content, causing its temperature to fall. In these circumstances, the *measured* water temperature differs from the water's original temperature and depends on the relative thermal capacities of the thermometer and the water in which it has been placed. This implies that both the thermometer's size and the volume of water to which it is exposed must be "standardized" and held constant if water temperature measurements are to be meaningful. This strategy is closely analogous to that which has long been used in conventional ligand assays.

Following exposure to an analyte-containing medium, the occupancy of binding sites can, in principle be determined in various ways, of which the simplest (Fig. 8) is by exposure of the microspot to a solution of a labeled binding agent that recognizes occupied (the noncompetitive approach) or unoccupied sites (competitive approach).[*]

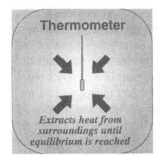 

**Figure 7** A thermometer and microspot are closely analogous. Note that a thermometer should be of such size that it does not disturb the ambient temperature and it should be readable.

---

[*] To clarify these concepts these steps are portrayed as sequential, they may be combined as is commonly done in conventional ligand assay protocols.

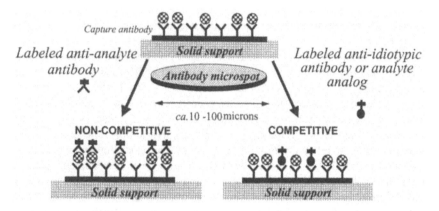

**Figure 8** The occupancy of capture binding sites located within a microspot may be determined using competitive and noncompetitive strategies.

Furthermore, the perception that, under ambient analyte assay conditions, the ratio of occupied (or unoccupied) to total binding sites is solely dependent on the ambient analyte concentration leads to the concept of a dual-label, "ratiometric," microspot assay (see Fig. 9), a simple (albeit not the only nor an obligatory) approach to the determination of binding-site fractional occupancy. This approach involves the labeling of the binding (or "capture) agent located within the microspot (or an inert "tracer" material added thereto) as well as (labeled with a second, distinguishable label) the second binding agent. The ratio of the signals emitted by the two labels reveals the capture binding agents fractional occupancy (Fig. 10). Among several advantages, this strategy implies that the influence on assay results of variations in the surface density or number of binding sites within the microspot area (such as caused by variations in spotting volume, or attachment of sites to the solid support) is minimized. In short, it obviates some of the problems that may arise in the course of microarray production. Alternatively, labeling of the binding (or "capture) agent provides the basis of a quality control system enabling the presence and quality of each individual microspot in a microarray to be monitored in the course of manufacture. Nevertheless, the use of this approach is not mandatory and can be disregarded if microspots can be reliably constructed of such constant shape, size, and binding agent content that assay results can be guaranteed to be of acceptable precision.

## C. Microspot Assay Sensitivity

The proposition that miniaturized microspot-based ligand assays not only provide the basis of multianalyte testing but are potentially more sensitive and rapid than conventional systems challenged long accepted ideas in the 1980s and constituted

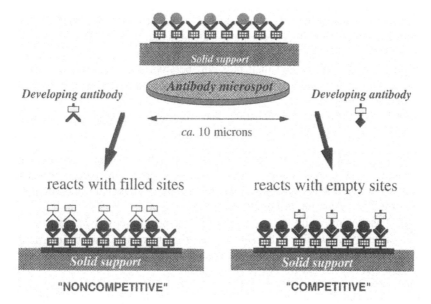

**Figure 9**  Capture binding site occupancy may be determined by labelling the capture reagent (e.g., antibody) with a label distinguishable from that used to label the ''developing'' antibody and determining the ratio of the signals emitted by the two labels.

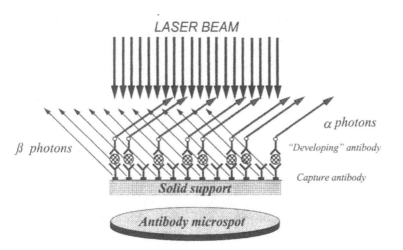

**Figure 10**  Dual fluorescent labels are especially convenient in the context of the ratiometric approach portrayed in Fig. 9.

the principal, arguably the only, barrier to their earlier development. However, the attainment of high sensitivity necessitates an insight into the principles underlying microarray methods, and even at present few reports exist (aside from those published by the authors and their past collaborators (e.g., Ref. 43)) claiming equal or increased sensitivity as compared with that yielded by ligand assays of conventional design.

The basis of the above proposition and the practical achievement of high microspot sensitivity rely on two principal concepts:

1. That the signal–noise ratio observed in an assay based on the use of a binding agent located within a defined area of a solid support as a microspot increases to a limiting maximum value as the area of the spot decreases but declines as the area of the microspot closely approaches zero
2. That as the area of the spot is decreased, the specific activity of labels used in the assay and visible to the signal measuring instrument must be increased provided the volume of the analyte-containing solution remains constant, see page 104

The specific activities of labels commonly used in the ligand assay field are shown in Table 1. Note that the specific activities of fluorescent labels are orders of magnitude greater than those of radioisotopic labels and are, therefore, a potential choice for use in microarray-based assay technologies. However, many materials (such as the glass used in microscopes, etc.) also fluoresce, causing high backgrounds, and a consequent loss of sensitivity unless means are found to circumvent this problem (such as the use of time-resolving fluorescence measurement methods).

The first of these concepts is illustrated in Fig. 11. This shows the fall in the signal–background ratio (expressed as a percentage of the ratio when the "binder-coated" area is zero) that accompanies an increase in the area of solid support over which the binding agent is distributed at a constant and uniform surface density (an increase in area implying a corresponding increase in binding-agent concentration in the system). This curve is based on the simplifying assumption that the instrument background is zero and, thus, that the background signal

**Table 1** Specific Activities of Some Commonly Used Ligand Assay Reagent Labels

| | |
|---|---|
| $^{125}I$ | 1 detectable event/s/$7.5 \times 10^6$ labeled molecules |
| Enzyme labels | Dependent on enzyme "amplification factor" and detectability of reaction product |
| Chemiluminescent labels | Total of one detectable event/labeled molecule (or less) |
| Fluorescent labels | Many detectable events/sec/labeled molecule |

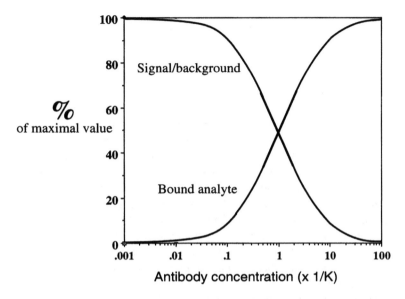

**Figure 11**   The signal–background ratio falls with increase in the amount of antibody located (as a monomolecular layer) on a spot, albeit the total amount of analyte bound increases.

observed in the system derives entirely from, and is proportional to, the area on which the binding agent is located. Also plotted is the fraction (%) of the total analyte present that is bound to the solid binding agent at equilibrium (assuming the total analyte concentration is small; i.e., $< 0.01/K$). Clearly, as the coated area is increased, both the percentages of the total analyte present "captured" by the binding agent and the signal increase, yet the signal–background ratio decreases. For example, if the spot is of such large area that $\sim$ 100% of target analyte molecules are captured thereon, further increase in spot size solely increases background noise, with no increase in the analyte-generated signal. Assuming that the statistical variation in the background (i.e., the background "noise") is (approximately) proportional to the background (i.e., that $\sigma_{R_0}/R_0$ is approximately constant), the signal–noise ratio (and hence sensitivity) are greatest when the coated area tends toward zero.

Obviously, were the interrogated area (i.e., the microspot) actually reduced to zero, both the specific signal and background noise would likewise both fall to zero, implying that, in the limit, no signal of any kind would be recorded and the system would thus be totally insensitive.* In practice, certain statistical consid-

---

\* Just as a thermometer of zero dimensions would be unable to measure temperature.

erations come into play prohibiting reduction of spot size (and hence of the binding-agent concentration of in the system) beyond a certain limit without loss in sensitivity.

One such factor is the ability to observe the specific signal generated from the spot, this depending on the specific activity of the labeled binding agent used. In short, the greater the label's specific activity, the smaller the spot size that is permissible without such loss of signal that the precision of the signal measurement deteriorates, ultimately resulting in an unacceptable reduction in sensitivity.

However, even if the label were of infinite specific activity, a second consideration comes into play as the spot size is reduced (i.e., the statistical variation in the number of target analyte molecules captured within the microspot area). As the spot size is reduced, increasing statistical variations in this number will eventually cause an unacceptable loss in precision and sensitivity.

Nevertheless, it is evident that, given very high-specific-activity labels, circumstances can be envisioned in which, even in a ''noncompetitive'' system, the optimal size of the spot and, hence, concentration of the capture binding agent may be extremely low

These concepts are further illustrated in Fig. 12. This represents areas of differing diameter, *each assumed to be coated with a binding agent at the same surface density*. When exposed to equal volumes of an analyte-containing solution, the resulting capture binding site concentrations will differ as shown. Meanwhile, the fractional occupancy of binding sites on the surface increases as the microspot area decreases, reaching a plateau when the binding-site concentration falls below $0.01/K$, the analyte surface density reaching a maximum in this circumstance. Assuming adoption of a noncompetitive strategy (i.e., observation of the signal generated by occupied sites), the signal–background ratio will be greatest when the antibody coated area is very small and the antibody concentration is below $0.01/K$. Further reduction in spot size reduces the signal, but does not further improve the ''visibility'' of the microspot against the background.

A more detailed theoretical consideration [44][*] of (noncompetitive) microspot ligand assay (e.g., immunometric assay) sensitivity suggests that

$$[An]_{min} = S^*_{min} \times (6 \times 10^{20}) \frac{(1+K^*[S^*])}{S\,KK^*[S^*]} \tag{5}$$

---

[*] This early analysis disregards errors arising in the case of extremely small spots where the statistical variation in the number of molecules captured on the spot may—using very high specific activity labels—become the principal contributor to the overall variation in the signal measurement and hence to a loss in sensitivity. In other words, the statistical errors incurred in the measurement of the signal *per se* (e.g., photons emitted by a fluorophor) are assumed in the analysis to constitute the dominant source of error, implying that statistical errors deriving from other sources can be disregarded. Bearing in mind that methods based on microspots in the order of 1 μm in diameter are currently under development, this assumption can no longer be considered to be invariably valid.

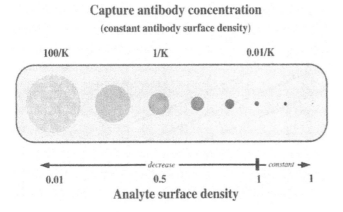

**Figure 12** The "visibility" of the signal emitted from a microspot increases as the spot size decreases (see Fig. 11).

where $S$ is the surface density of the capture binding agent (binding sites/$\mu m^2$), $K$ is the capture agent affinity constant (l/mol), [S*] is the labeled analyte binding-agent concentration (mol/L), $K^*$ is the affinity constant of a labeled analyte-binding agent (l/mol), $S_{min}^*$ is the minimum detectable surface density of labeled analyte-binding agent (molecules/$\mu m^2$), and $[An]_{min}$ is the assay detection limit (molecules/mL). For example, if $K^*[S^*] = 1$, $S = 10^5$ binding sites/$\mu m^2$, $K = 10^{11}$ L/mol, and $S_{min}^* = 20$ molecules/$\mu m^2$, then $[An]_{min} = 2.4 \times 10^6$ molecules/ mL $= 4 \times 10^{-15}$ mol/L, the fractional occupancy of sensor antibody-binding sites by the minimum detectable analyte concentration being 0.04%. Note in particular that $S_{min}^*$ constitutes an important determinant of microspot assay sensitivity. Figure 13 shows theoretical sensitivities attainable using sensor-binding agents of varying affinities, plotted as a function of $S_{min}^*$.

In summary, these considerations confirm, *inter alia*, that the attainment of high microspot assay sensitivity requires the use of very high-specific-activity labels and an instrument capable of accurately measuring low surface densities of labeled binding agents. Close packing of capture binding agent molecules within the microspot area, by maximizing the analyte-generated signal emitted per unit area minimizes the relative effects of background signals generated from the support, contributes to the fulfillment of this requirement. Thus, the packing density of binder molecules is an important parameter. (Note, however, that the surface density of the capture binding agent is of lesser importance if its occupancy is determined by numerical counting of occupied sites rather than by measurement of the integrated signal generated from the interrogated microspot segment.) They also suggest (1) that microspot immunoassay sensitivities higher

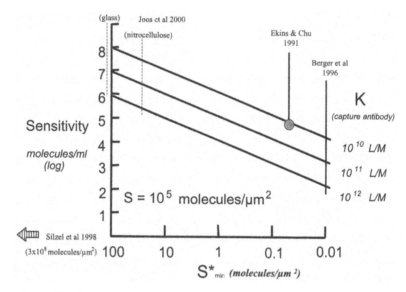

**Figure 13** Theoretically predicted assay sensitivities plotted as a function of the minimum detectable surface density ($S^*_{min}$) of the labeled binding agent. Note the values reported by the authors and Boehringer Mannheim's microspot research team as compared with more recent reports by others.

than those obtainable, for example, by conventional isotopically based immunoassays are potentially achievable and (2) that—assuming the use of high-specific-activity nonisotopic labels—sensitivities yielded by microspot assays are unlikely to be inferior, and (depending on the instrumentation used) may be considerably superior, to the sensitivities achievable with assays of conventional design.

It should perhaps be noted in this context that surface densities (i.e., values of $S^*_{min}$) of labeled analyte-binding antibodies in the order of 0.1 molecules/$\mu m^2$ or less were detected in experimental studies in the authors' laboratory in the late 1980s. Later improvements in instrumentation, and so forth by the authors' industrial collaborators (Boehringer Mannheim GmbH) in the mid-1990s reduced this value to 0.01 molecules/$\mu m^2$ using spots ~5000 $\mu m^2$ in area (see Section IV.B), implying the detection of ~50 analyte molecules on the spot. These observations indicated that detection limits in the order of $10^{-17}$ mol/L (i.e., $10^3$–$10^4$ molecules/mL) were achievable using the microspot approach (see Fig. 13), albeit the attainment of such sensitivities would have required longer incubation times than those (<15 min) being used at the time (1996) at which the Boehringer Mannheim team reported their observations.[*]

## D. Microspot Assay Kinetics

The theoretical considerations summarized earlier are largely based on the assumption that measurements are carried out following sufficient time for thermodynamic equilibrium to have been essentially established in the assay system. However, termination of the binding reaction significantly before the attainment of equilibrium has long been known to result in a reduction of the apparent affinity constant (i.e., the affinity constant as determined by Scatchard analysis[†]) and hence to some loss of assay sensitivity. Nevertheless the demand for rapid results has resulted in early termination becoming the rule rather than the exception. The conclusions drawn earlier may, albeit with some loss in theoretical rigor, be extrapolated to this situation (in accordance with normal practice) by relying on the value of the capture-agent affinity constant as determined by conventional methods at the time at which the assay is terminated.

Meanwhile, it is readily demonstrable that a microspot format (conforming to ambient analyte conditions) is, also counterintuitively, capable of yielding assays that are more rapid than many conventional methodologies. It should be noted in this context that it has long been recognized, in conformity with the mass action laws, that the greater the concentration of the binding agent in a homogeneous liquid-phase assay system, the greater the velocity of the binding reaction, and the sooner equilibrium is reached.

However, a different picture emerges if results are expressed in terms of the increase in the fractional occupancy of the binding agent with time. In short, fractional occupancy is at all times greater, the lower the binding agent concentration (Fig. 14). Thus, if—provisionally ignoring the diffusion constraints that reduce reaction velocities in assay systems in which the binding agent is linked to a solid support—we consider the kinetics of two microspot-based assays each relying on spots on which binding agent is coated at the same surface density, one microspot being 100-fold greater in diameter (i.e., 10,000-fold greater in area) than the other, the *analyte* surface density (and hence the signal per unit area) will at all times be greater in the case of the smaller spot. Thus, the signal–background ratio is always higher in the latter case despite the larger spot's more rapid attainment of equilibrium. In other words, contrary to general expecta-

---

[*] Reported at the 1996 Oak Ridge Conference by Dr. Hans Berger, leader of the Boehringer Mannheim Microspot® project.

[†] This involves the determination of bound and free analyte fractions and the plotting of the bound–free ratio as a function of the bound analyte concentration. The slope of the resulting curve yields a measure of the "apparent" or "effective" affinity constant governing the reaction.

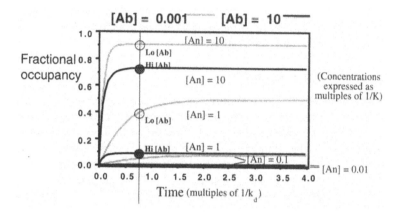

**Figure 14**  Kinetics of binding reactions between binding reagent (e.g., antibody) and target analyte. Reagent concentrations expressed in terms of the reciprocal of the affinity constant and time in terms of the reciprocal of the dissociation constant. Note that, at all times, the fractional occupancy of binding sites of the binding reagent when at a lower concentration is greater.

tions and assuming, as earlier, that background noise is proportional to the background itself, a higher sensitivity is reached in a shorter time when the lower amount of antibody located on the smaller spot is used.

Moreover, as indicated earlier, this simplified analysis disregards the diffusion constraints on the rate of the binding reaction if binding agent molecules are linked to a solid support. These constraints on the migration rates of analyte molecules to and from the solid support reduce both the effective association and dissociation rates of the reaction prior to the attainment of equilibrium, albeit (assuming that linking the capture agent to the support neither alters its structure nor affects its microenvironment and physicochemical properties) the final equilibrium state is unaffected. These constraints normally depend, *inter alia*, on sample volume and the area on which the binding agent is coated; thus, following any defined incubation time, the apparent or effective affinity constant may be considerably reduced as compared with that observed were the binding agent distributed homogeneously in solution. Nevertheless, in the limiting case of a microspot containing only a single binding-agent molecule, the kinetics of the binding reaction with analyte are essentially identical to those that would be observed were the molecule to be moving freely. In other words, the smaller the microspot area, the closer will the kinetics of the binding reaction approximate to those that would obtain were the reagents mixed in a homogeneous single-phase system.

This conclusion is supported by a more detailed consideration of the rate at which analyte molecules migrate and bind to a binding-agent microspot and reveals that the (initial) occupancy rate (OR) per unit microspot area is given by [45]

$$OR = 4r_m k_a D[An]S/(\pi r_m^2 k_a S + 4Dr_m) \text{ molecules/second/cm}^2 \tag{6}$$

where $D$ is the the the analyte diffusion coefficient ($cm^2/s$), $r_m$ is the microspot radius, $k_a$ is the association rate constant ($cm^3$/molecule/s), $S$ is the binding agent surface density (molecules/$cm^2$), and [An] is the ambient analyte concentration (molecules/$cm^3$). This expression reveals that as $r_m$ tends to zero, the term $\pi r_m^2 k_a S$ becomes negligible compared with $4Dr_m$ and the occupancy rate approximates $k_a[An]S$ molecules/s/$cm^2$. In other words, the kinetics of the reaction increase with reduction in $r_m$, ultimately approximating those observed in solution.

The conclusions deriving from this analysis may perhaps be more readily appreciated by again invoking the thermometer analogy referred to earlier. In short, the smaller the thermometer inserted into a volume of water, the less it affects, and the more rapidly it measures, the water's temperature.

Computer studies in the authors' laboratory reveal the sequence of events following the introduction of binding-agent microspots of varying diameters into an analyte-containing solution and embrace the kinetics of the antibody-analyte reaction per se, the establishment of concentration gradients within the solution, analyte diffusion, and so forth. These have confirmed that the smaller the microspot area, the lower the diffusion constraints on the rate of analyte binding to the antibody and the more closely the kinetics of the reaction approximate those seen in a homogeneous liquid-phase system (see Figs. 15 and 16).

Such studies, although instructive, might be said to be of limited practical value because they assume an unstirred sample. Moreover, a number of factors (such as the extent and influence of mixing on layers close to the solid support surface and hence on reaction kinetics) are difficult to model. Nevertheless, the studies predict that higher signal–noise ratios are likely to be attained in a shorter time using a microspot format, implying that microspot assays are likely to prove at least as rapid as assays of conventional macroscopic design.

## E.  Microspot Assay Specificity

The specificity problems caused in ligand assays by the presence in test samples of structurally different analytes that nonetheless react—albeit generally with lesser affinity than the target analyte—with the binding agent against which the latter is directed are broadly similar in the case of nucleic acid and antibody-based ligand assays. Nevertheless, they differ insofar as reactions between complementary polynucleotide sequences involve the entire sequence, whereas reactions between an antigen and an antibody generally involve only small zones of the

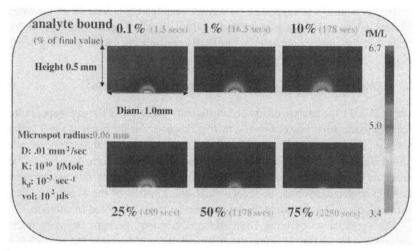

**Figure 15** Computer studies by Edwards embracing analyte diffusion and binding reaction rates showing the analyte concentration gradients established immediately following exposure of a binding-agent microspot to an analyte-containing medium.

two molecules [i.e., the antibody binding site (or ''paratope'') and a corresponding ''epitope'' within the antigen's three-dimensional molecular structure].

Of particular relevance in a publication relating to proteomics are the effects of cross-reactive antigens in microspot assays. If the epitope is identical in both target and cross-reactive antigens, both will react with identical potency if the

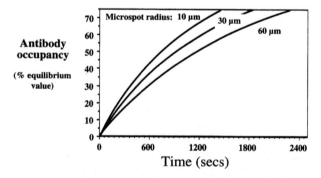

**Figure 16** Studies of the kind illustrated in Fig. 15 reveal that, at all times, reactions using microspots of smaller size reach equilibrium more rapidly.

binding reactions are permitted to proceed to near or quasiequilibrium. However, they may display differing potencies if the reactions are terminated earlier, essentially because differences in their overall structure may affect their diffusion characteristics and, hence, their apparent or effective affinity constants. A difference in the affinity constants of target and cross-reacting antigens is likewise likely to occur if the epitopes in target and cross-reacting antigens to which the antibody binds are of different (albeit similar) structure.

The complex quantitative effects of differences in the binding affinity in target and cross-reactive antigens were addressed by one of the present authors many years ago (and confirmed by others; see, e.g., Ref. 46), albeit many practitioners in the ligand assay field have, perhaps due to ignorance, adopted a simpler and more pragmatic (albeit incorrect) view of this issue.

Theoretical analysis revealed that, in a ligand assay, the target ligand reacts with the binding agent with a potency (RP) relative to a cross-reactive ligand given by

$$RP = (f \times K_T/K_C) + b$$
in a competitive (labeled analyte analog) assay [17]

$$RP = K_T (K_C [\text{fAb}] + 1]) / K_C (K_T [\text{fAb}] + 1)$$
in a noncompetitive (labeled antibody) assay    (7)

where $f$ and $b$ represent the free and bound labeled analyte fractions; [fAb] is the unoccupied capture antibody binding site concentration, and $K_T$ and $K_C$ are antibody affinity constants vis-à-vis the target analyte and cross-reactant, respectively.

The key conclusion deriving from these considerations is that the relative potency of a cross-reacting antigen is not constant (as assumed in many representations of cross-reaction effects), but that cross-reacting and target ligands are equipotent when $f = 0$ in a competitive (labeled analyte analog) assay, and when $[\text{fAb}] >> 1/K_T$ in a noncompetitive (labeled antibody) assay.

Conversely, the relative potency of the target antigen vis-à-vis a cross-reactive antigen tends to $K_T/K_C$ as $f$ tends to 1 (i.e., as $b$ tends to 0) in a competitive assay, and as the capture antibody concentration (and hence [fAb]) tend to 0.

It is, of course, of paramount importance, when designing an antibody microarray (and analogously any other type of microarray), to attempt to identify antibodies that are specific to epitopes exclusively possessed by target antigens. However, the existence of such epitopes cannot be guaranteed and the presence of cross-reactive antigens must always be suspected. Fortunately, the microspot approach reduces the effects of cross-reactants by its reliance on the use of capture binding site concentrations that are much lower than have been conventional. Moreover, it is in principal possible to include in a microarray microspots directed against suspected cross-reactants and to correct for their effects.

In general, sandwich (i.e., two-site) assays are of greater specificity than single-site assays and have for this reason been invariably employed for the assay of proteins (i.e., for analytes of relatively large molecular size) by the authors and (later by Boehringer Mannheim) in experimental studies. Note, however, that high concentrations of cross-reactants may cause negative bias in such assay formats by competing with target analytes for capture antibody binding sites.

## IV.  MICROARRAYS

The primary aim of the preceding sections has been to clarify the basic principles governing ligand assays and their design and the supposedly counterintuitive concepts that revealed the possibility of developing microspot-based assays enabling the simultaneous and sensitive assay of thousands of substances in a small sample. Even though purpose-designed instruments (such as microarrayers and laser-based scanners) were not commercially available in the mid-1980s[*], many experimental studies—using manual methods and rudimentary equipment—were slowly and laboriously performed by the authors during this period to confirm the feasibility of microarray-based assay technology and the validity of the concepts on which it was based. However, following an agreement with Boehringer Mannheim in 1991 and this company's subsequent production of a prototype array–construction facility and prototype analyzers, the rate of progress substantially increased (the principal role of the authors subsequently being confined to the exploration of ancillary ideas, not of performing assays in their entirety). Although little useful practical purpose would be served by describing these early studies in detail, a brief account may be instructive in illustrating the experimental strategies adopted during this period, many of which underlie methodologies now used by manufacturers in this field.

### A.  Early Experimental Developments in the Authors' Laboratory

As indicated earlier, the proposition that miniaturized "ambient analyte" ligand assays using vanishingly small amounts of a binding agent located on a microspot could prove at least as sensitive as conventional ligand assay systems contradicted long accepted ideas in the field, this perception (rather than difficulties in depositing microspots of binding agents on a solid support or in constructing arrays of such microspots) constituting the principal barrier to the development of microar-

---

[*] Fortuitously, the author became aware of the development (at the MRC Molecular Biology Laboratory at Cambridge, UK) of a prototype confocal microscope, which was generously made available to the author and his colleagues for occasional array-scanning purposes.

ray-based assay methods. The authors' original experimental efforts were therefore directed toward verifying this proposition.

In practice, the construction of an array of microspots on a solid support did not constitute a major difficulty and was readily, albeit arduously, achieved using simple manual methods. In general, initial studies centered on the construction and use of small antibody arrays, because the requisite reagents were freely available to the authors and assay results could be readily confirmed using conventional methods. Moreover, it was recognized that the measurement of multiple antigens using antibody arrays required particularly high sensitivity, and the demonstration that this objective was achievable represented the basic prerequisite to the commercial development of a truly ubiquitous microarray-based ligand assay technology.

Spotting of antibodies onto solid supports was effected using a variety of devices such as very fine stainless-steel rods (ranging in diameter from 100 to 300 $\mu$m available from Laboratory Systems Ltd, Southampton, UK), Rotring drawing pens (the smallest being nominally 180 $\mu$m in diameter), fine glass capillaries, and so forth. Spotting procedures generally depended on very brief ($\sim$1 s) exposure of solid supports to minute droplets of high concentration antibody-containing solutions followed rapidly by conventional washing and protein-blocking steps prior to storage of the resulting array. Clearly, such laborious manual methods of microarray production would not represent the basis of a viable technology, although it was anticipated that mechanical arrayers would ultimately be developed based on similar spotting techniques, as indeed proved to be the case.

Solid supports used in preliminary studies were essentially selected on the basis on the theoretical considerations discussed above, the basic requirements being that they should display the capacity to retain a high surface density of the binder combined with low intrinsic signal-generating properties (e.g., low-intrinsic fluorescence), thus minimizing background "noise." A wide variety of support materials were examined with these criteria in mind, including polypropylene, Teflon, cellulose and nitrocellulose membranes, and microtiter plates (clear polystyrene plates from Nunc; black, white and clear polystyrene plates from Dynatech). Glass slides and quartz optical fibers (diameter-40 $\mu$m) coated with a solution of 3-(amino propyl) triethoxy silane and treated with glutaraldehyde, (the antibody being attached thereto via a layer of protein A covalently linked to the glutaraldehyde) were also tested. White Dynatech Microfluor microtiter plates formulated specially for the detection of low-fluorescence signals proved to yield high signal-to-noise ratios and were initially used, albeit they were subsequently replaced by black Dynatech microtiter-well polystyrene strips (to which antibodies could be attached using well-known adsorption methods).

Using such methods, surface densities in the order of $5 \times 10^4$ IgG molecules/$\mu$m$^2$ or greater were achieved, a high proportion of the antibodies retaining immunological activity. Lower coating densities were observed using the protein A method, albeit the proportion of functionally active antibody molecules was somewhat higher.

Although a few preliminary experiments carried during this period on the construction of oligonucleotide microspots revealed that, using simple adsorption methods, only low surface densities of functionally active molecules could be obtained, earlier publications (e.g., Refs. 47 and 48) had described other techniques for the creation of "solid-phase" nucleic acid assays (closely analogous to "solid-phase" immunoassays) using a variety of solid supports. It was thus evident that array-based nucleic acid microarrays could also be readily developed at a later stage (as proved to be the case).

In recognition of the theoretical prediction that, in order to obtain the high sensitivities required for protein assays, reagent labels would need to possess specific activities very much higher than radioisotopes and other conventional labels, initial studies in the authors' laboratory relied on the use of (conventional) fluorophors (essentially fluorescein and Texas Red or similar). Microspots deposited on the base of sample containers were scanned using a prototype laser-scanning confocal microscope then under development by White et al. [49] at the MRC Molecular Biology Laboratory at Cambridge* (the first such instrument to become commercially available and kindly occasionally made available to the authors). Nevertheless, initial expectations of this approach were low because the measurement of fluorescence typically results in high-fluorescence backgrounds deriving from the measuring instrument itself and other sources. It was therefore fully anticipated at this stage that a requirement would ultimately arise to construct a time-resolving, laser-scanning, confocal microscope (using rare-earth cryptates as fluorescent labels) to achieve the highest sensitivity.

Nevertheless, by a combination of minor improvements to the commercial version of the Cambridge MRC confocal microscope that was ultimately purchased† (when this became available), significantly increased signal–noise ratios were achieved, resulting in the acceptable although not outstanding, sensitivities reported at the 1991 Oak Ridge Conference at St Louis [50]. Meanwhile, experiments using commercially available fluorescent microspheres as high-specific-activity reagent labels were proving encouraging, such microspheres acting as powerful signal amplifiers. Provided the tendency of certain microspheres of this type to stick nonspecifically to solid supports could be overcome, these were anticipated to further enhance sensitivity. These experiments ultimately proved successful, and by mid-1991, microspot assay sensitivities were consistently achieved for proteins (e.g., thyroid-stimulating hormone) considerably higher than

---

* Subsequently manufactured by Lasersharp Ltd as the MRC 500.
† From a grant generously provided by the Wolfson Foundation in 1986.

obtained using the best, conventionally formatted, immunoassay methodologies (see, e.g., Ref. 43). Indeed, on the basis of preliminary experiments using these techniques, the authors perceived that highly sensitive assays using microspots as low as 50 $\mu m^2$ in area could be developed, and thus, as first reported in 1989 [51], that high-density arrays comprising up to 2 $\times$ $10^6$ microspots/cm$^2$ were feasible.*

In summary, these preliminary studies (albeit restricted by the limited financial resources available to a small academic laboratory and by the use of relatively rudimentary methods and equipment) nevertheless demonstrated the technology's feasibility. Moreover, assay sensitivities achieved by 1991 were not only superior to those yielded by conventional assay methodologies but have not since been matched by others working in the field.

## B. Subsequent Developments in Collaboration with Boehringer Mannheim

An approach by Boehringer Mannheim GmbH in 1991 led (following the company's verification of the authors' experimental results) to an agreement to develop microarray-based methods for the assay of both proteins (and other antigenic analytes) and nucleic acids. The ensuing collaborative studies were inevitably dominated by the much larger research team and greater financial resources devoted to the project by Boehringer Mannheim. These studies nevertheless largely conformed (with certain small technical differences) to the above-described ideas and experimental techniques, being primarily devoted to placing them on an industrial footing.

Despite the unexpected termination of the project (because of a "strategic change in priorities" [52]) by Hoffmann–la Roche following its acquisition of Boehringer Mannheim in 1998, the technology developed at considerable cost over the preceding 7 years was in many important respects the most advanced in the world. A brief overview of the project's technical achievements is therefore germane to the main thrust of this chapter.

The project's primary objective was the development of a microarray-based diagnostic technology yielding multianalyte assay results of higher sensitivity than conventional methods in a total processing time per microarray of 15 min.

The company immediately commenced development of prototype "chips" and microarraying machinery (see Figs. 17 and 18) using, in the case of the latter, piezo-electric ink-jet technology. Thus, by 1995–1996, the prototype arraying equipment developed by the company was producing ~5000 individually quality-

---

* Note that the minimum number of target molecules detectable on a spot of these dimensions, using the equipment and methods then being used, was five.

**Figure 17**   Black plastic chip developed by Boehringer Mannheim GmbH.

controlled* chips per hour (later increased to 10,000). The chips themselves comprised small, black, injection-molded plastic wells with an optically flat basal surface 3 mm in diameter, on which ~200 microspots (each ~80 μm in diameter and spaced ~40 μm apart) could be deposited (Fig. 19). Careful attention was paid to the materials used in their manufacture to minimize fluorescent background.

Prototype analyzers developed by the company incorporated a purpose-built confocal laser scanner (which proved superior in sensitivity to charge-coupled device (CCD)-based devices, interrogating each chip in a total scanning time of 10 s. Initial prototypes processed chips at the rate of 1 chip min; later versions under development in 1997–1998 processed chips at a rate of one every 15 s [53] (Fig. 20).

An initial problem encountered in these analyzers' design was the unexpected observation of differences in response between identical microspots located at

---

* Using a noninvasive process verifying spot size and position, imperfect chips are automatically discarded.

**Figure 18** Prototype microarray production line developed by Boehringer Mannheim GmbH.

different positions within the microarray, apparently caused by variations in diffusion and mixing effects at certain points. Considerable efforts were therefore directed toward the elimination, *inter alia*, of standing waves seen within test solutions using certain mixing protocols. These efforts finally proved successful.

Another important issue addressed in the course of these studies was the development of fluorescent microspheres displaying minimal nonspecific binding to solid supports. Again, a reliable method of synthesis was developed that achieved this objective.

These technical improvements implied that, by 1996, Boehringer Mannheim's Microspot® project leader Dr. Hans Berger was able to report a 10-fold improvement in sensitivity over that achieved by the present authors in their earlier studies such that a value for $S^*_{min}$ of 0.01 labeled molecules/$\mu m^2$ had been achieved.

A more detailed report of the instrumentation and methods used by Boehringer Mannheim is given by Finckh et al. [53] in what proved to be one of the few written accounts published by the company's research team of its pioneering activities in the microarray field. This publication reports on some of the analyte

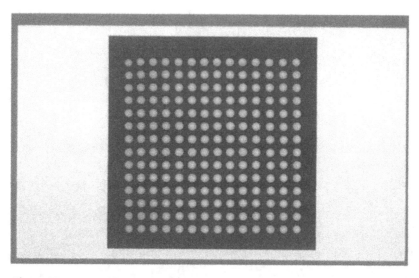

**Figure 19** Array of fluorescent microspots deposited (using piezoelectric ink-jet technology) on base of chip shown in Fig. 17.

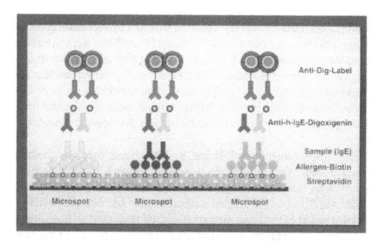

**Figure 20** Assay format and protocol used by Boehringer Mannheim for protein assays (see Ref. 53).

classes to which the technology had been successfully applied. These fell broadly within the fields of endocrinology, allergy, and infectious disease but similar techniques were also employed for the screening of a number of therapeutic drugs. The basic protocols used for protein and nucleic acid arrays are shown in Figs. 21 and 22. Analytes included a wide variety of antigens (such as hormones), antibodies for the detection, for example, of a variety of allergies and infectious diseases (e.g., acquired immunodeficiency syndrome, hepatitis B and C, rubella) and DNA, the latter being illustrated by the detection of rifampicin-resistant forms of *Mycobacterium tuberculosis*. [*Note*: Rifampicin inhibits the RNA polymerase of the bacterium gene by binding to its β-subunit (rpo-β); however, various single-base transitions clustered in a 27-codon segment of the gene cause resistance. This project was selected because of the technical challenges it posed (e.g., the occurrence of single-point mutations, formation of strong intrastrand secondary structures, extremely GC-rich segments, etc.), in addition to its clinical relevance. A study on 80 selected samples from two clinical centers specializing in tuberculosis diagnosis showed a high degree of concordance with the reference (culture) method.]

In its conclusion, this report [53] stressed the diagnostic advantages of the microarray-panel approach, claiming that "the ability to divide the analyte specificities conventionally embodied in a single test into separate assays resulted in an increase in both sensitivity and specificity."*

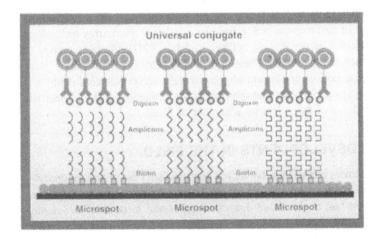

**Figure 21** Assay format and protocol used by Boehringer Mannheim for nucleic acid assays (see Ref. 53).

---

* The terms "sensitivity" and "specificity" here are employed in the diagnostic sense used by clinicians.

Oligo-tagged antibodies

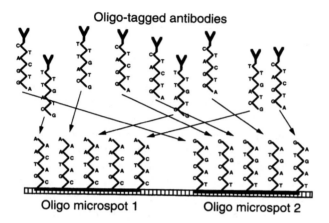

Oligo microspot 1          Oligo microspot 2

**Figure 22** "In-house" method of array construction developed by present authors based on the use of a oligonucleotide array template.

Boehringer Mannheim's studies embraced a variety of formats falling within both noncompetitive and competitive assay designs, the latter being employed for the assay of analytes of small molecular size not permitting the use of the labeled antibody "sandwich" approach. Regrettably, however, details of these pioneering studies have neither been, nor are ever likely to be, published. In summary, many of the concepts and techniques relating to microarray technology originally developed by the present authors had, by 1997-1998, been carried to the point at which the commercial launch of highly sensitive multianalyte assay methods based on both antibody and oligonucleotide arrays was confidently expected by the end of 1998. Regrettably Roche's termination of the project has resulted in the delay of this event by a number of years.

## V. OTHER DEVELOPMENTS IN THE FIELD

Although the primary aim of this chapter has been to clarify the principles underlying high-sensitivity microarray technology, certain, more recent developments by others are relevant to the above-discussed issues and merit brief discussion.

It has, for example, become commonplace in the DNA field to use two fluorescent labels (Cy3 and Cy5) to label nucleic acids present in different samples, which are then exposed to the same (polynucleotide) array (see Ref. 54). The relative amounts of individual polynucleotides deriving from the two sources can be compared using a two-color detection system. A principal advantage of this approach is a saving in expensive microarrays; also, the effects of interarray variations are thereby obviated. Emulating this approach, certain investigators (e.g.,

Haab et al. [55]) have reported its adoption to label antibodies and antigens deriving from different test samples. However, not only does such a strategy imply exposure to the danger of numerous protein-labeling artifacts, but by implicitly relying on a single-site approach to molecular recognition, it disregards the advantages (regarding assay specificity and sensitivity) of two-site assay designs. Also, high-specific-activity labeling with fluorescent microspheres in such a manner might prove difficult if not impossible.

Moreover, the dual-label approach developed by the authors and described earlier (whereby fluorescent signals generated by target analytes are compared with those generated by the capture binding agents located within microspots) obviates problems arising from interarray variability. Thus, the advantages of dual labeling with Cy3 and Cy5 as conventionally practiced in the nucleic acid field appears to offer only a putative economic advantage, at the expense of losses of assay sensitivity and specificity. Thus, it is thus not an approach that is likely to prove relevant to protein arrays.

Another recent and widely reported development is the creation of arrays of microspots of micron or submicron diameter, the occupancy of target analyte thereon being determined by atomic force microscopy [56]. Although uses for such "nanoarrays" may emerge, the claim that they offer greater sensitivity must be viewed with caution. Although it is arguable that this claim may be valid if a nanospot is exposed to a sample of comparable dimensions (such that the capture binding agent binds all, or most, of the target analyte it contains, it loses validity if the sample size is such that it requires an array of such dimensions that ambient analyte assay conditions prevail. Because in these circumstances sensitivity is governed by the value of $S^*_{min}$, the use of nanospots of areas in the order of 1 $\mu m^2$ implies that, at best, the value of this parameter must equal, or be greater than, 1 molecule/$\mu m^2$. This value is some 100-fold less than that reported by Boehringer Mannheim using larger areas, disregarding the lower probability of analyte capture on spots of such a small area.

These two examples are cited to illustrate that certain of the technologies currently under development as a result of the explosive interest in microarray technology should be critically examined before some of the claims made for them are accepted. It is hoped that this chapter has helped readers to do this.

Methods of array construction have been reviewed by a number of authors, including Schena et al. [2]. Among other methods of note are the electronic oligonucleotide-localization techniques developed by Sosnowski et al. [57], which depend on positioning presynthesized oligonucleotides at known locations within the array. Another method evidently now being used by a number of US companies relies on a method originally developed by the present authors–that is, of attaching antibodies or other binding agents (to which oligonucleotides have been attached, see Fig. 22) to an array template made up of an array of complementary oligonucleotides [58].

A number of groups have developed microarray-based immunoassays for simultaneous determination of multiple analyte concentrations (see e.g., Refs. 59–62), albeit generally relying on methods considerably less sensitive than those originally devised by the present authors and subsequently improved upon by Boehringer Mannheim.

## VI. FUTURE DEVELOPMENTS AND SUMMARY

This chapter has been primarily intended to clarify the basic principles underlying microarray technology and its development, with particular emphasis on the concepts that relate to the possible use of ultrasensitive microarrays in the proteomics field.

Clearly, the principal technical problems to be solved in this area (aside from the development of techniques of adequate sensitivity) are the identification of the proteins to be assayed and that of specific binding agents to be located on the arrays. A variety of methods have been described (see Refs 4, 12, and 63–69). How successful recombinant proteins will prove in this context is, however, a matter of conjecture.

Whether antibodies will prove to be the specific "protein-recognition" reagents of choice is also uncertain. Because these bind to epitopes of limited molecular size rather than to protein molecules in their entirety, they are not overly specific and are known often to bind to different isoforms of target ligands of heterogeneous molecular composition, generally arising as a result of posttranslational modifications. Whether a combination of antibody arrays with other molecular separation (e.g., electrophoretic methods [70]), or the use of aptamers or other more specific binding agents remains speculative.

Microarray-based methods are not, of course, restricted to the observation of binding reactions solely between antibodies and antigens or complementary polynucleotides, and a number of groups have reported their use to examine protein–protein, enzyme–substrate, and protein–nucleic acid interactions (e.g., Refs. 71–75). Other possibilities are obvious.

However, although these techniques are likely to prove of enormous importance in research, their major impact is likely ultimately to be felt in the field of clinical diagnostics, as the authors predicted more than a decade ago [76,77]).

## REFERENCES

1. Fodor, S.P.A.; Read, J.L.; Pirrung, M.C.; Stryer, L.; Lu, A.T.; Solas, D. Science. 1991, 251, 767–773.
2. Schena, S.; Heller, R.A.; Theriault, T.P.; Konrad, K.; Lachenmeier, E.; Davis, R.W. Trends Biotechnol. 1998, 16, 301–306.

3.  Ekins, R.; Chu, W.F. Trends Biotechnol. 1999, 17, 217–218.
4.  Albala, J.S.; Humphery-Smith, I. Curr Opin Mol Ther. 1999, 1, 680–684.
5.  Yalow, R.S.; Berson, S.A. J Clin Invest. 1960, 39, 1157–1175.
6.  Ekins, R.P. Clin Chim Acta. 1960, 5, 453–459.
7.  Köhler, G.; Milstein, C. Nature. 1975, 256, 495–497.
8.  Goletz, S.; Christensen, P.A.; Kristensen, P.; Blohm, D.; Tomlinson, I.; Winter, G.; Karsten, U. J Mol Biol. 2002, 315, 1087–1097.
9.  Gao, C.; Mao, S.; Lo, C.H.; Wirsching, P.; Lerner, R.A.; Janda, K.D. Proc Natl Acad Sci USA. 1999, 96, 6025–6030.
10. Hesselberth, J.; Robertson, M.P.; Jhaveri, S.; Ellington, A.D. J Biotechnol. 2000, 74, 15–25.
11. Brody, E.N.; Gold, L. J Biotechnol. 2000, 74, 5–13.
12. Wilson, D.S.; Keefe, A.D.; Szostak, J.W. Proc Natl Acad Sci USA. 2001, 98, 3750–3755.
13. Haupt, K.; Mosbach, K. Trends Biotechnol. 1998, 16, 468–475.
14. Litman, D.J.; Ullman, E.F. Preferential signal production on a surface in immunoassays. US patent N. 4,299,916, 1981.
15. Hess, P. J Clin Ligand Assay. 1996, 19, 11–15.
16. Yalow, R.S.; Berson, S.A. In Radioisotopes in Medicine: In Vitro Studies Hayes RL, Goswitz FA, Pearson Murphy BE, eds; US Atomic Energy Commission: Oak Ridge TN, 1968, 7–39.
17. Ekins, R.P.; Newman, B.; O'Riordan, J.L.H. In Radioisotopes in Medicine: In Vitro Studies Hayes RL, Goswitz FA, eds; US Atomic Energy Commission: Oak Ridge, TN, 1968, 59–100.
18. Yalow, R.S.; Berson, S.A. In Statistics in Endocrinology McArthur JW, eds; MIT Press: Cambridge, MP, 1970, 327–344.
19. Ekins, R.P.; Newman, G.B.; O'Riordan, J.L.H. In Statistics in Endocrinology McArthur JW, eds; MIT Press: Cambridge, MA, 1970, 345–378.
20. Ekins, R.; Edwards, P. Clin Chem. 1997, 43, 1824–1831.
21. Pardue, H.L. Clin Chem. 1997, 43, 1831–1837.
22. Ekins, R.; Edwards, P. Clin Chem. 1998, 44, 1773–1776.
23. Pardue, H.L. Clin Chem. 1998, 44, 1776–1778.
24. Freiser, H.; Nancollas, G.H. International Union of Pure and Applied Chemistry, Analytical Division, Compendium of Analytical Nomenclature, Definitive Rules 1987. 2nd ed; Blackwell Scientific: Oxford, 1987, 5–6, 35–36, 206, 242.
25. International Vocabulary of Basic and General Terms in Metrology. 2nd ed; International Organization for Standardization: Genevea, 1993.
26. Macurdy, L.B.; Alber, H.K.; Benedetti-Pichler, A.A.; Carmichael, H.; Corwin, A.H.; Fowler, R.M.; Huffman, E.W.D.; Kirk, P.L.; Lashof, T.W. Anal Chem. 1954, 26, 1190–1193.
27. Berson, S.A.; Yalow, R.S. In Methods in Investigative and Diagnostic Endocrinology Berson SA, Yalow RS, eds; North-Holland/Elsevier: Amsterdam, 1973; Vol. 2A, 84–135.
28. Ekins, R.; Newman, B. In Steroid Assay by Protein Binding. Karolinska Symposia on Research Methods in Reproductive Endocrinology, ed; WHO/Karolinska Sjukhuset: Stockholm, 1970, 11–30.

29. Ekins, R.P. In Protein and Polypeptide Hormones, Part 3 (Discussions) Margoulies M, ed; Excerpta Medica: Amsterdam, 1968, 612–616, 672–682.

30. Larsen, P.R.; Alexander, N.M.; Chopra, I.J.; Hay, I.D.; Hershman, J.M.; Kaplan, M.M.; Mariash, C.N.; Nicoloff, J.T.; Oppenheimer, J.H.; Solomon, D.H.; Surks, M.I. J Clin Endocrinol Metab. 1987, 64, 1089–1094.

31. Wide, L.; Bennich, H.; Johansson, S.G.O. Lancet. 1967, ii, 1105–1107.

32. Miles, L.E.H.; Hales, C.N. Nature. 1968, 219, 186–189.

33. Wide, L. In Radioimmunoassay Methods Hunter WM, eds; Churchill Livingstone: Edinburgh, 1971, 405–412.

34. Miles, L.E.H.; Hales, C.N. In Protein and Polypeptide Hormones, Part 1, ed; Excerpta Medica: Amsterdam, 1968, 61–70.

35. Rodbard, D.; Weiss, G.H. Anal Biochem. 1973, 52, 10–44.

36. Yorde, D.E.; Sasse, E.A.; Wang, T.Y.; Hussa, O.O.; Garancis, J.C. Clin Chem. 1976, 22, 1372–1377.

37. Ekins, R.; Chu, F.; Biggart, E. Clin Chim Acta. 1990, 194, 91–114.

38. Marshall, N.J.; Dakubu, S.; Jackson, T.; Ekins, R.P. In Monoclonal Antibodies and Developments in Immunoassay Albertini A, eds; Elsevier/North-Holland: Amsterdam, 1981, 101–108.

39. Ekins, R.P. Measurement of analyte concentration. British Patent. 8224600, 1983.

40. Chang, T. Matrix of antibody-coated spots for determination of antigens. US patent. 4,591,570, 1986.

41. Silzel, J.W.; Cercek, B.; Dodson, C.; Tsay, T.; Obremski, R.J. Clin Chem. 1998, 44, 2036–2043.

42. Obremski, R.J.; Silzel, J.W.; Tsay, T.; Cercek, B.; Dodson, C.L.; Wang, T.R.; Liu, Y.; Zhou, S. Detection of very low quantities of an analyte bound to a solid phase. US patent application. 09/063,978, filed 21 April 1998.

43. Ekins, R.; Chu, F. Ann Biol Clin. 1992, 50, 337–353.

44. Ekins, R.P.; Chu, F.; Biggart, E. Anal Chim Acta. 1990, 227, 73–96.

45. Ekins, R.P. In Principles of Nuclear Medicine Wagner HN, Szabo Z, eds; WB Saunders: Philadelphia, 1995, 247–66.

46. Rodbard, D.; Lewald, J.E. In Steroid Assay by Protein Binding. Karolinska Symposia on Research Methods in Reproductive Endocrinology, ed; WHO/Karolinska Sjukhuset: Stockholm, 1970, 79–103.

47. Polsky-Cynkin, R.; Parsons, G.H.; Allerdt, L.; Landes, G.; Davis, G.; Rashtchian, A. Clin Chem. 1985, 31, 1438–1443.

48. Wolf, S.F.; Haines, L.; Fisch, J.; Kremsky, N.; Dougherty, J.P.; Jacobs, K. Nucleic Acid Res. 1987, 15, 2911–2925.

49. White, J.G.; Amos, W.B.; Fordham, M. J Cell Biol. 1987, 105, 41–48.

50. Ekins, R.P.; Chu, F.W. Clin Chem. 1987, 37, 1955–1967.

51. Ekins, R.P.; Chu, F.; Micallef, J. J Biolumin Chemilumin. 1989, 4, 59–78.

52. Press release formerly agreed between Roche Diagnostics GmbH and Multilyte (spin off from University College London) announcing its termination of license agreement of microarray patents.

53. Finckh, P.; Berger, H.; Karl, J.; Eichenlaub, U.; Weindel, K.; Hornauer, H.; Lenz, H.; Sluka, P.; Weinreich, G.E.; Chu, F.; Ekins, R.P. Proc UK NEQAS Meeting. 1998, 3, 155–165.

54. Hacia, J.G.; Brody, L.C.; Chee, M.S.; Fodor, S.P.A.; Collins, F.S. Nature Genet. 1996, 14, 441–447.

55. Haab, B.; Dunham, M.; Brown, P. Protein microarrays for highly parallel detection and quantitation of specific proteins and antibodies in complex solutions. Genome Biology. 2001, 2, 2 (www.genomebiology.com. Access date 22 January 2001).

56. Limansky, A.; Shlyakhtenko, L.S.; Schaus, S.; Henderson, E.; Lyubchenko, Y.L. Amino modified probes for atomic force microscopy. Probe Microscopy. 2002, 2(3–4), 227–234.

57. Sosnowski, R.G.; Tu, R.G.E.; Butler, W.F.; O'Connell, J.P.; Heller, M.J. Proc Natl Acad Sci USA. 1997, 94, 1119–1123.

58. Ekins, R.P. Binding assay using binding agents with tail groups. European patent 0 749 581, 1998.

59. Huang, R.-.P.; Huang, R.; Fan, Y.; Lin, Y. Anal Biochem. 2001, 294, 55–62.

60. Belov, L.; de la Vega, O.; dos Remedios, C.G.; Mulligan, S.P.; Christopherson, R. Cancer Res. 2001, 61, 4483–4489.

61. Joos, T.O.; Shrenk, M.; Hopfl, P.; Kroger, K.; Chowdery, U.; Stoll, D.; Schorner, D.; Durr, M.; Herick, K.; Rupp, S.; Sohn, K.; Hammerle, H. Electrophoresis. 2000, 21, 2641–2641.

62. Mendoza, L.G.; McQuary, P.; Mongan, A.; Gangadharan, R.; Brignac, S.; Eggers, M. BioTechniques. 1999, 27, 778–788.

63. Anderson, N.L.; Anderson, N.G. Electrophoresis. 1998, 19, 1853–1861.

64. Emili, A.Q.; Gagney, G. Nature Biotechnol. 2000, 18, 393–397.

65. Cahill, D.J. TIBTECH. Proteomics. 2000, 18(7), 47–51.

66. Holt, L.J.; Bussow, K.; Walter, G.; Tomlinson, I.M. Nucleic Acids Res. 2000, 28(15), E72.

67. Lueking, A. Anal Biochem. 1999, 270, 103–111.

68. De Wildt, R.M.; Mundy, C.R.; Gorick, B.D.; Tomlinson, I.M. Nature Biotechnol. 2000, 18, 989–994.

69. Humphery-Smith, I. High density arrays for proteome analysis and methods and compositions therefor. International patent WO 99/39210.

70. Kaufmann, H.; Bailey, J.E.; Fussenegger, M. Proteomics. 2001, 1, 194–199.

71. Arenkov, P.; Kukhtin, A.; Gemmell, A.; Voloshchuk, S.; Chupeeva, V.; Mirzabekov, A. Anal Biochemm. 2000, 278, 123–131.

72. Zhu, H.; Klemic, J.F.; Chang, S.; Bertone, P.; Casamayor, A.; Klemic, K.G.; Smith, D.; Gerstein, M.; Reed, M.A.; Snyder, M. Nature Genet. 2000, 26, 283–289.

73. Ge, H. Nucleic Acids Res. 2000, 28, e3.

74. MacBeath, G.; Koehler, A.N.; Schreiber, S.L. J Am Chem Soc. 1999, 121, 7967–7968.

75. MacBeath, G.; Schreiber, S.L. Science. 2000, 289, 1760–1763.

76. Ekins, R. J Pharm Biomed Anal. 1989, 7, 155–168.

77. Ekins, R.; Chu, F.; Biggart, E. Ann Biol Clin (Paris). 1990, 48, 655–666.

# 3
# Practical Approaches to Protein Microarrays

**BRIAN HAAB**
*Van Andel Research Institute*
*Grand Rapids, Michigan, U.S.A.*

## I. INTRODUCTION

High-throughput protein analysis methods have advanced scientific research in areas such as the identification of disease-related proteins or the study of protein function. For example, the efficient testing of binding interactions between proteins by high-throughput yeast two-hybrid technology enabled the generation of comprehensive protein–protein interaction maps of yeast [1] and *Helicobacter pylori* [2]. Similarly, the capability of two-dimensional (2D) electrophoresis to separate and quantify thousands of proteins has facilitated the identification of proteins that are implicated in disease, as in a study that found differentially expressed proteins in squamous cell carcinoma [3]. Much research has been devoted to the development of additional methods that would enhance the current high-throughput protein analysis art. Protein microarrays have great potential to meet this demand, shown by recent work demonstrating the feasibility of this tool for the highly parallel and sensitive quantitation of proteins in complex biological samples [4]. The aim of this chapter is to survey current approaches to protein microarray technology and to focus on methods that are the most practical to implement, with the broad goal of promoting the wider use of the technology for biological studies.

## II. PROTEIN DETECTION USING MICROARRAYS

Protein microarrays are analogous to DNA microarrays, comprising microscale spots of unique antibody, protein, or peptide samples arrayed on a surface. The widespread use of DNA microarrays testifies to the many advantages of the microarray format. Primary experimental advantages are the high information density (thousands of unique analytes can by quantified in a single experiment) and low sample consumption, both in the amount of material spotted onto the microarray and in the amount of sample to be analyzed. Low consumption is important for limiting reagent costs and preserving valuable experimental samples. Additional positive experimental features of microarrays include quantitative accuracy, sensitivity, and speed. A recent characterization of DNA microarray performance showed a linear signal response over three orders of magnitude of analyte concentration, with a low coefficient of variation (12–14%) and a low limit of detection (2 pg of mRNA) [5]. A similar characterization of protein microarray performance also showed linearity in response over three orders of magnitude with low detection limits ($<$ 0.3 ng/ mL) [4]. The rapidity and ease of use of microarray experiments enable large studies to be performed, such as the validation of tumor markers from hundreds of patients. Such studies would not be practical using other proteomics tools such as 2D gel electrophoresis.

Although many significant biological studies have contributed to the scientific literature by use of the DNA microarray [6–10], the application of microarrays to the study of proteins has been slower to develop due to the increased complexity of proteins as compared to nucleic acids. For example, the sequence of a nucleic acid can be readily converted to the sequence of its complementary binding partner, and the strength of the interaction can be calculated for lengths up to ~50 base pairs. No corresponding formulas exist for protein sequences. Binding partners and binding strengths are determined empirically. Furthermore, nucleic acids are easily synthesized and replicated, and the similarities in structure and chemistry allow common methods of manipulation and surface attachment. In contrast, peptides are costly to synthesize, and expressed proteins can be difficult to purify and replicate. Because proteins have widely variant structures and chemistries, it can be difficult to find experimental protocols that work well for all proteins.

Because of the significant advantages of the microarray format noted earlier, there has been much motivation for addressing these experimental challenges to develop practical protein microarray methods. Several reports have demonstrated the feasibility of using microarrays for protein analysis. An early demonstration of a parallel approach to immunoassays was the antigen spot or immuno-dot-blot assay [11], in which multiple samples were spotted (~ 1 μL per spot) onto a membrane and probed with a single antibody. Arrays of peptides were synthesized on cellulose supports by spot synthesis [12] and on polyethylene supports by multipin synthesis [13]. Reports of arrays of spotted proteins [14–19], antibodies [4,20–22] and phage display clones [23] appeared later, with various uses and technological implementations.

## III.  PRACTICAL APPROACHES TO PROTEIN MICROARRAYS

This section focuses on practical approaches to implementing and using protein microarrays without the need for the development or acquisition of new equipment or special and difficult attachment or labeling chemistries. In particular, this chapter looks at ways to use microarrays that are manufactured using commonly available microarray printing robots (see http://cmgm.stanford.edu/pbrown/). By building on the established technology of DNA microarrays, one may take advantage of the existing array manufacturing and detection technologies and the supporting software. As many labs already have invested in cDNA microarray platforms, the extension of this infrastructure to protein microarray experiments should be readily achievable.

The technical platform that provides the basis for most experimental variants described in this chapter is described in Ref. 24. A robotic printer transfers tiny amounts of protein solutions from the wells of 96- or 384-well microtiter plates into ordered arrays on the surfaces of derivatized microscope slides or membranes. Up to 40,000 unique protein spots 150–250 μm in diameter can be printed on one microscope slide. Protein solutions are incubated on the arrays, and specific binding (e.g., antibody–antigen) interactions localize specific components of the complex mixtures to defined cognate spots in the array. The bound proteins are detected through an attached luminescent or radioactive label.

Although much of the technology is common between protein and DNA microarrays, including the spotting and scanning methods and the fluorescent dyes used for detection, certain challenges unique to protein microarrays require special consideration. These challenges include attachment of capture proteins, blocking the surface, detection of bound proteins, and the collection of capture proteins. These aspects of the technology are discussed in the following subsections.

### A.  Attachment and Blocking

Whereas nucleic acids share four basic chemical units and have a limited folding repertoire, proteins have an almost immeasurable variety of polarities, hydrophobicities, charges, sizes, and structures. This variety in chemistries presents a challenge for protein microarrays: to attach all types of proteins while repelling nonspecific binding. Surfaces that readily bind spotted proteins, such as polystyrene and certain membranes, also readily bind background proteins nonspecifically. However, surfaces that are effective in repelling nonspecific binding present difficulties in attaching the spotted favored proteins.

Table 1 outlines the protein attachment methods presented in this section. The main protein attachment strategies are adsorption, affinity binding, and covalent binding. Adsorption is the easiest—spotted proteins attach by electrostatic or

**Table 1**  Summary of Published Protein Array Attachment Methods

| Method | Ref. |
|---|---|
| Poly(vinylidene fluoride) (PVDF) | 15, 16, 25 |
| Nitrocellulose | 17, 19, 23 |
| Poly-l-lysine coated glass | 4, 19 |
| Polystyrene | 20 |
| Specific binding | |
| Binding of biotinylated capture antibody to avidin adsorbed on polystyrene | 20 |
| Binding of biotinylated antibodies to avidin attached to silanized slides through a cross-linker | 26 |
| Adsorption of biotinylated bovine serum albumin to hydrophobic silanes; attachment of a biotinylated antibody through streptavidin | 27 |
| Covalent binding | |
| Polymerization of gel matrices to silinated glass; reaction of protein amine groups with a glutaraldehyde crosslinker | 14, 21 |
| Reaction of protein amine groups to silanated slides through an aldehyde-containig cross-linker | 22 |
| Reaction of protein amine groups to silanated slides through a succinimide cross-linker | 18 |
| Oxidation of carbohydrate groups on an agarose film; reaction of protein amine groups with the resulting aldehyde groups | 28 |

hydrophobic forces. Surfaces such as Poly(vinylidene fluoride) (PVDF), nitrocellulose, polystyrene, or poly-L-lysine-coated glass adsorb proteins well, and there have been several demonstrations of protein arrays using these surfaces [4,15,17,23]. A potential drawback is the difficulty in preventing nonspecific binding when using a surface that indiscriminately binds all proteins. Comparisons of various blocking agents [29,30] have found milk proteins to effectively saturate the free binding sites on membranes, thus reducing nonspecific binding. Others have found that for specific applications, the use of high-salt conditions [31], high detergent concentrations [32], or the cocoating of surfaces with anionic proteins [33] or denatured proteins [34] reduced nonspecific binding while maintaining specific binding. Nevertheless, despite rigorous blocking, it seems that the level of nonspecific binding still depends somewhat on the intrinsic protein-binding capability of a surface, perhaps reflecting the fact that it is impossible to completely block a surface. Another potential drawback in using adsorptive attachment is that some proteins may not bind, depending on the chemistry of the protein. A weakly bound protein may not stay attached during the washing of the surface, especially if stringent washes are used, such as those with high salt or high detergent concentrations.

Affinity binding can provide a strong and highly selective attachment through the use of specific biological interactions, such as between biotin and avidin or between protein A and IgG. Both the surface and the protein to be attached may need to be derivatized with the components of the interaction. If the derivatized surface is resistant to nonspecific protein adsorption, a lower background is achieved while maintaining specific binding. For example, a polyethylene glycol (PEG)-coated surface has been developed that is functionalized with biotin and streptavidin [35]. Spotted antibodies containing biotin attach to the surface, but nonspecific background proteins are repelled by the PEG, which is highly resistant to protein binding [36] (see also Chap. 4). Others have attached biotinylated proteins to avidin that was covalently attached [26] or adsorbed [20,27] to the solid phase. A drawback of attachment by specific binding is that proteins have to be modified before spotting, adding steps to protein preparation and potentially altering the protein structure. The affinity binding strategy may be worth the extra preparation effort to achieve a lower background level of nonspecific binding or to orient capture antibodies to expose the active site of the antibody (see following discussion).

Covalent binding provides the strongest attachment of proteins to surfaces, allowing high-stringency binding conditions and washes that remove weakly bound proteins. Reaction of the amine groups on proteins, from either lysine residues or terminal amines, with immobilized functional groups such as aldehydes or succinimides is the most straightforward and common strategy. Methods to form aldehyde groups on surfaces include oxidation of the carbohydrate groups of adsorbed polyacrylamide [21] or agarose [28] and cross-linking gluteraldehyde to immobilized polyacrylamide [14] or aminosilanes [37]. In addition, commercially available "silynated" glass slides with terminal aldehyde groups have been used to make protein microarrays [19,22]. Similarly, succinimide groups were formed on glass surfaces using a cross-linker attached to aminosilanes [18] or a bovine serum albumin (BSA) coating [22]. The requirement of accessible amine groups on the spotted proteins may, in some cases, limit this approach. Other covalent attachment schemes that have been used for immunosensors and chromatography may be useful for protein microarrays. IgG molecules were covalently bound by oxidizing the carbohydrate group on the Fc region of the molecule, creating an aldehyde, and reacting the aldehyde with a hydrazide-activated surface [38]. However, oxidation of the antibody may sometimes damage the antigen-binding site [39]. For general use, the best covalent attachment strategy currently may be the reaction of protein amine groups with an immobilized amine-reactive cross-linker.

A general concern when using either specific binding or covalent binding is drying of the spotted protein. Bulk liquid conditions are necessary for efficient chemical reactions, which may be difficult to maintain when spotting miniscule volumes (slotted-pin spotters deliver about 5 nL per spot). Strategies to slow the evaporation of the spots on microarrays include the addition of glycerol [22] or betain [40] to the spotting solution and the maintenance of a high ambient humidity during the spotting process [41]. An additional motivation to maintain the hydration of the spotted proteins is to prevent denaturation.

Certain covalent and specific binding strategies enable the orientation of antibodies so that the antigen binding site is facing outward from the point of attachment, perhaps yielding an increased antigen binding capacity [42]. Protein A or protein G is used in affinity chromatography to bind the Fc region of IgG, leaving the active region of the IgG free [43]. Another technique oxidizes the carbohydrate moiety on the Fc region of IgG molecules to an aldehyde group, enabling oriented covalent attachment to hydrazide-derivatized surfaces or hydrazide-derivatized biotin [44]. Additionally, the sulfhydryl group in the C-terminal region of antibody Fab fragments can be reacted with the appropriate immobilized reactive groups (e.g., maleimide) to achieve proper orientation [45]. The measured increase in the capacity to bind antigens has ranged from threefold [38] to no increase [46]. Considering the difficulty in preparing oriented antibodies and the modest increase in antigen-binding capacity, for many applications randomly immobilized antibodies should be sufficient, especially if many antibodies are to be tested and used, as with protein microarrays. In conclusion, each attachment method (adsorption, affinity binding, and covalent) has features that must be evaluated with regard to ease of use, versatility, and the number of antibodies or antigens to be used.

## B.  Detection

Protein microarrays can be detected using established protein detection methods such as enzyme-linked immunosorbent assay (ELISA), radioisotope detection, and fluorescence, either by direct or secondary detection. Direct detection is the direct labeling and detection of analyte proteins, and secondary detection is the detection of the analyte proteins through a second labeled antibody that binds to captured analyte proteins after an initial incubation.

Secondary detection is often used in ELISA assays, with labeled secondary antibodies directed against a captured antigen (''sandwich ELISA'') or the Fc region of a primary antibody [47]. Applied to protein microarrays, secondary detection is most practical when only one type of species is to be detected, such as a single antigen or one type of immunoglobulin. Epitope mapping by incubating an antibody on a peptide array [48] and the characterization of immune responses by incubating blood sera on arrays of potential autoantigens [19,49] are microarray applications well suited to secondary detection. The secondary antibody can be labeled either with a fluorophore [49] or with an enzyme that produces a detectable luminescent product [18] (i.e., ELISA).

Direct detection of protein analytes is useful for protein microarray applications that measure multiple nonantibody proteins. Fluorescence is the most widely used and versatile detection method, but radioisotope labeling is feasible if a limited number of proteins are to be incubated on an array [17]. A significant advantage of radioisotope labeling is that the labeled protein is not altered in structure, as it might be with the addition of a fluorescent or enzymatic tag. Direct labeling by the addition of a fluorescent tag is easier to perform than radioisotope labeling and is usually achieved by the coupling of protein amine groups to fluorophore-conjugated reactive groups such as N-hydroxysuccinimide. Fluorescent tagging was used for the direct labeling of individual proteins [22] and the bulk labeling of complex pools of proteins [4] to be detected by protein microarrays.

A useful feature of fluorescence is the capability to detect in multiple wavelength regions, enabling the simultaneous detection and quantitative comparison of analytes from separate sample pools. Comparative fluorescence has been widely used with DNA microarrays and was recently demonstrated with protein microarrays [4], using the labeling and detection scheme in Fig. 1. Two complex protein samples, one serving as a standard for comparative quantitation (the reference), and the other representing an experimental sample in which the protein quantities are to be measured, are labeled by covalent attachment of spectrally resolvable fluorescent dyes (here shown as Cy3 and Cy5). The unreacted dye is removed, and the samples are incubated together on a protein microarray. After the solution proteins have bound to cognate spots on the array according to specific binding interactions, the relative intensity of the fluorescent signal representing the experimental sample and the reference standard provides a measure of each protein's abundance. Quantitation by comparative fluorescence is highly accurate because the reference sample provides an internal control and a basis for normalization and comparison between arrays. Certain commercially available microarray readers can detect fluorescence from commonly used labeling dyes in up to five different wavelength regions (e.g., from Perkin Elmer). Because of the multicolor capability and ease of use, direct fluorescence labeling is currently the best method to measure the relative abundances of multiple proteins from multiple sample pools.

Other detection methods have been demonstrated (e.g., by atomic force microscopy [50] or mass spectrometry [51]) but are not as practical for labs wanting to develop their own protein microarray applications. Showing particular promise for improving the sensitivity of protein microarray detection is rolling circle amplification (RCA), which was demonstrated to lower detection limits in solid-phase immunoassays by over 100-fold [52]. A secondary detection antibody was tagged with a DNA primer, from which a lengthy repetitive DNA strand was

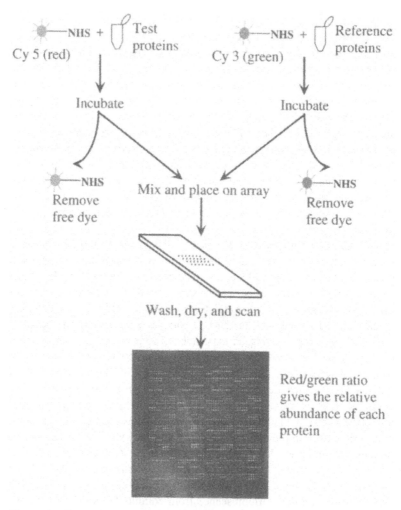

**Figure 1**  Two-color fluorescent labeling and detection of protein microarrays. A test protein solution and a reference protein solution are mixed with two different *N*-hydroxy-succinimide (NHS)-conjugated fluorescent dyes. The NHS group reacts with amine groups on the proteins. Free dye is removed, and the solutions are mixed and placed on an array. The array is read by scanning fluorescence microscopy in two color channels specific to the conjugated dyes. The fluorescence ratio between the color channels reflects the relative protein concentration between the two solutions. (See the color plate.) (Courtesy of Current Drugs; B Haab. Curr Opin Drug Dis Dev 4:116–123, 2001.)

extended using RCA after the antibody bound its target. Fluorescently labeled oligonucleotides complementary to the repetitive sequence were hybridized to the extended DNA strand, providing the means for the amplified detection signal. Efforts are underway to apply RCA to microarray assays (Lizardi, personal communication, 2001).

## C. Collection of Binding Reagents

The usefulness of protein microarray experiments depends on having a set of reagents to spot on the arrays that bind biologically interesting proteins. The acquisition of protein sets will be greatly facilitated by the recently developed high-throughput protein expression and purification methods, based on recombinant baculoviruses [53] or the Gateway™ recombinant cloning system [54]. The proteins are produced in 96-well microtiter plates and efficiently purified through the amino- or carboxyl-terminal attachment of an epitope tag, such as poly-histidine or Glu-Glu. An efficient way to test for the expression of recombinant proteins is based on a method for arraying individual bacterial colonies of a cDNA library onto membranes [15,16,25]. The arrayed colonies were induced for protein expression, and after lysing the cells on the membrane, the expressed proteins were tested for proper expression, folding, and antibody specificity.

The acquisition of large antibody sets that are useful for protein microarrays is difficult because of the cost of antibody production and the requirement for the characterization of each antibody. Therefore, comprehensive protein expression studies analogous to cDNA microarray whole-genome mRNA expression profiling may not be feasible using antibody microarrays. A more practical experimental design will be to direct studies to particular biological problems, such as a particular signaling pathway, for which one could construct microarrays using an appropriate and manageable set of antibodies. The limited procurement of antibodies for microarrays could also be directed by information from cDNA microarrays, by generating antibodies to protein products that are identified as upregulated in global mRNA expression profiles.

Another approach to the collection of binders for protein microarrays is the selection of reagents from protein display libraries, such as phage display [55], yeast display [56], or ribosome display [57]. Each member of a display library contains a different piece of genetic information that is expressed as an accessible protein product (e.g., on the coat of a phage particle). These accessible protein products can be screened for binding to an immobilized protein of interest. After unbound library members are washed away, the bound members are eluted and amplified by growing more of the display particles. This "biopanning" process could rapidly generate protein capture reagents for protein microarrays, as recently demonstrated by a high-throughput antibody screening assay using arrays of phage-display-generated scFv antibodies [23].

There are several potential advantages to using display libraries to generate binders for protein microarrays. Most significantly, both the cost and time required to generate a binder to a particular protein could be greatly reduced as compared to conventional antibody production methods. Once a display library is produced, the selection process takes about a week, with minimal reagent costs. An additional benefit may be the simultaneous production of binders to multiple proteins in complex mixtures, as demonstrated by the selection of scFv clones specific to multiple antigens in dilute mixtures [23]. This demonstration is promising for protein discovery using protein microarrays, because binders to uncharacterized proteins can be generated by panning complex mixtures (see discussion of applications in Section V).

In summary, efficient methods to generate peptides and proteins for microarrays exist, but the collection of antibodies for protein microarrays should be directed to particular biological questions. However, phage display technology shows promise for rapidly generating binding reagents to arbitrary protein targets in complex environments, thereby broadening the utility of antibody and protein microarrays.

## IV.  FEASIBILITY STUDIES

Recent work showed the feasibility of the practical application of protein microarrays by characterizing the detection limits, specificity, and accuracy of quantitation for a set of 115 antibody/antigen pairs [4]. Six different mixtures of the 115 antibodies and 6 different mixtures of 115 antigens were prepared so that the concentration of each species varied in a unique pattern across the protein mixtures over a range of three orders of magnitude. Using a comparative fluorescence assay as depicted in Fig. 1, each of the six test protein mixtures was labeled with Cy5 dye (red fluorescence) and then mixed with a Cy3-labeled (green fluorescence) "reference" mixture containing each of the same 115 proteins at a constant concentration. The variation across the six microarrays in the red-to-green (R/G) ratio measured for each antibody or antigen spot should reflect the variation in the concentration of the corresponding binding partner in the set of mixes. By comparing the observed variation in the R/G ratios with the known variation in the concentrations, the performance of each antibody/antigen pair was assayed. Figure 2 presents the measured R/G ratios plotted as a function of the cognate analyte solution concentrations for 12 spotted antigens. The dashed line in each graph represents the ideal linear relationship, and the solid line is the average red-to-green ratio of 6–12 replicate spots (with the error bar representing the standard deviation between the replicates). Many of the tested antigens showed near-ideal response over three-orders-of-magnitude change in analyte concentration, such as AIM-1 and TEF-1. In some cases, the standard deviation increased (e.g., ARNT1) or the red-to-green ratio deviated from linearity (e.g., MST3) at

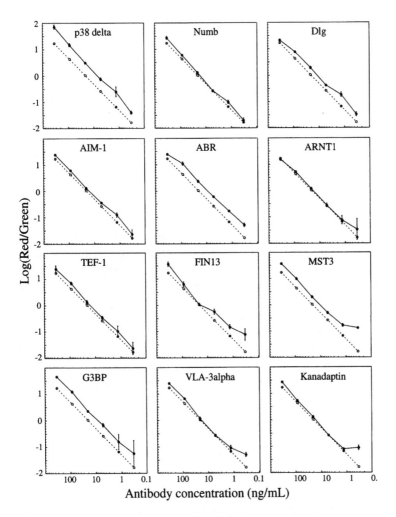

**Figure 2** Relationship between the Cy5/Cy3 fluorescence ratios measured using antigen microarrays and the concentration ratio of the cognate antibodies. The solid line represents the median of the $\log_{10}$- transformed Cy5/Cy3 fluorescence ratios from 6–12 replicate antigen spots, with the error bars representing the standard deviation between the replicate spots. The dashed line represents the ideal linear relationship between the R/G ratio and the concentration ratio. (From Ref. 4.)

low analyte concentrations, indicating the limit of detection. In other cases, the plot was still linear at the lowest concentration of 0.3 ng/mL (e.g., NUMB), indicating that the detection limit is below this value.

The observed detection sensitivities in these control experiments were similar to those required by several established clinical tests. For example, the prostate cancer marker, prostate-specific antigen (PSA), is clinically significant at 4 ng/mL and above [58], and the breast cancer markers c-erbB-2, CEA, and CA 15.3 are considered clinically significant at concentrations of 15, 5, and 35 μg/mL, respectively [59]. These concentrations for biomarker identification are within the observed detection limits of the protein microarray. These experiments have established the use of protein microarrays as a sensitive, quantitative, and highly parallel tool to measure protein concentrations in complex backgrounds.

Not all of the antibodies and antigens that were tested performed well on the microarrays, although all performed well in other assays such as Western blot or using immunohistochemistry. Fifty percent of the arrayed antigens and 20% of the arrayed antibodies provided specific and accurate measurements of their cognate ligands at or below concentrations of 0.34 and 1.6 μg/mL, respectively. The failures might have been due to degradation of the capture proteins on the surface or the inability to correctly label some proteins in solution. Areas identified in this study for further development are better preparation of the array surface against nonspecific binding, optimization of fluorescent labeling across protein species, and better stabilization of the proteins on the array. However, the existing methodology has already enabled biological study.

## V.  APPLICATIONS OF PROTEIN MICROARRAYS

Many valuable protein microarray applications have been demonstrated or are foreseen. Antigen, protein, and peptide arrays are valuable for protein functional studies, by measuring protein–protein, protein–nucleic acid, protein–small molecule, and enzyme–substrate interactions. Synthesized peptide arrays, because they were much earlier in development, have long been used for applications such as the characterization of antibody epitopes [13], protein–protein contact sites [60], and DNA–metal binding sites [61]. Peptide and protein arrays also can be used as a highly parallel screen for antibodies in blood serum to type an immune response or to characterize the autoantigen in an autoimmune disease. For example, Robinson et al. have collected and arrayed hundreds of peptides spanning several candidate autoantigens in systemic lupus erythematosus and rheumatoid arthritis [49]. After incubation of the blood sera of autoimmune patients on the arrays, the binding of antibodies identified immunological reactivity to self proteins. The use of peptides instead of proteins allows fine mapping of autoreactive sites and the observation of epitope spreading [62]. In addition, the individual patterns of antigen reactivity may be used to tailor treatments to specific patients.

Antibody arrays will have significant applications in basic and applied biological research, particularly because of the rapidity of the experiments. For example, a detailed study on the changes in protein expression patterns in cells that have been perturbed could give great insight into protein function and coordinately co-regulated pathways. Such analysis is only possible if many experiments with highly parallel, quantitative information can be performed. Applications in disease marker discovery and diagnostics could also be furthered by antibody microarrays. For example, the serum protein expression patterns of hundreds of cancer patients could be correlated with clinical information to assess the clinical value of multiple proteins or sets of proteins. The microarray format facilitates not only the rapid evaluation of many proteins individually but also the evaluation of coordinate patterns of expression.

The use of phage display with protein microarrays also has potential for biomarker discovery. A limitation of using only known antibodies for protein discovery is that the candidate marker proteins must be known prior to experimentation. Figure 3 depicts a strategy to address this limitation using phage display technology. A phage display library is mixed with a complex pool of proteins from disease tissue, and all of the members of the library that bind proteins are plated out and spotted onto microarrays. The protein expression levels from individual disease protein samples and normal protein samples are compared on these microarrays using two-color comparative fluorescence. After the collection of data from many samples, the patterns of fluorescence intensities from spotted phage display clones could be correlated with clinical information about the samples. Clones that yield valuable information could be followed up with further studies. In this way, the microarray provides a high-throughput screening tool for a selected subset of the phage display library, and aberrantly expressed proteins in disease tissue can be identified without knowing the candidates in advance. Although the production of high-quality, high-affinity single-chain antibody (scFv) libraries can be very time-consuming, further advances in the technology and commercially available kits (e.g., from New England Biolabs and Novagen) are making the use of phage display more accessible to researchers.

The above is only a sampling of the potential applications for peptide, protein, and antibody microarrays. As the continued developments in the technology lead to more sensitive and accurate detection, and with increased availability of sets of proteins, peptides, antibodies, and display libraries, the usefulness of these methods will be furthered for use in many areas of biological discovery.

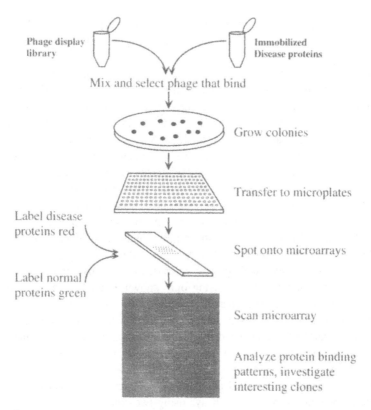

**Figure 3** Strategy for protein discovery using phage display libraries and protein microarrays. Members of a phage display library that bind to immobilized disease proteins are plated out and spotted onto microarrays. The pattern of binding to the spotted clones from individual patient samples is assessed using two-color comparative fluorescence. Clones that provide informative patterns of binding are investigated further. (See the color plate.)

## REFERENCES

1.  Uetz, P.; Giot, L.; Cagney, G.; Mansfield, T.A.; Judson, R.S.; Knight, J.R.; Lockshon, D.; Narayan, V.; Srinivasan, M.; Pochart, P.; Qureshi-Emili, A.; Li, Y.; Godwin, B.; Conover, D.; Kalbfleisch, T.; Vijayadamodar, G.; Yang, M.; Johnston, M.; Fields, S.; Rothberg, J.M. Nature. 2000, 403, 623–627.
2.  Rain, J.-.C.; Selig, L.; De Reuse, H.; Battaglia, V.; Reverdy, C.; Simon, S.; Lenzen, G.; Petel, F.; Wojcik, J.; Schaechter, V.; Chemama, Y.; Labigne, A.; Legrain, P. Nature. 2001, 409, 211–215.

3.  Celis, J.E.; Celis, P.; Ostergaard, M.; Basse, B.; Lauridsen, J.B.; Ratz, G.; Rasmussen, H.H.; Orntoft, T.F.; Hein, B.; Wolf, H.; Celis, A. Cancer Res. 1999, 59, 3003–3009.
4.  Haab, B.B.; Dunham, M.J.; Brown, P.O. Genome Biol. 2001, 2, 1–13.
5.  Yue, H.; Eastman, P.S.; Wang, B.B.; Minor, J.; Doctolero, M.H.; Nuttall, R.L.; Stack, R.; Becker, J.W.; Montgomery, J.R.; Vainer, M.; Johnston, R. Nucleic Acids Res. 2001, 29, e41.
6.  Perou, C.M.; Sorlie, T.; Eisen, M.B.; van de Rijn, M.; Jeffrey, S.S.; Rees, C.A.; Pollack, J.R.; Ross, D.T.; Johnsen, H.; Akslen, L.A.; Fluge, O.; Pergamenschikov, A.; Williams, C.; Zhu, S.X.; Lonning, P.E.; Borresen-Dale, A.L.; Brown, P.O.; Botstein, D. Nature. 2000, 406, 747–752.
7.  Golub, T.R.; Slonim, D.K.; Tamayo, P.; Huard, C.; Gaasenbeek, M.; Mesirov, J.P.; Coller, H.; Loh, M.L.; Downing, J.R.; Caligiuri, M.A.; Bloomfield, C.D.; Lander, E.S. Science. 1999, 286, 531–537.
8.  Alizadeh, A.A. Nature. 2000, 403, 503–511.
9.  Bittner, M.; Meltzer, P.; Chen, Y.; Jiang, Y.; Seftor, E.; Hendrix, M.; Radmacher, M.; Simon, R.; Yakhini, Z.; Ben-Dor, A.; Sampas, N.; Dougherty, E.; Wang, E.; Marincola, F.; Gooden, C.; Lueders, J.; Glatfelter, A.; Pollock, P.; Carpten, J.; Gillanders, E.; Leja, D.; Dietrich, K.; Beaudry, C.; Berens, M.; Alberts, D.; Sondak, V. Nature. 2000, 406, 536–540.
10. Hedenfalk, I.; Duggan, D.; Chen, Y.; Radmacher, M.; Bittner, M.; Simon, R.; Meltzer, P.; Gusterson, B.; Esteller, M.; Kallioniemi, O.P.; Wilfond, B.; Borg, A.; Trent, J. N Engl J Med. 2001, 344, 539–548.
11. Herbrink, P.; Van Bussel, F.J.; Warnaar, S.O. J Immunol Methods. 1982, 48, 293–298.
12. Frank, R. Tetrahedron. 1992, 48, 9217–9232.
13. Geysen, H.M.; Meloen, R.H.; Barteling, S.J. Proc Natl Acad Sci USA. 1984, 81, 3998–4002.
14. Guschin, D.; Yershov, G.; Zaslavsky, A.; Gemmell, A.; Shick, V.; Proudnikov, D.; Arenkov, P.; Mirzabekov, A. Anal Biochem. 1997, 250, 203–211.
15. Buessow, K.; Cahill, D.; Nietfeld, W.; Bancroft, D.; Scherzinger, E.; Lehrach, H.; Walter, G. Nucleic Acids Res. 1998, 26, 5007–5008.
16. Lueking, A.; Horn, M.; Eickhoff, H.; Buessow, K.; Lehrach, H.; Walter, G. Anal Biochem. 1999, 270, 103–111.
17. Ge, H. Nucleic Acids Res. 2000, 28, e3.
18. Mendoza, L.G.; McQuary, P.; Mongan, A.; Gangadharan, R.; Brignac, S.; Eggers, M. Biotechniques. 1999, 27, 778–788.
19. Joos, T.O.; Schrenk, M.; Hopfl, P.; Kroger, K.; Chowdhury, U.; Chowdhury, S. .D..; Schorner, D.; Durr, M.; Herick, K.; Herick, R.; Herick, S.; Sohn, K.; Hammerle, H. Electrophoresis. 2000, 21, 2641–2650.
20. Silzel, J.W.; Cercek, B.; Dodson, C.; Tsay, T.; Obremski, R.J. Clin Chem. 1998, 44, 2036–2043.
21. Arenkov, P.; Kukhtin, A.; Gemmell, A.; Voloshchuk, S.; Chupeeva, V.; Mirzabekov, A. Anal Biochem. 2000, 278, 123–131.
22. MacBeath, G.; Schreiber, S.L. Science. 2000, 289, 1760–1763.
23. deWildt, R.M.T.; Mundy, C.R.; Gorick, B.D.; Tomlinson, I.M. Nature Biotechnol. 2000, 18, 989–994.

24. Eisen, M.B.; Brown, P.O. Methods Enzymol. 1999, 303, 179–205.
25. Buessow, K.; Nordhoff, E.; Luebbert, C.; Lehrach, H.; Walter, G. Genomics. 2000, 65, 1–8.
26. Rowe, C.A.; Scruggs, S.B.; Feldstein, M.J.; Golden, J.P.; Ligler, F.S. Anal Chem. 1999, 71, 433–439.
27. Mooney, J.F.; Hunt, A.J.; McIntosh, J.R.; Liberko, C.A.; Walba, D.M.; Rogers, C.T. Proc Natl Acad Sci USA. 1996, 93, 12,287–12,291.
28. Afanassiev, V.; Hanemann, V.; Woelfl, S. Nucleic Acids Res. 2000, 28, e66.
29. Kenna, J.G.; Major, G.N.; Williams, R.S. J Immunol Methods. 1985, 85, 409–419.
30. Vogt, R.F.J.; Phillips, D.L.; Henderson, O.; Whitfield, W.; Spierto, F.W. J Immunol Methods. 1987, 101, 43–50.
31. Emmrich, F. J Immunol Methods. 1984, 72, 501–503.
32. Robertson, P.W.; Whybin, L.R.; Cox, J. J Immunol Methods. 1985, 76, 195–197.
33. Graves, H.C.B. J Immunol Methods. 1988, 111, 157–166.
34. Mauracher, C.A.; Mitchell, L.A.; Tingle, A.J. J Immunol Methods. 1991, 145, 251–254.
35. Ruiz-Taylor, L.A.; Martin, T.L.; Zaugg, F.G.; Witte, K.; Indermuhle, P.; Nock, S.; Wagner, P. Proc Natl Acad Sci USA. 2001, 98, 852–857.
36. Prime, K.L.; Whitesides, G.M. Science. 1991, 252, 1164–1167.
37. Dalluge, J.J.; Sander, L.C. Anal Chem. 1998, 70, 5339–5343.
38. Hoffman, W.L.; O'Shannessy, D.J. J Immunol Methods. 1988, 112, 113–120.
39. Matson, R.S.; Little, M.C. J Chromatogr. 1988, 458, 67–77.
40. Diehl, F.; Grahlmann, S.; Beier, M.; Hoheisel, J.D. Nucleic Acids Res. 2001, 29, 38.
41. Hegde, P.; Qi, R.; Abernathy, K.; Gay, C.; Dharap, S.; Gaspard, R.; Hughes, J.E.; Snesrud, E.; Lee, N.; Quackenbush, J. Bio Techniques. 2000, 29, 548–562.
42. Lu, B.; Smyth, M.R.; O'Kennedy, R. Analyst. 1996, 121, 29r–32r.
43. Widjojoatmodjo, M.N.; Fluit, A.C.; Torensma, R.; Verhoef, J. J Immunol Methods. 1993, 165, 11–19.
44. Shannessy, D.J. .O.; Quarles, R.H. J Appl Biochem. 1985, 7, 347–355.
45. Prisyazhnoy, V.S.; Fusek, M.; alakhov, Y.B. J Chromatogr. 1988, 424, 243–253.
46. Vankova, R.; Gaudinova, A.; Suessenbekova, H.; Dobrev, P.; Strnad, M.; Holik, J.; Lenfeld, J. J Chromatogr A. 1998, 811, 77–84.
47. Harlow, E.; Lane, D. Using Antibodies. Cold Spring Harbor; Cold Spring Harbor Laboratory Press: 0, 1999, 81–84.
48. Frank, R.; Overwin, H. Methods Mol Biol. 1996, 66, 149–169.
49. Robinson, W.H.; DiGennaro, C.; Hueber, W.; Haab, B.B.; Kamachi, M.; Dean, E.J.; Fournel, S.; Fong, D.; Genovese, M.C.; Neuman de Vegvar, H.E.; Steiner, G.; Hirschberg, D.L.; Muller, S.; Pruijn, G.J.; vanVenrooij, W.J.; Smolen, J.S.; Brown, P.O.; Steinman, L.; Utz, P. Nat Med. 2002, 8, 295–301.
50. Jones, V.W.; Kenseth, J.R.; Porter, M.D.; Mosher, C.L.; Henderson, E. Anal Chem. 1998, 70, 1233–1241.
51. Davies, H.; Lomas, L.; Austen, B. Biotechniques. 1999, 27, 1258–1261.
52. Schweitzer, B.; Wiltshire, S.; Lambert, J.; O'Malley, S.; Kukanskis, K.; Zhu, Z.; Kingsmore, S.F.; Lizardi, P.M.; Ward, D.C. Proc Natl Acad Sci USA. 2000, 97, 10,113–10,119.

53. Albala, J.S.; Franke, K.; McConnell, I.R.; Pak, K.L.; Folta, P.A.; Rubinfeld, B.; Davies, A.H.; Lennon, G.G.; Clark, R. J Cell Biochem. 2000, 80, 187–191.
54. Walhout, A.J.; Temple, G.F.; Brasch, M.A.; Hartley, J.L.; Lorson, M.A.; van den Heuvel, S.; Vidal, M. Methods Enzymol. 2000, 328, 575–592.
55. Scott, J.K.; Smith, G.P. Science. 1990, 249, 386–390.
56. Boder, E.T.; Wittrup, K.D. Nature Biotechnol. 1997, 15, 553–557.
57. Roberts, R.W.; Szostak, J.W. Proc Natl Acad Sci USA. 1997, 94, 12,297–12,302.
58. Kardamakis, D. Anticancer Res. 1996, 16, 2285–2288.
59. Molina, R.; Jo, J.; Filella, X.; Zanon, G.; Pahisa, J.; Munoz, M.; Farrus, B.; Latre, M.L.; Escriche, C.; Estape, J.; Ballesta, A.M. Breast Cancer Res Treat. 1998, 51, 109–119.
60. Reineke, U.; Sabat, R.; Kramer, A.; Stigler, R.D.; Seifert, M.; Michel, T.; Volk, H.D.; Schneider-Mergener, J. Mol Divers. 1995, 1, 141–148.
61. Kramer, A.; Volkmer-Engert, R.; Malin, R.; Reineke, U.; Schneider-Mergener, J. Peptide Res. 1993, 6, 314–319.
62. James, J.A.; Harley, J.B. Immunol Rev. 1998, 164, 185–200.

# 4

# Protein Biochips: Powerful New Tools to Unravel the Complexity of Proteomics?

**STEFFEN NOCK and PETER WAGNER**
*Zyomyx, Inc.*
*Hayward, California, U.S.A.*

## I. INTRODUCTION

The determination of the human genome sequence coupled with a highly parallel analysis of gene expression patterns have produced an avalanche of genomic data with the promise to unravel complex biological processes at the molecular level. To realize this promise, a comprehensive approach to data collection must be extended beyond nucleic acids to proteins—the biomolecules that are responsible for virtually all biological structures and processes.

The explosion in our understanding of DNA sequence and expression patterns resulted from technical innovations, including automated high-throughput DNA sequencing machines and miniaturized DNA biochips. Comparable advances in the technology of protein analysis, combining miniaturization with increased throughput, will be complicated by the diversity and more delicate nature of protein structures.

One technical approach that will have a profound impact on protein analysis is the development of miniaturized protein biochips containing high-density arrays of proteins. Protein biochips can be developed to monitor protein expression patterns, protein structures, and protein activities and to identify protein–protein, protein–nucleic acid, and protein–small molecule interactions in a highly parallel manner. The successful development of these biochips is being catalyzed and enabled by new technological advances in material science and the integration of these innovations with other disciplines such as microdevice technology, detection physics, and bioinformatics.

## II.  PROTEOMICS: THE POSTGENOMIC REVOLUTION

The precise definition of a biological system by quantitative measurements of its individual components is an essential but largely unexplored area of biology. DNA microarrays in combination with techniques such as the serial analysis of gene expression (SAGE) are providing global, relatively quantitative profiles of mRNA in different cells and organisms at specific times. The correlation between mRNA expression and abundance of the encoded protein is, however, very poor [1,2]. Furthermore, mRNA expression data provide no information about post-translational modifications or protein activities, interactions, and localization. Proteome analysis or "proteomics," defined as the analysis of the protein complement expressed by a genome [3], is required to accurately complete the quantitative description of the precise state of a complex biological system [4].

## III.  PROTEOMIC TECHNOLOGIES: THE NEED FOR INNOVATION!

Some of the currently available protein discovery and analysis methods are two-dimensional polyacrylamide gel electrophoresis (2D PAGE), new chromatographic techniques, metabolic labeling and enrichment of isotopically labeled proteins, and isotope-coded affinity tag (ICAT). Although each of these techniques has proven quite useful under certain circumstances, none are able to thoroughly characterize the complex proteome of an organism.

Since O'Farrell [5] and Klose [6] demonstrated that it was possible to separate proteins according to their charge and molecular weight by electrophoresis, 2D PAGE has become the method of choice for the analysis of complex protein mixtures. High-resolution 2D PAGE has been combined with extremely sensitive fluorescence-based detection systems and mass spectroscopy to permit the identification of individual proteins in complex mixtures.

Nevertheless, as a protein discovery tool, 2D PAGE is limited by its overall throughput, resolution, and sensitivity [7]. Over the last couple of years, methods designed to increase the sensitivity of 2D PAGE have emerged along with new technologies that avoid electrophoresis entirely (for review, see Ref. 8).

Procedures that enrich samples for low-abundance proteins by using chromatographic methods like hydroxylapatite chromatography have been successful in special cases but are certainly not generally applicable [9]. A technique to facilitate quantitative comparisons has also been developed that relies on the metabolic labeling of proteins with either $^{15}N$ or $^{14}N$ as the sole nitrogen source. Two differently labeled protein sources are then combined and separated by 2D PAGE. Spots of interest are excised, digested with trypsin, and the resulting peptides analyzed by mass spectrometry. The mass difference of identical peptides from each sample (attributable to the isotopic labeling) allows for the determination of relative amounts of several proteins in both samples [10,11]. Clearly, this technology is limited to biological systems that can be grown in special media in order to incorporate an isotopic label.

A slightly similar but more broadly applicable technique that avoids 2D PAGE, and therefore its limitations, has been developed based on a class of reagents termed isotope-coded affinity tags (ICATs) [12]. Cysteine residues in proteins are labeled after cell growth with either a "heavy" or "light" reagent created using isotopes of different mass. After the tagging, the two fractions are mixed, digested, affinity purified, and analyzed by mass spectrometry. Ratios between "heavy" and "light" peptides allow relative quantification of the same protein in the different samples. By using this approach, no electrophoretic separation step is involved and less protein is required.

Isotope-coded affinity tags and metabolic labeling, as described, can be applied to define and quantify the levels of proteins in individual samples but only under very specific or ideal conditions. 2D PAGE can also be used to identify and quantify proteins, but with limited resolution, sensitivity, and denaturation of the protein. Because proteomics is aimed at defining the complete protein status of a cell, tissue, or organism, including the abundance, structural status, activities, and interactions of the proteins, these current methods need to be complemented by new technologies.

Miniaturized protein biochips are one of the emerging technologies that will have the ability to expand our techniques for quantitative analysis of the proteome. This technology should work under physiological conditions to assure that the proteins maintain their native form. Furthermore, protein biochips will allow the analysis of small samples (micrograms), unlike 2D PAGE, which requires milligram amounts of protein. The remainder of this chapter will address the issues and solutions for the development of such protein biochips.

## IV. FROM DNA BIOCHIPS TO PROTEIN BIOCHIPS: A DIFFICULT TASK

The success of the DNA biochip technology is mainly due to the integration and miniaturization of new developments in material science and nucleic acid biochemistry. Macroscopic assays such as DNA sequencing and hybridization have been scaled down to microscopic dimensions and combined with new, highly sensitive detection methodologies. The spatial resolution of single spots in DNA microarrays has allowed hundreds of thousands of assays to be conducted in a 1-cm$^2$ area. The possibility of analyzing the expression patterns of all genes of a cell or organism by DNA microarrays together with the outcome of the Human Genome Project has revolutionized the thinking of biologists. Hypothesis-driven science will evolve to real discovery science.

The rigid and stable properties of DNA molecules permitted the straightforward development of DNA microarrays. These microarrays currently contain either intact cDNAs or short oligonucleotides. The immobilization of the DNA relies either on physisorption of the negatively charged DNA on adhesive supports like poly-lysine or nitrocellulose [13] or on coupling through amino or thiol groups added in the form of modified nucleotides during the synthesis process.

## V.  CAN THE SAME APPROACH BE USED FOR THE DEVELOPMENT OF PROTEIN BIOCHIPS?

As previously mentioned, the delicate nature of protein structures requires a more sophisticated approach for the construction of protein biochips. Retention of protein activity is required to maximize sensitivity and reproducibility. Protein activity is typically dependent on conformation. Upon adsorption to a surface, interaction forces at the solid–liquid interface generally alter protein conformation and frequently cause complete denaturation. Paramount to the successful development of protein biochips is the careful and specific surface engineering, designed to avoid denaturing effects at each step of both manufacturing and use. The diversity and complexity of proteins tend to obfuscate a general strategy for the development of a protein biochip technology. However, careful consideration for the design of the biochip, including chip material, architecture, surface coating, the mode of protein immobilization, the type of protein dispenser used to transfer proteins to discrete areas, and the mode of detection will result in a broadly applicable protein-based biochip.

Unfortunately there is no one key innovation that will fully enable protein biochips. What is needed is an optimization of specific protein–surface interactions coupled with a minimum amount of nonspecific binding and protein denaturation. In addition, the parallel development of protein-compatible dispensers and sample label-independent detection methods are essential to take full advantage of the versatility and flexibility of the protein chip and to implement high-throughput protein analysis and discovery.

## VI.  PROTEIN IMMOBILIZATION

A major bottleneck in the construction of high-density protein microarrays is the limited understanding of protein interactions with engineered materials. To date, proteins have been immobilized on solid supports to develop enzymatic and immunoassays, as well as affinity chromatography techniques. These "macro" devices function reasonably well, even though a substantial portion of the immobilized protein is either inactive or inaccessible as a result of unfavorable interactions with the solid support. Enzymatic and immunological assays amplify the detection signal to compensate for the limited fraction of protein that contributes to the assay. In fact, the fraction of protein molecules participating in a standard enzyme-linked immunosorbent assay (ELISA) is often below 5% with respect to the total number of immobilized species. The development of miniaturized, high-density protein microarrays will require significant advances in protein immobilization because the amounts of such proteins will be limiting ($10^7$ molecules per microspot).

Protein immobilization techniques have evolved over the years from ill-defined, physical adsorption onto highly porous materials of macroscopic devices, to site-specific immobilization of proteins onto fairly well-defined interfaces [14]. Poly-lysine, nitrocellulose, and polystyrene are high-affinity protein-binding materials and are used routinely in diagnostic assays [blotting techniques, ELISA, or radio immunoassay (RIA)]. High-throughput protein detection using both of these systems has been investigated with limited success [15]. Alternatively, proteins have been covalently immobilized directly onto inorganic surfaces by using coupling reagents that react with either glass or gold surfaces, or by taking advantage of natural functionalities on proteins such as cysteine thiols that have a high affinity for noble metal substrates. This immobilization method is ineffective for device fabrication because protein denaturation at the inorganic surface interface makes assays unpredictable or even impossible at the micron scale [16].

The first real improvement in protein immobilization came with the discovery of methods for creating thin films of self-assembled monolayers (SAMs) on gold or glass with functional groups that can be conjugated to proteins [14]. These layers are often terminated with either amine or carboxylic acid groups that can be coupled to the respective carboxyl or amine groups in the proteins, using *N*-ethyl-*N* (3-dimethylaminopropyl)-carbodiimide (EDC) as a coupling agent. Unfortunately, because the surface-active groups, either carboxyl or amino, are also present within the proteins, polymerization is difficult to avoid during the activation of the surface species. Regardless, many useful assay techniques are based on this technology, including bead-based or resin-based assays, and the Biacore surface plasmon resonance assays [17].

The introduction of protein reactive ω-functionalization in organic thin films was the next significant improvement in protein immobilization [18]. For example, monolayers derived from dithio-bis(succinimidylundecanoate) (DSU) react directly with exposed amine groups on proteins. Working with protein-reactive surfaces removes the need for the potentially destructive coupling agent. In addition, it guarantees a specific molecular density of end groups, which is important for obtaining specific molecular densities of immobilized proteins. This is, in fact, a very important feature if spot-to-spot reactivity is to be analyzed quantitatively. Bioreactive surfaces prepared in this way will react indiscriminately with amine groups on proteins, resulting in multiple points of attachment and random orientation of immobilized species [19].

The problems of multiple sites of attachment and random orientation are solved by site-specific immobilization techniques using functionalized surfaces that react specifically with tags fused to target proteins [20–22]. These techniques provide a means for immobilizing a protein through a single engineered position within that protein. However, the attachment is not through a covalent bond. Strong, covalent attachment is essential for applications in proteomics or diagnostics where the protein array is exposed to heterogeneous mixtures of analytes that might disrupt noncovalent interactions. Optimal protein immobilization will merge both covalent and site-specific immobilization methods to optimize surface coverage, molecular density, and protein activity.

In addition to maximizing the activity of the bound protein, the biochip surface must be resistant to nonspecific protein binding. Reduced nonspecific binding is an extremely important feature because proteomics applications often involve exposing the chip to complex biological samples. New developments in biomaterials engineering and implant integration might help to address these issues if implemented in a biochip environment.

Mixed monolayers are easily integrated with the site-specific immobilization methods. They consist of one component resistant to protein adsorption such as polyethylene glycol chains [23] and a second component with functional groups that are reactive to specific tags in proteins [24]. One possibility is to take advantage of the specific interaction between an exposed hydrazine on a chip surface and an introduced keto group at a specific site in an engineered protein. Incorporating orthogonal chemistry into well-chosen sites in proteins will allow oriented, covalent binding of the protein via a single attachment site. This will assure maximal interaction with the binding partner from the solution phase. At the same time, the protein-resistant moieties of the surface will prevent nonspecific protein adsorption.

## VII.  SURFACE TOPOGRAPHY

Another important aspect of protein immobilization lies in the topography of the chip surface. Both the presence of micron-sized irregularities on the surface and the nanotopography and texture at the molecular level will have an impact on the performance of assays in high-density arrays. Deep cavities in surfaces lead to protein entrapment and protein inactivation. Extremely flat surfaces, in contrast, have been shown to improve the orientation and activity of the immobilized protein [25]. Figure 1 depicts scanning tunneling microscopy images of both a rough and an atomically flat surface. The gold surface in Fig. 1a shows a high density of macroscopic defects and is therefore a poor substrate for protein immobilization. In contrast, the surface shown in Fig. 1b is free of large defects and atomically flat. The extremely flat nature of the substrate helps to minimize the interaction of a bound protein with the surface and therefore reduces surface-

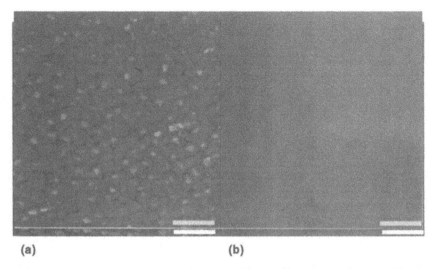

(a)                                            (b)

**Figure 1**   Scanning transmission micrographs of gold surfaces. Scale bars are 300 nm. Panel a shows a 100-nm-thick gold film evaporated onto a silicon substrate at room temperature. Panel b shows a Template Stripped Gold (TSG) surface optimized for topography. Both images are displayed with the same Z-scale for comparison. TSG exhibits subnanometer roughness over micron-sized areas. Roughness of regular gold surfaces is on the order of magnitude of protein dimensions. (See the color plate.)

induced inactivation. It also increases the accessibility of surface-immobilized proteins to other proteins in the sample.

## VIII.   PROTEIN DISPENSERS

The dispensing of the proteins onto the chip surface is another critical step in the development of a successful protein biochip technology. Due to the fragile nature of proteins, standard capillary-based spotters (e.g., common DNA arrayers) are often unsuitable and their contact printing mode is generally destructive to the chip surface. Sequential spotting of proteins via glass capillaries would lead to protein inactivation at several steps during the spotting event. First, proteins would adsorb to the glass capillary and become inactivated. Droplet formation during the dispensing event is also unfavorable due to denaturing interactions of the protein with the air–liquid interface. Finally, the entire biochip surface needs to be kept moist to prevent protein inactivation.

## IX.   ZYOMYX PROTEIN BIOCHIPS

One concept, taking all the mentioned issues into account, is being developed in our laboratories. A prototype of a Zyomyx protein chip is shown in Fig. 2. The

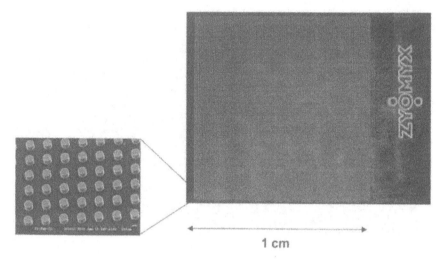

1 cm

**Figure 2**   Example of a possible protein biochip design. In a 1-cm × 1-cm field, 10,000 proteins can be immobilized on the top of small posts. In combination with a specially designed dispenser, proteins can be deposited in parallel onto the posts.

chip consists of an array of 100 × 100 posts in a 1-cm² area. The posts are etched into a silicon substrate using deep-reactive ion etch technology. The posts create defined areas for protein immobilization and allow a contact-free dispensing process. The top surfaces of the posts are then modified using different thin films to make them protein reactive. Figure 3 shows different protein immobilization approaches. On the left side, the posts are coated with nitrocellulose, on the right side, the posts are coated with a monomolecular thin film reactive towards amino groups. Physisorption of a fluorescently labeled antibody on the nitrocellulose-coated posts leads to very heterogeneous, nonreproducible protein adsorption with a high background. Defined covalent interaction of the same antibody with the monomolecular thin-film, however, shows a very homogeneous and reproducible protein distribution.

Figure 4 shows a top view of an array of 13 different monoclonal antibodies dispensed onto 25 posts on the chip. The diameter of each post is only 15 μm. In one experiment, the chip was exposed to a mixture of fluorescently labeled antigens, specific for two of the immobilized antibodies. Specific interaction of the antibodies with the cognate antigen is shown in Fig. 4b. This approach demonstrates a miniaturized version of an ELISA with, to date, unsurpassed density that can be used for the detection and quantification of proteins.

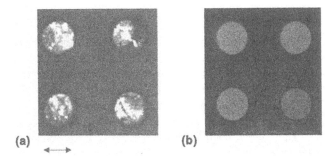

(a)

(b)

50 μm

**Figure 3** Example of protein immobilization on the posts described in Fig. 2. A fluorescently labeled antibody was immobilized on posts coated with nitrocellulose (a) or a bioreactive monolayer (b). The defined approach using monomolecular thin films as a surface coating shows reproducible and homogeneous protein immobilization, whereas physisorption on nitrocellulose is very heterogeneous.

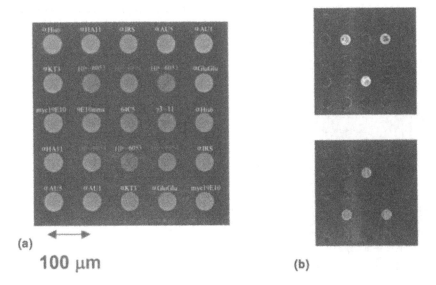

(a)

100 μm

(b)

**Figure 4** Example of a protein array experiment. Twenty-five post surfaces (Fig. 3) contain 1 of 13 immobilized antibodies raised against peptides and proteins (a). The array was then incubated with a mixture of two fluorescently labeled antigens. Fluorescence microscopy shows specific interaction of the antigens with the corresponding antibodies (b). (See the color plate.)

## X.  DETECTION

Using fluorescence for detection for ultrahigh-throughput analysis is impractical because the technique relies either on sandwich-type assays or the ability to incorporate labels into the analyte. For complex protein mixtures such as cell lysates or blood with fluorescent markers, it is very unlikely that labeling would be successful. Additionally, the use of sandwich assays becomes impractical due to the fact that specific secondary antibodies have to be generated for each spot.

The optimal readout for the described protein biochip will be label independent. Technologies such as imaging surface plasmon resonance and imaging ellipsometry are very promising in that respect. In Fig. 5, different densities of an immobilized antibody on the top of the posts are visualized using label-independent imaging ellipsometry.

## XI.  DOWNSCALED IMMUNOASSAYS AS A PROTEOMICS TOOL?

DNA arrays are widely used to monitor expression levels of mRNAs. A similar device to detect changes in expression levels of proteins in serum or cell lysates would revolutionize the understanding of the status and progression of diseases.

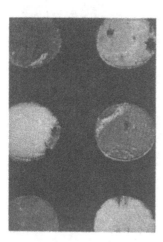

## 50 μm

**Figure 5**  Imaging ellipsometry is one possibility for label-independent detection of different amounts of protein bound to the post surface from Fig. 2. (See the color plate.)

A protein biochip containing capture agents like antibodies or fragments thereof could be used as an expression-profiling tool for proteins in complex samples.

Antibody-based immunoassays are the most commonly used type of diagnostics assay and still one of the fastest growing technologies for the analysis of biomolecules, especially proteins [26]. These assays are usually performed in 96-well microtiter plates, which limits the throughput to a few thousand assays per day. The volume per assay is usually between 50 and 100 µL and therefore consumes relatively larger volumes of reagents. Ekins [27] described downscaled immunoassays on microspots of immobilized antibodies. Recent attempts to transfer immunoassays onto filter membranes or glass slides to increase the density and reduce the sample volume have shown some promising results. Mendoza et al. [28] describe a protein biochip to perform high-throughput ELISAs using antigen arrays in a miniaturized 96-well format on a glass substrate. With marker antigens, they demonstrate the feasibility of multiplexed detection of arrayed antigens. In a similar approach, Joos et al. [29] explored the use of a microarray ELISA for autoimmune diagnostics. The data based on a microarray ELISA of several autoantigens were comparable with standard ELISA tests. Mirzabekov and co-workers have used 3D polyacryalmide gel patches on glass slides [30] for immobilization of antibodies. A recent publication of MacBeath and Schreiber [31] showed interaction of a small set of surface immobilized proteins with other proteins and small molecules in solution. There are several other approaches to protein microarrays using either photolithography on silane monolayers [32], combining microwells with microsphere sensors [33], or ink-jetting onto polystyrene film [34].

In most of the above cases, fluorescently labeled detection antibodies are used in a sandwich-type assay to monitor the binding of an analyte. Low- density, macroscopic arrays have been shown to be interfaceable with SELDI (surface-enhanced laser desorption/ionization) mass spectrometry [35].

## XII. OTHER PROTEIN BIOCHIP APPLICATIONS

Various embodiments of the protein biochips can be developed for a wide range of uses, including protein discovery, profiling, structure determination, and the high-throughput assessment of protein–protein and protein–small molecule interactions. The specific application for which a protein biochip is to be used will dictate the biochip design. In combination with phage display or other in vitro display technologies, protein chips could also be used as a protein discovery tool. Phage display is a method whereby bacteriophage particles are made to express libraries of proteins or peptides as a fusion to their coat proteins. Libraries of active antibody fragments on phage could be selected against proteins in tumor cells for example. Protein biochips containing thousands of randomly selected antibody fragments in combination with mass spectrometry as readout could then be used as a discovery tool for these antigens.

Another very important application of protein biochips is in the field of protein–protein interaction studies. Current technologies like the yeast two-hybrid systems are based on in vivo protein interactions and often suffer from high false-positive rates. Another attractive way to study protein–protein interactions is to purify the entire multiprotein complex by affinity-based methods. Usually, one protein is bound onto a matrix (e.g., a bead) and then used as bait for the interaction partners. Even low-density arrays of bait proteins on planar surfaces have shown successful results [35,36]. The use of high-density protein arrays in combination with mass spectrometry will have a huge impact on the characterization of, for example, signal transduction pathways and their cross-talk.

Further applications are in the field of protein–small molecule interactions. This platform can also be integrated into a microfluidics system to perform micro-scale functional assays on complex enzyme systems. These enzyme biochips will allow drug discovery researchers to rapidly detect activities from an array of enzymes and to conduct high-throughput screening of substrate and inhibitor libraries in nanoliter volumes.

## XIII.  CONCLUSION

It is evident that recent years have seen advances in 2D gel electrophoresis and methods to characterize complex protein mixtures. However, none of the methods will allow the comprehensive screening of proteins in an organism. The majority of current technologies are only capable of revealing proteins of high-to-moderate abundance. In the DNA field, the development of high-density nucleic acids arrays has become a tool of choice to analyze complex biological systems. With the development of similar microanalytical tools for proteins, scientists will be armed with the ideal tool to better understand the complex interplay of proteins in an organism.

## REFERENCES

1. Anderson, L.; Seilhamer, J. Electrophoresis. 1997, 18, 533–537.
2. Gygi, S.; Rochon, Y.; Franza, B. R.; Aebersold, R. Mol. Cell. Biol. 1999, 19, 1720–1730.
3. Pennington, S. R.; Wilkins, R. W.; Hochstrasser, D. F.; Dunn, M. J. Trends Cell Biol. 1997, 7, 168–173.
4. Wilkins, M. R.; Williams, K. L.; Appel, R. D.; Hochstrasser, D. F. In Proteome Research: New Frontiers in Functional Genomics; Springer-Verlag: Berlin, 1997.
5. O'Farrell, P. H. J. Biol. Chem. 1975, 250, 4007–4021.

6. Klose, J. Humangenetik. 1975, 26, 231–243.
7. Gygi, S. P.; Corthals, G. L.; Zhang, Y.; Rochon, Y.; Aebersold, R. Proc Natl. Acad. Sci. USA. 2000, 97, 9390–9395.
8. Washburn, M. P.; Yates, J. R. Curr. Opin. Microbiol. 2000, 3, 292–297.
9. Fountoulakis, M.; Takacs, M. F.; Berndt, P.; Langen, H.; Takacs, B. Electrophoresis. 1999, 20, 2181–2195.
10. Oda, Y.; Huang, K.; Cross, F. R.; Cowburn, D.; Chait, B. T. Proc. Natl. Acad. Sci. USA. 1999, 96, 6591–6596.
11. Pasa-Tolic, L.; Jensen, P. K.; Anderson, G. A.; Lipton, M. S.; Peden, K. K.; Martinovic, S.; Tolic, N.; Bruce, J. E.; Smith, R. D J. Am. Chem. Soc. 1999, 121, 7949–7950.
12. Gygi, S. P.; Rist, B.; Gerber, S. A.; Turecek, F.; Gelb, M. H.; Aebersold, R. Nature Biotechnol. 1999, 17, 994–999.
13. Schena, M.; Shalon, D.; Davis, R. W.; Brown, P. O. Science. 1995, 270, 467–470.
14. Taylor, R. F. In Protein Immobilization: Fundamentals and Applications; Marcel Dekker: 0, 1991.
15. Hage, D. S. Anal. Chem. 1999, 71, 294–304.
16. Blawas, A. S.; Reichert, W. M. Biomaterials. 1998, 19, 595–609.
17. Johnsson, B.; Lofas, S.; Lindquist, G.; Edstrom, A.; Muller, A.; Hillgren, R. M.; Hansson, A. J. Mol. Recogn. 1995, 8, 125–131.
18. Wagner, P.; Hegner, M.; Kernen, P.; Zaugg, F.; Semenza, G. Biophys. J. 1996, 70, 2052–2066.
19. Wagner, P.; Zaugg, F.; Kernen, F.; Hegner, M.; Semenza, G. J. Vac. Sci. Technol. B. 1996, 14, 1466–1471.
20. Sigal, G. B.; Bamdad, C.; Barberis, A.; Strominger, J.; Whitesides, G. M. Anal. Chem. 1996, 68, 490–497.
21. Keller, T. A.; Duschl, C.; Kroger, D.; Sevin-Landais, A. F.; Vogel, H. Supramol. Sci. 1995, 2, 155–160.
22. Nock, S.; Spudich, J. A.; Wagner, P. FEBs Lett. 1997, 414, 233–238.
23. Jeon, S. I.; Lee, J. H.; Andrade, J. D.; de Gennes, P. G. J. Colloid Interf. Sci. 1991, 142, 149–158.
24. Ruiz-Taylor, L. A.; Martin, T. L.; Zaugg, F. G.; Witte, K.; Indermuhle, P.; Nock, S.; Wagner, P. Proc. Natl. Acad. Sci. USA. 2001, 89, 852–857.
25. Wagner, P.; Hegner, M.; Guentherodt, H.-J.; Semenza, G. Langmuir. 1995, 11, 3867–3875.
26. Borrebaeck, C. A. Immunol. Today. 2000, 21, 379–382.
27. Ekins, R. P. J. Pharm. Biomed. Anal, 7, 155–168.
28. Mendoza, L. G.; McQuary, P.; Mongan, A.; Gangadharan, R.; Brignac, S.; Eggers, M. Biotechniques. 1999, 27, 778–788.
29. Joos, T. O.; Schrenk, M.; Hopfl, P.; Kroger, K.; Chowdhury, U.; Stoll, D.; Schorner, D.; Durr, M.; Herick, K.; Rupp, S.; Sohn, K.; Hammerle, H. Electrophoresis. 2000, 21, 2641–2650.
30. Arenkov, P.; Kukhtin, A.; Gemmell, A.; Voloshchuk, S.; Chupeeva, V.; Mirzabekov, A. Anal. Biochem. 2000, 278, 123–131.
31. MacBeath, G.; Schreiber, S. L Science. 2000, 289, 1760–1763.
32. Mooney, J. F.; Hunt, A. J.; McIntosh, J. R.; Liberko, C. A.; Walba, D. M.; Rogers, C. T. Proc. Natl. Acad. Sci. USA. 1996, 93, 12,287–12,291.

33. Michael, K. L.; Taylor, L. C.; Schultz, S. L.; Walt, D. R. Anal. Chem. 1998, 70, 1242–1248.
34. Brizzolara, R. A. Biosens. Bioelectron. 2000, 15, 63–68.
35. Davies, H.; Lomas, L.; Austen, B. BioTechniques. 1999, 27, 1258–1261.
36. Lueking, A.; Horn, M.; Eickhoff, H.; Bussow, K.; Lehrach, H.; Walter, G. Anal. Biochem. 1999, 270, 103–111.

# 5

# Functionalized Surfaces for Protein Microarrays: State of the Art, Challenges, and Perspectives

**ERIK WISCHERHOFF**
*Utrecht University*
*Utrecht, The Netherlands*

## I. REQUIREMENTS FOR PROTEIN MICROARRAY SURFACES

Protein microarrays have attracted considerable attention as an alternative and complement to conventional proteomics. With the related technologies rapidly advancing, they promise to provide a more comprehensive picture of an organism's proteome. Moreover, they will significantly acclerate proteomic discovery because of their distinct high-throughput capability. Protein microarrays will permit the screening of very high numbers of protein–protein or compound–protein interactions simultaneously with a high dynamic range and low detection limits, thus enabling one to create protein abundance profiles from complex biological samples, as extreme as an entire organism, with high sensitivity, giving rise to good proteomic coverage, and in short time. However, in order to achieve these desirable features, functionalized surfaces which fulfill several prerequisites are needed.

High-throughput capability is linked to a sufficiently high spot density [i.e., the ability to immobilize a certain number of spots with different receptors (proteins, peptides, or haptens which function as capture probes that bind target molecules from the fluidic sample) on a limited area]. Technological solutions supplying high spot densities do exist, such as piezo-electric, pin, or ring and pin arrayers. A more critical issue seems to be an intelligent interface design, because using proteomic analysis means handling very complex biological solutions. To cope with this, the design of the array surface has to fulfill some requirements which are similar to the specifications for surfaces of affinity interaction biosensors:

- High immobilization density of receptors
- Capability of keeping immobilized biomolecules in an active state
- Capability of inhibiting nonspecific adsorption, because no detector technology to date can discriminate between biologically relevant binding events and undesired physical adsorption of sample material, being responsible for false positives
- Capability of suppressing or at least limiting matrix effects of complex biological solutions on the qualitative and quantitative specific detection of the one component of interest

In the following, existing solutions will be presented. Then, improvements relevant for array surface technology will be discussed.

## II. STATE OF THE ART: ACHIEVEMENTS AND PROBLEMS

A variety of differently functionalized surfaces and immobilization strategies have been used to date to couple proteins to surfaces in array formats. Recent examples are found in the work of MacBeath et al. [1], Brown et al. [2], and Snyder et al. [3]

The approach of MacBeath et al. [1] is based on the classical procedure to physically adsorb proteins onto a surface and then block with albumin or milk powder, as used in enzyme-linked immunosorbent assays (ELISAs). However, to avoid the problem of sterically shielding small proteins and peptides when performing a bovine serum albumin (BSA) blocking step, it was attempted to improve the method by covalently immobilizing BSA on an activated surface and then covalently bind proteins and peptides *on top* of the BSA layer. This kept the small proteins and peptides accessible for analytes in solution (Fig. 1). Among other experiments, a microarray with 10,800 spots was prepared with this immobilization strategy; 10,799 of these spots were prepared with protein G [which interacts specifically with immunoglobulin G (IgG)] and 1 was prepared with FKBP12-rapamycin binding domain (FRB) of FKBP-rapamycin-associated protein, interacting specifically with the human immunophilin FKBP12 in the presence of rapamycin. This microarray was then probed with a mixture of BODIPY-FL-IgG and Cy5-FKBP12, with 100 n$M$ rapamycin included in the buffer. The single FRB spot with Cy5-FKBP12 bound could be clearly identified among the 10,800 spots, 10,799 of them having specifically bound BODIPY-FL-IgG. The experiment highlights the potential of protein microarrays; however, the study lacked a test for resistance

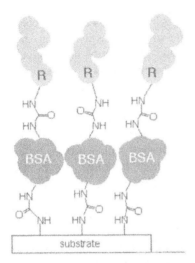

**Figure 1** Schematic representation of the immobilization strategy used by MacBeath et al. [1]. A layer of BSA was immobilized to a surface via disuccinimidyl carbonate coupling chemistry. Then, small proteins or peptides (represented by R) were coupled on top of this BSA layer, again utilizing disuccinimidyl carbonate. Thus, immobilized small biomolecules remain accessible for the analyte.

to nonspecific adsorption (i.e., incubation with a non-interacting, "sticky" protein in large excess and subsequent measurement of the resulting background signal); thus, the suitability of this approach for analysis of complex biological samples remains to be demonstrated.

Brown et al. [2] immobilized proteins via electrostatic interactions. On a glass slide, they deposited a poly(lysine) layer. Then, in a second step, antigens or antibodies were immobilized on this positively charged surface. One hundred fifteen antigen–antibody pairs were tested in this study. Although the antigen–antibody interaction could be monitored precisely, for some interacting pairs with remarkable sensitivity, the effects of nonspecific adsorption were not investigated extensively. As a test for the influence of complex sample compositions on the analytical result, some arrays were incubated with a mixture of specific binding antigen and fetal calf serum, increasing the total protein concentration by 10-fold and 100-fold, respectively. Then, the effects of the added calf serum on the detection limits were investigated. A higher background at higher protein concentrations is supposed to be the major cause of the diminished analytical performance encountered. However, no clear statement is given about abundance and quantity of false positives in this experimental setting (i.e., the occurrence of signal on spots of proteins not interacting specifically with the analyte). Although not addressing the problem experimentally, the authors seem to be aware of the necessity of improved surface properties, concluding with the statement: "A reduction in background through improved blocking of non-specific adsorption should further lower the detection limits."

In another investigation, Zhu and Snyder [3] covalently coupled proteins to a glass surface via low-molar-mass silane linkers. Proteins were coupled via 3-glycidoxypropyltrimethoxysilane in the microwells of a microwell chip. Then, analogous to ELISAs, the surfaces were blocked with 1% BSA. Subsequently, these protein-modified surfaces were used in kinase assays with radiolabeled $^{33}P$-$\gamma$-ATP. Because only the phosphorylation of the attached proteins is monitored, nonspecific adsorption is not a major concern in this particular experiment. Consequently, the authors did not address their attention to this aspect in their investigation. However, because a BSA blocking strategy is chosen, it is unlikely that this immobilization protocol can be successfully adopted to protein–protein interaction measurements in complex biological samples which are the focus of investigations for proteomic discovery.

In summary, all of the discussed coupling methods used for the protein microarrays result in surfaces with some limitations, especially lacking the ability to prevent nonspecific adsorption to an extent mandatory for array-based proteomics. Improvements in this property are crucial to enhance sensitivity and to avoid false positives.

## III. SUPPRESSION OF NONSPECIFIC ADSORPTION: A COMPLEX PROBLEM

Over the last decade, surface-modification methods to improve the biocompatibility of surfaces and to enhance the performance of various types of biosensors has evolved. Improvements are still under way, many of them related to the covalent attachment of hydrophilic polymer or oligomer layers to a substrate. In the following, the achievements in this sector will be discussed in detail, and in the next section, some applications to microarrays will be shown.

There are two principal strategies to attach a polymer or oligomer to a surface: grafting from and grafting onto. Grafting from means the *in situ* polymerization from a surface, whereas grafting onto involves the covalent attachment of previously formed polymer chains to a substrate. Grafting from is believed to posess an inherent advantage: Theoretically, it is supposed to result in higher densities of polymer chains covering the surface. When the polymer chains are growing at the interface, only monomers of small to moderate size have to access the ends of the growing chains, a process which takes place with high diffusion rates and which does not imply steric hinderance. On the contrary, in the grafting onto scenario, entire polymer chains need to access a surface and react there. Diffusion of polymers is generally slower, and especially when some polymer chains are already attached, the access of further polymer chains to a reaction site on the surface will be sterically limited. However, experimental proof of superior performance of surfaces modified by grafting from is lacking. This may be a reason why most of the work related to biocompatible surfaces and biosensors was done with polymeric materials grafted onto the substrate.

In 1990, Löfås and Johnsson [4] introduced a method to modify noble metal surfaces in an effective manner for biosensor applications. Hydrogel matrices composed of polysaccharides (dextran) were utilized to immobilize biomolecules analogous to procedures in affinity chromatography. Noticing the importance of a barrier between the original noble metal surface and the hydrogel itself, they used a self assembled monolayer (SAM) of long-chain ω-functionalized alkyl thiols for primary surface functionalization, because it forms a layer much less prone to defects than shorter-chain variants (Fig. 2). The combination of this dense SAM and the dextran covalently-coupled is rather efficient and became the "standard" sensor surface for the surface-plasmon-resonance (SPR)-based biosensor system commercialized by Biacore AB (Sweden). It still enjoys much popularity among the users of such biosensors today. However, this specific solution is only applicable to noble metal surfaces. The presence of a thin noble metal layer is essential when SPR is employed as a sensing technique, but may be undesirable or even prohibitive for other principles of detection. Furthermore, the study of Löfås and Johnsson is rather limited concerning the parameters of the surface modification. Taking a closer look, one recognizes that a biocompatible surface suitable for biosensing is a complex system. Many parameters have the potential to influence the final performance of the biocompatible layer: chemical composition and morphology of the substrate, thickness, density and chemical composition of barrier and linker layers, immobilization density, average molar mass and chemical composition, as well as substitution pattern of the biocompatible polymer or oligomer layer. Consequently, in the following years, more research was performed investigating these different aspects, some of the work being related only to the prevention of nonspecific adsorption and some also extended to actual biosensing applications.

In 1991, Gombotz et al. [5] performed a study on protein adsorption to poly(ethylene glycol) (PEG) layers bound to poly(ethylene terephthalate) (PET) surfaces. After introducing amino groups on the substrate by plasma polymerization of allylamine, $\alpha,\omega$-amino functionalized PEG was immobilized via cyanuric chloride activation. Then, the surface coverage with PEG and the adsorption of fibrinogen and albumin were investigated as a function of the average molar mass of the PEG. Average molar masses ranged from 200 to 20,000. The adsorption studies showed a clear trend: The higher the average molar

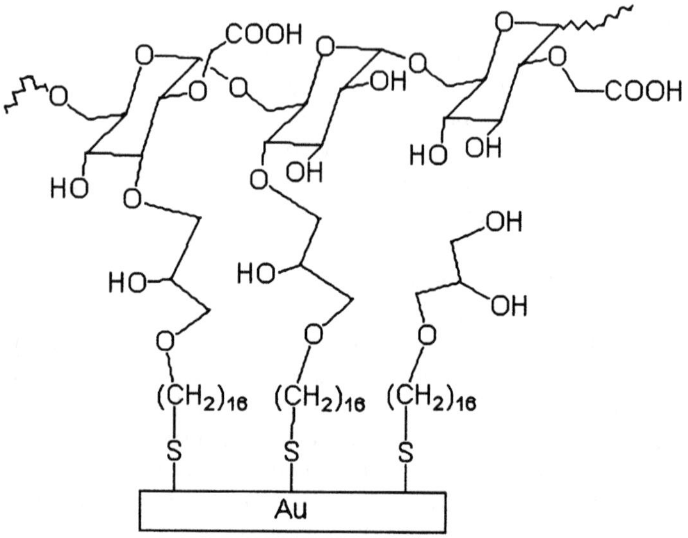

**Figure 2**  Schematic representation of the immobilization matrix developed by Löfås and Johnsson [4]. The self-assembled monolayer serves as a barrier to the gold surface and simultaneously provides the functional groups for the covalent coupling of the dextran. Receptor biomolecules can be immobilized via the carboxymethyl groups on the polysaccharide chains.

mass of the PEG, the more effectively nonspecific adsorption of both albumin and fibrinogen were suppressed. Typically, remaining nonspecific adsorption was in the range of some 100 ng/cm$^2$.

Subsequently, other substrates, linker chemistries, polymers of different chemical compositions and experimental conditions were investigated by others. In their study in 1996, Gauglitz et al. [6] coated surfaces with various amino- and carboxy-substituted polymers. The polymers used were branched poly(ethyleneimine), $\alpha,\omega$-amino functionalized PEG, chitosan, poly(acrylamide-*co*-acrylic acid), and an amino-modified dextran. The amino-substituted polymers were immobilized on glass via an aminosilane/succinic anhydride/*N*-hydroxysuccinimide linker chemistry, whereas the poly(acrylamide-*co*-acrylic acid) was directly coupled to an aminosilanized surface. When probed with a 1 mg/mL ovalbumin solution, nonspecific adsorption was lowest for the dextran derivative. Notably, nonspecific adsorption increased in most cases when a hydrophobic hapten (atrazine) was coupled to the polymer-modified surface. Although the choice of compounds is limited, this investigation gives some insight into the influence of chemical composition on nonspecific adsorption: It demonstrates that charged as well as hydrophobic groups are undesirable.

A problem associated with this study relates to the average molar masses of the polymers. Only one average molar mass per polymer was investigated and the masses varied significantly from 2000 for the PEG to 300,000 for the amino dextran. Therefore, influences of the chemical composition and the molar mass of the compounds are probably superposed. Further complicating the matter, a more recent article of Gauglitz et al. [7] strongly suggests that the chemical composition of the linker layer and especially the exact experimental conditions of the assembly also have a dramatic influence on the performance. In contrast to the work of 1996, particularly dense layers of PEG were obtained by applying the polymers solvent free at 75°C. Thus, the nonspecific adsorption from a 1-mg/mL ovalbumin solution to a PEG surface is brought down from 10 ng/cm$^2$ to below 1 ng/cm$^2$.

Schacht et al. [8] investigated the influence of degree of substitution and of average molar mass of dextrans on the nonspecific adsorption. One percent to 4% of the glucose repeat units of the polymers were modified with thiol groups, which were then used to directly bind the polymers to noble metal surfaces. The molar masses were 5000, 70,000 with two different degrees of substitution, and 500,000. The differences observed in nonspecific adsorption when incubating with a 0.5-mg/mL BSA solution were relatively small for these systems. Absolute values are in the range of some 10 ng/cm$^2$, the smallest molar mass giving rise to the least nonspecific binding (0.01° SPR angle shift, corresponding to ~10 ng/cm$^2$). When these findings are compared to the study of Gombotz et al. [5] on PEGs, one must remark there is no general rule for the dependence of nonspecific adsorption on the average molar mass of the hydrogel.

To date, the most comprehensive investigations addressing the influence of the chemical composition of the interface on the nonspecific adsorption were performed by Whitesides et al., first for monolayers on noble metals modified with low molar mass or oligomeric compounds [9] and later with polymers [10]. On a self-assembled monolayer exhibiting carboxylic anhydride groups, a variety of compounds were immobilized via amino groups. This study was then extended to polymers being immobilized via differently structured interlayers and having different average molar masses. Among other findings, the authors came to several conclusions for the chemical structure of surfaces resisting nonspecific adsorption: According to them, the surface has to be (1) hydrophilic, (2) overall electrically neutral, (3) a hydrogen-bond acceptor, but (4) not a hydrogen-bond donor. Although statements 1–3 are fully consistent with all investigations in the field, statement 4 is in contradiction to many other results and not even fully justified by the experimental findings of the authors' own studies. Many of the surfaces showing very low nonspecific adsorption contain hydrogen-bond donors (all dextrans and also many of the PEGs, among these are the surfaces of Gauglitz et al. [7] with the extremely low nonspecific adsorption of less than 1 ng/cm$^2$). Moreover, even the compound exhibiting the least nonspecific binding in Whitesides' monolayer study [9], an oligo(ethylene glycol), has a terminal hydroxy group and therefore does not fit well into the explanation scheme of the authors.

In conclusion, to date, many questions about the influence of the surface design on nonspecific adsorption remain unresolved and a complete picture is not at hand: even worse, some of the results contradict each other. Successes often seem to be rather a result of trial and error paired with profound experimental skills of the researchers than a consequence of systematic approaches. In all of the studies mentioned, either rather limited choices of compounds were used and/or several parameters were varied simultaneously, leaving uncertainties about the exact influence of one specific parameter. This is not astonishing, as a systematic multiparameter investigation would consume enormous resources.

Furthermore, the physical methods for detection of nonspecific binding and the experimental conditions for nonspecific binding tests varied. This makes it hard to directly compare the different investigations and to draw generally valid conclusions for the rational design of biosensor surfaces. However, some rules can be deduced:

• Surfaces must be rendered hydrophilic and noncharged.
• Coatings comprising oligomers or polymers seem to be most efficient.
• An effective barrier between the original surface and the biocompatible layer seems to be helpful not only for noble metal substrates.

## IV. PROTEIN ARRAYS BASED ON HYDROGEL COATINGS

Although the potential of hydrogel coatings may not yet be fully exploited, it can be beneficial to apply them to protein microarrays. Even if more in-depth characterizations seem to be desirable in all cases, encouraging results can be shown.

Mirzabekov et al. [11] prepared arrays of 100-μm × 100- μm × 20-μm gel pads composed of acrylamide units by photopolymerization from an activated surface. This investigation is one of the few examples of the application of the grafting method for hydrogel immobilization at interfaces. Here, in contrast to other hydrogel coatings consisting of defined molecular layers with thicknesses in the nanometer range, thick, cross-linked structures are generated, because bifunctional monomers are present in the reaction mixture. Different proteins such as antibodies, antigens, and enzymes were immobilized by utilizing either aldehyde or hydrazide groups. Protein microchips generated this way were used in immunoassays for the detection of antigens or antibodies, as well as to carry out enzymatic reactions. Although the protein-binding capacity of a gel pad array must be extremely high compared to molecular-scale layers, the sensitivity reported does not exceed typical values found elsewhere. Tests for nonspecific adsorption were made using a 1-mg/mL BSA solution. Only specific interactions were reported. However, because a statement about the background noise is missing, the ability of the system to suppress nonspecific absorption is still difficult to judge. Moreover, there are no experiments performed with protein mixtures; therefore, the susceptibility to matrix effects remains to be investigated.

Gauglitz et al. recently introduced a method to prepare microarrays with haptens immobilized to dextran [12]. In a first step, a hapten (atrazine) was covalently bound to an amino-derivatized dextran in solution. Then, the dextran–hapten conjugate was immobilized on a surface. A solution containing this conjugate was applied in microdrops to functionalized surfaces with a piezo printer. Several coupling chemistries were tried: azide photolinker, activated ester and epoxide. Although the photolinkers gave rise to problems, the latter two chemistries provided satisfactory results. As an example for specific interactions on such surfaces, a 50-μg/ml solution of anti-atrazine IgG was detected specifically. Nonspecific adsorption was tested with 1 mg/mL ovalbumin and was found to be insignificant. The system may be interesting not only for immobilization of small organic molecules but also for peptides; it therefore possesses some potential for application in microarrays for proteomics. However, its suitability for higher-molar-mass proteins or antibodies remains to be proven. Furthermore, the characterization will have to be extended to lower analyte concentrations and the influence of complex sample matrices will have to be investigated.

Wagner et al. [13] electrostatically adsorbed poly(L-lysine)-graft-PEG to negatively charged surfaces. Some of the grafted PEG chains were end-functionalized with biotin (Fig. 3). These biotin moieties were then used to perform biomolecular recognition experiments. Immobilized poly(L-lysine)-graft-PEGs with different degrees of biotinylation were incubated with solutions of monoclonal mouse antibody, streptavidin, streptavidin plus excess of biotin, and streptavidin with *Escherichia coli* cytoplasmic fraction. Within the limits of accuracy, no binding was found on the sample without biotin moities, increasing amounts of streptavidin were detected with increasing degree of biotinylation, and the adsorption of monoclonal mouse antibody and biotin-saturated streptavidin were constantly insignificant. Moreover, no significant differences in streptavidin binding were detected when using pure streptavidin solutions or streptavidin solutions containing a 200- or a 1000-fold excess of *E. coli* cytoplasmic fractions, respectively. This indicates that nonspecific

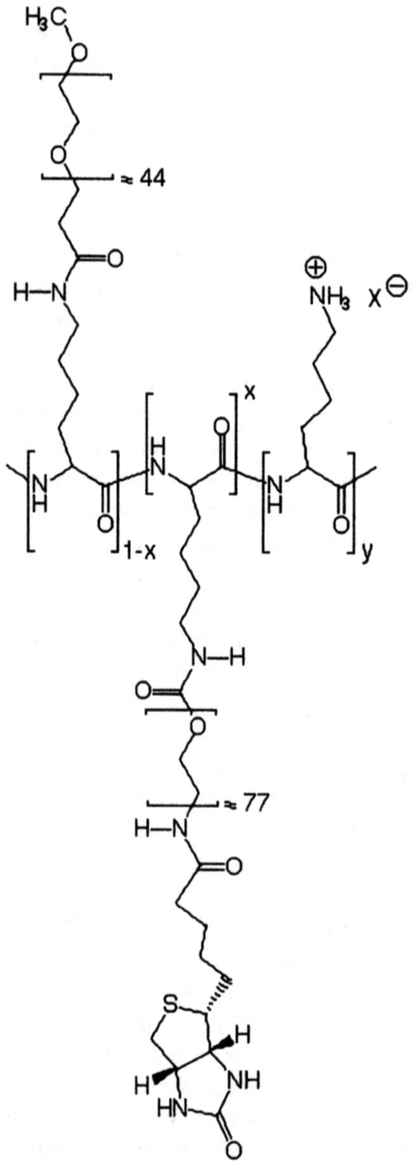

**Figure 3** Chemical structure of the PEG-modified poly-L-(lysine) carrying variable amounts of biotin, which was used by Wagner et. al. for surface modification. (Adapted from Ref. 13 © National Academy of Sciences.)

adsorption is suppressed by the PEG layer and that matrix effects do not influence the detection. As a first demonstration, a microarray with spots measuring 50 μm in diameter was prepared. In this case, fluorescent biotinylated phycoerythrin was bound to streptavidin on poly(L-lysine)-grafted PEG with 30% of biotin units. However, the experimental conditions chosen in this investigation do not yet resemble realistic conditions as encountered in biological samples. The high conformity in results from streptavidin solutions with and without cytoplasmic fractions from *E. coli* is encouraging, but it may be a consequence of the extremely high affinity of the biotin–streptavidin system. Furthermore, the conditions to test nonspecific adsorption are, by far, less demanding compared to other investigations (~30 μg/mL of monoclonal mouse antibody compared to milligrams per milliliter of serum/ovalbumin/BSA). Noteworthy, similar graft copolymers without biotin moities had previously been subjected to much more difficult conditions (pure serum, 1 mg/mL human serum albumin, or 1 mg/mL fibrinogen) [14]. Their performance was reported to be good to excellent depending on the chemical composition of the substrate and on the exact structure and molar mass of the polymers (e.g., adsorption of ~20 ng/cm$^2$ down to lower than 1 ng/cm$^2$ from a 1-mg/mL human serum albumin solution). It remains to be verified whether this very good resistance to nonspecific adsorption can also be obtained for similar systems carrying biofunctional moieties, especially proteins.

Glaucus Proteomics has recently developed a hydrogel interface for microarray applications based on an immobilization strategy via covalently linked polymeric interlayers [15]. Receptor–analyte interactions were investigated with the model system protein A (from *Staphylococcus aureus*)/rabbit–anti-mouse IgG. When characterized with SPR, specific interactions could be detected at least in a concentration range from 100 down to 0.1 μg/mL.

Limitations of this experimental design did not allow for characterization with lower concentrations. Tests for nonspecific adsorption were performed with BSA solutions. When incubating with 150 μg/mL, no adsorption can be detected within the limits of accuracy. Increasing the BSA concentration to the very high level of 4 mg/mL, nonspecific adsorption is detectable, but still very low with a value of 5 ng/cm$^2$. Furthermore, the influence of matrix effects was investigated with the protein A/IgG system. The hydrogel surface with immoblized protein A was incubated with a mixture of 4 μg of IgG and 4 mg BSA in 1 mL of buffer. The IgG binding signal dropped to about 60% of the one expected for a 4 μg/mL IgG solution without additional protein, but in spite of the 1000-fold excess of matrix protein, the specific binding could be easily monitored (Fig. 4).

In conclusion, first results demonstrate that hydrogels have the potential to improve detection in protein arrays, especially for proteomic applications, where complex biological samples with highly differential concentrations need to be

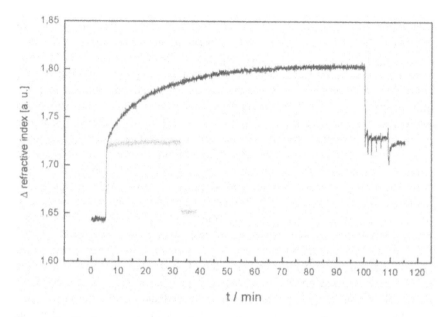

**Figure 4** Monitoring of nonspecific adsorption of BSA (4 mg/mL) to a dextran-modified surface prepared according to a procedure recently developed by Glaucus Proteomics (gray curve). The same surface was also used for a specific interaction experiment: Rabbit–anti-mouse IgG (4 µg/mL) was mixed with 4 mg/mL BSA. The 1000-fold excess of matrix protein did not inhibit the specific binding of the IgG to immobilized protein A (black curve).

analyzed. To date, this potential is not yet fully exploited. Both more comprehensive characterization of existing hydrogel interfaces and optimization of surface modification procedures need to follow.

## REFERENCES

1.  MacBeath, G.; Schreiber, S.L. Science. 2000, 289, 1760–1763.
2.  Haab, B.B.; Dunham, M.J.; Brown, P.O. http://genomebiology.com/2000/1/6/pre-print/0001.1, 2000.
3.  Zhu, H.; Snyder, M. Curr Opin Chem Biol. 2001, 5, 40–45.
4.  Löfås, S.; Johnsson, B. J. Chem. Soc. Chem. Commun. 1990, 1526–1528.
5.  Gombotz, W.R.; Guanghui, W.; Horbett, T.A.; Hoffman, A.S. J Biomed Mater Res. 1991, 25, 1547–1562.
6.  Piehler, J.; Brecht, A.; Geckeler, K.E.; Gauglitz, G. Biosensors Bioelectron. 1996, 11(6/7), 579–590.

7. Piehler, J.; Brecht, A.; Valiokas, R.; Liedberg, B.; Gauglitz, G. Biosensors Bioelectron. 2000, 15, 473–481.

8. Frazier, R.A.; Matthijs, G.; Davies, M.C.; Roberts, C.J.; Schacht, E.; Tendler, S.J.B. Biomaterials. 2000, 21, 957–966.

9. Chapman, R.G.; Ostuni, E.; Takayama, S.; Holmlin, R.E.; Yan, L.; Whitesides, G.M. J Am Chem Soc. 2000, 122(34), 8303–8304.

10. Chapman, R.G.; Ostuni, E.; Liang, M.N.; Meluleni, G.; Kim, E.; Yan, L.; Pier, G.; Warren, H.S.; Whitesides, G.M. Langmuir. 2001, 17, 1225–1233.

11. Arenkov, P.; Kukhtin, A.; Gemmell, A.; Voloshchuk, S.; Chupeeva, V.; Mirzabekov, A. Anal Biochem. 2000, 278, 123–131.

12. Gauglitz, G. http://barolo.ipc.unituebingen.de/infomat/microdrop/index_en.html.

13. Martin, T.L.; Zaugg, F.G.; Witte, K.; Indermuhle, P.; Nock, S.; Wagner, P. Proc Natl Acad Sci USA. 2001, 98(3), 852–857.

14. Kenausis, G.L.; Vörös, J.; Elbert, D.L.; Huang, N.; Hofer, R.; Ruiz-Taylor, L.; Textor, M.; Hubbell, J.A.; Spencer, N.D. J Phys Chem B. 2000, 104, 3298–3309.

15. Wischerhoff, E. European patent application 00203767.9, 2000.

# 6

# High-Throughput Protein Expression, Purification, and Characterization Technologies

**STEFAN R. SCHMIDT**
*AstraZeneca*
*Södertälje, Sweden*

## I. INTRODUCTION

In the postgenomic era, in an age in which whole genomes are completely sequenced within months, a wealth of DNA sequence information is available.
Genome sequences have been decoded from simple bacteria, like *Escherichia
coli*, *Bacillus subtilis*, and pathogenic strains, to model organisms, *Saccharomyces
cerevisiae*, *Caenorhabditis elegans*, and *Drosophila melanogaster*, up the evolutionary tree to *Homo sapiens*. Bioinformatic analysis usually reveals an enormous
number of putative open reading frames (ORFs), whose functions most often are
unknown. Generally, sequence and structural homology to previously characterized genes is the key element to identify the function of novel ORFs. A severe
limitation lies in the fact that no putative function can be assigned to genes that
do not share homology with other known genes, as homology is the tool to assign
and predict function of an identified gene. Between 31% (*Helicobacter pylori*)
and 50% (*Saccharomyces cerevisiae*) of putative ORFs for these model organisms
are of unidentified function as determined using homology-based approaches
[1,2].

The functional characterisation of novel genes is central to a genomic approach
for drug discovery. Novel targets, once identified, offer the possibility for drug
development or other pharmaceutical therapies. Pharmacologically, however,
only a very limited number of gene classes are considered suitable targets for
drug discovery programs [3]. Moreover, functional analysis should not focus on
pharmacological targets alone. Therefore, methods need to be established that
are applicable for a more global approach. Traditional methods employing fractionation, purification, and biochemical analysis are technically difficult, time-

consuming, and typically require a large amount of starting material. This approach has been adapted and modified in order to allow a higher throughput by the use of expression libraries and pooling of clones [4]. Expression libraries are important tools for an approach aimed toward the functional characterization of large numbers of novel genes or putative ORFs. There have been a number of different technologies like expression cloning [5,6], surface display [7] or phage display [8] described, targeting the functional analysis in libraries. Another indispensable prerequisite are purified recombinant proteins derived from those libraries. Once large numbers of novel purified proteins are available, many possibilities emerge for the functional analysis of the identified proteins. Those proteins can then be used for structural crystallization studies, for the generation of antibodies, and for a completely new technological field—the analysis of proteins on microarrays (Fig. 1).

However, all of these interesting applications suffer from several impediments. First, methods need to be developed that allow for the selection of clones which truly express from expression libraries, because only 33% of all clones in an oligo-dT-primed cDNA library will contain cDNA inserts in a proper reading frame. Second, the purification must be optimized in order to achieve a throughput in the range of at least several hundred purified proteins per day with a yield sufficient for several different assays or structural analysis [9]. Third, techniques have to be established allowing the simultaneous functional analysis of hundreds or thousands of proteins in parallel. Ideally, the analysis is performed in a format that minimizes sample consumption and is compatible with different assay conditions. Interesting assays are those interrogating any kind of interaction with other molecules. These may be drugs, other proteins, antibodies, DNA, or assays deducing biochemical functions which are particularly suitable for enzymes of the pharmacologically relevant classes like kinases, phosphotases, and proteases or enzymes used in biotechnological processes. An analysis system containing all of these properties can be found by use of protein microarrays, which consist of a solid support and immobilized purified proteins. The first example of such a global approach was a biochemical screen of all yeast open reading frames (ORFs) [10,11] followed by a universal protein array [12]. Until now, no commercial solution to all these problems is available. Therefore, methods for optimized expression libraries, improved purification technologies, and the generation and analysis of protein microarrays will be described in this chapter.

## II. EXPRESSION LIBRARIES

High-throughput protein expression and purification plays a central role in a series of functional analytical and structural biological approaches, but it is critically dependent on the reliable generation of expression libraries. One specific bottleneck in the process of expression library production is the identification of clones

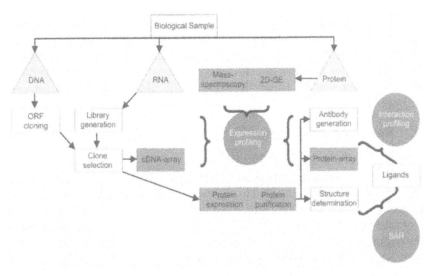

**Figure 1** Work flow integrating genomics and proteomics. Biological samples in the form of extracted DNA or RNA are split into individual clones that can be the basis for a cDNA array or the production of recombinant proteins. The purified protein serves as a resource for the construction of protein arrays, the generation of antibodies, and structural determinations. A central goal of proteomics, expression profiling, can either be achieved by using well-established methods like RNA profiling or the separation and identification of proteins by two-dimensional gel electrophoresis (2D-GE). Protein or antibody arrays will complement these approaches and probably allow a refined analysis of the proteome. However, protein arrays are a multifunctional tool which can also be utilized together with a broad range of ligands (inhibitors, small compounds, substrates, etc.) for interaction profiling studies. The availability of large numbers of three-dimensional structures from purified proteins will also contribute to the rational design of drugs and improved structure–activity relationships (SARs) between targets and drugs.

containing an in-frame fusion of the cDNA insert with affinity tags that are needed for purification purposes. In order to distinguish between expressing and nonexpressing clones, several procedures like trapping of ORFs with intein [13], the selection with antibiotic resistance markers [14], or the fusion to green fluorescent protein [15] have been described. Most of these approaches rely on the construction of an integrated selection mechanism. The selection mechanism must be amenable to a high-throughput approach able to process large numbers of clones for complete coverage of all genes expressed in the specific tissue or cell type from which the library was derived. Generally, there are several possibilities to accomplish this goal. First, the selection procedure can be performed by automated devices, which are capable of distinguishing between expressing and nonexpressing clones by physical means. Second, the vector can contain a genetic switch that interferes with the growth or survival of clones not containing proper in-frame fusions. A third approach relying on the selection of clones that express by analysis of the presence of affinity tags is a less favourable method because it is time-consuming and requires massive robotic infrastructure to perform rearraying [16].

## A. Selection Procedure

### 1. Physical Selection Approaches

The physical parameters, which are assessed mostly when characterizing cells, are those which can be measured by optical devices. Usually, characteristics like fluorescence, luminescence, color, refraction, or optical density are determined. Unfortunately, not all of the methods mentioned allow for an automated, high-throughput approach. For the last decade, instruments have been available that were specifically constructed for the purpose of measuring and selecting cells based on their optical parameters, the so-called fluorescence-activated cell sorters (FACSs). Those devices measure fluorescent signals emitted by the cells or attached molecules and they sort those cells accordingly into individual, optically identical populations. State-of-the-art units are even capable of selecting and plating single cells (Cytomation, Fort Collins, CO, USA; Becton Dickinson, Erembodegem-Aalst, Belgium) [17]. Clones can be selected using charge-coupled device (CCD) cameras that distinguish between differential staining of cells. These cameras are mounted onto picking robots (Genetix, New Milton, UK; Autogen, Framingham, MA, USA, Biorobotics, Cambridge, UK) taking images of agar trays containing the plated library and select cells based on different colors, stains, or contrast for picking [18].

### 2. Genetic Selection Approaches

Genetic selection usually relies on specific factors, which are transferred into the host cells by the vector containing the cDNA insert. The absence or presence of these factors promotes growth or survival on selective medium for all those cells in which the gene coding for the factor is switched on or off. This leads to the positive selection of clones harboring the marker gene.

### 3. Reporter Constructs

Reporter systems have a long tradition in molecular biological methods. One of the first examples is the classical blue/white screening for recombinant clones. Here the alpha peptide complements a defect in β-galactosidase (β-GAL) resulting in blue colonies. When the peptide is disrupted by a cDNA insert, the result is the formation of white colonies, which contain the recombinant clones of interest [19]. This blue/white screening can also serve as one example for physical selection properties because the result can be distinguished optically. Using fluorescent substrates instead of the standard substrate X-gal (5-bromo-4-chloro-3-indolyl-β-D-galactopyranoside), the activity and presence of β-GAL can be monitored by FACS [20]. Recently, other optically active proteins have been discovered, the class of autofluorescent proteins [AFP, red fluorescent protein (RFP), green fluorescent protein (GFP)]. These proteins, mostly isolated form marine organisms, emit fluorescence when folded and excited properly [21]. Another extraordinary protein, the luciferase protein, isolated from beetles, generates light in the presence of a substrate and ATP and is also suitable for optical detection purposes. Other enzymes allowing photometric assays are the classical enzymes widely used in enzyme-linked immunosorbent assays (ELISA), alkaline phosphatase (ALP), and horseradish peroxidase (POD). However, all of these examples of reporters for optical detection share a common disadvantage—their size, which is in the range of ~20 kDa for AFPs up to 120 kDa in the case of β-GAL. This makes it very difficult to integrate them easily into expression vectors. Another disadvantage is their preferred selection for small cDNA inserts. In addition, the fusion between a cDNA and a large gene might lead to steric accession problems for the smaller recombinant protein encoded by the cDNA insert when functional assays are performed at a later time.

Enzymes, like chloramphenicol acetyltransferase (CAT), transferring antibiotic resistance, can also be used as genetic reporters [22]. The use of a suitable selection medium allows only growth of recombinant clones containing in-frame fusions with the resistance markers. For the selection of recombinant yeast clones, it is a widely used approach to perform a metabolic selection procedure. Genetic reporters outside the group of resistance and metabolic markers can be found in the class of toxic genes, thus allowing a negative selection. Now, only those clones having a disrupted gene are selected—in this case, a destroyed toxic gene, resulting in a positive signal and, consequently, survival of recombinant cells [23,24].

## 4. Practical Examples

The power of selecting clones according to their fluorescent properties can be demonstrated by separating RFP expressing cells from nonexpressing cells in a FACS (Fig. 2a). Both populations R1, having positive cells, and R2, having negative cells, can be clearly distinguished and separated in the sorting device.

An identical effect can be achieved using a fluorescent substrate, fluorescein di-βD-glucopyranoside (FDGlu) (Molecular Probes, Eugene, OR, USA) for β-GAL. The vector containing the alpha peptide fused to the carboxy-terminal of the cDNA complements the delta-lac mutation of β-GAL, thus reconstituting the enzymatic activity [25–28]. The signal intensity of FDGlu is similar to that of RFP (Fig. 2b), but the signal detection is faster because the slow folding of the RFP is not required for the alpha peptide. β-GAL signal selection can be performed in two steps to increase the selectivity of the whole process. After a

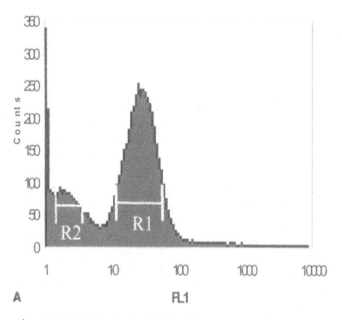

**Figure 2** FACS analysis of RFP clones. *E. coli* cells were mixed with RFP expressing clones. The population R1, being positive by red fluorescence, can be distinguished clearly from nonfluorescent cells.

crude prefractionation by FACS, which does not require single-cell sorting, the precleared population of β-GAL-positive cells can be fine sorted by plating the bacteria onto agar trays containing isopropyl-β-D-thiogaloctropyranoside (IPTG) and X-Gal, allowing a blue/white screening by the CCD camera of standard picking robots. A critical factor for all library-constructing processes is the removal of nonrecombinant clones that do not contain a cDNA insert. For that purpose, a toxic gene like *ccdb* can be integrated into the expression vector, which must be replaced by a novel cDNA insert in order to allow for the survival of the clone when incubated under inducing conditions [29,30].

Combining these two features, the optical reporter with a background suppressor gene, leads to the prototype of an ideal vector (Fig. 3). This vector contains standard cloning sites, such as *Sal*I and *Not*I, compatible to most other vectors used for library generation, like pSPORT (Invitrogen, Carlsbad, CA, USA). The affinity tag is fused to the amino-terminus of the recombinant protein, whereas the reporter construct is located at the carboxyl-terminus. The background suppressor is positioned between the cloning sites. The whole expression cassette creates one multifunctional polypeptide chain, consisting of three modules: affinity tag, suppressor gene, and reporter gene. A positive selection signal for either the FACS or the CCD camera will only be present in clones which are properly

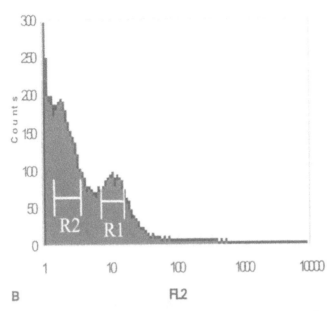

B                                    FL2

**Figure 2**  (*Continued*) FACS analysis of β-GAL complementation. The detection of β-GAL complementation with a fluorescent substrate (FDGlu) in a random mixture of *E. coli* cells. Fluorescence emitting cells in the population R1 can be separated from negative-stained cells R2.

read-through, meaning a proper in-frame fusion at both ends of the cDNA insert. This approach has several consequences, one being that the number of expected positive clones using one reading frame of the vector will result in maximally 17% positive clones. This figure is derived from the number of possible reading frames at the 3′ end (three) added to the number of possible reading frames at the 5′ end (three) resulting in six possibilities, i.e. as a consequence of positional uncertainties within the nucleic acid codon triplet in each coding direction. The positive clones will therefore be a population of 1/6 of all clones. This reduction in positive clones has been observed experimentally (data not shown). In order to circumvent this limitation, either the size of the library has to be increased by a factor of 6, which is not always possible, or the library is cloned into all nine possible variations of 5′ and 3′ fusion points. Another consequence is that oligo-dT-primed libraries can no longer be used because clones derived from that approach contain the full 3′-end sequence with the native stop signal and unpredictable stretches of untranslated regions (UTRs) at the 3′ end. This prohibits an in-frame read-through to the reporter gene at the 3′ end of the expression cassette. Therefore, only random, directionally cloned cDNAs can be transferred into this vector. These cDNAs have to be carefully size-selected, because the genetic selection process clearly

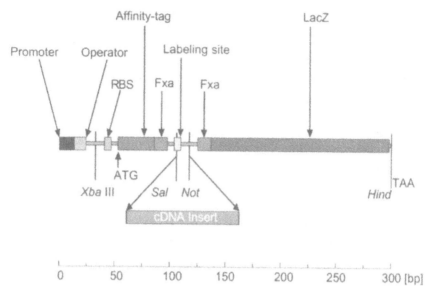

**Figure 3** Multifunctional expression cassette. The ideal expression cassette consists of a regulatory element containing promoter and operator sites and a series of functional modules. An important sequence, the ribosomal binding site (RBS), is located upstream of the first ATG. The translated sequence begins with an amino-terminal affinity tag followed by a protease cleavage site for the factor Xa protease (FXa). Recombinant cDNA inserts are ligated into unique restriction sites for *Sal*I and *Not*I, which are compatible to other standard vectors. The true native recombinant protein is flanked by protease recognition sequences at its amino- and carboxyl-terminus. The position of the *Sal*I site corresponds to a site which allows the radioactive labeling of the protein. The reporter construct located at the 3′ end of the cDNA codes for the LacZ alpha peptide. The whole construct can be transferred into other vectors containing different regulatory and resistance features by digests with *Xba*I and *Hind*III.

favors short cDNA inserts, which will not contain any stop signal in any frame, thus creating false-positive read-through clones. However, the use of random, directionally cloned cDNAs also has advantages. The greatest advantage results from dividing full-length proteins into fragments containing, ideally, only domains of proteins. In this specific type of library, one protein coded by a single gene will therefore be present as a number of potentially overlapping clones. Some of these clones will have a specific function depending on the cloned domain. This circumstance reflects nature. One example is the catalytic domain of receptor tyrosine kinases or phosphatases that can be functional even after separating these domains from their transmembrane regions. Protein domains, which are difficult to express (e.g., those with high hydrophobicity or transmembrane regions), will mostly be improperly folded and will consequently be underrepresented. However, all of the other domains, often exerting enzymatic activities, will be accessible for functional studies after expression and purification.

## B.  Further Improvements

This prototype vector with its three modules can be further improved by inserting protease cleavage sites allowing the removal of either the affinity tag or the reporter protein, or both, in order to obtain a native, unmodified protein. This can sometimes be quite useful for crystallization purposes. Another module useful for future interaction studies is the introduction of a labeling site. This site consists of a peptide sequence which is recognized in a highly specific manner by a kinase, can be used to create a radioactively labeled recombinant protein, and can be monitored in an in vitro interaction assay. The above-described vector is ideally suited for expression in bacterial cells, which are relatively easy to lyse and grow. The disadvantage of prokaryotic expression systems is the lack of posttranslational modifications, which are typical of eukaryotic and mammalian cells. In order to obtain those modifications, a vector allowing expression in yeast is probably better suited. Recently, a *Pichia/E. coli* shuttle vector has been described [31]. Another benefit of yeast expression is its powerful secretion system, transporting recombinant proteins very efficiently into the culture supernatant where the recombinant protein can easily be isolated. The use of secretion vectors has been recently demonstrated [32].

Techniques to improve cloning efficiency are available commercially. Use of homologous recombination to subclone whole libraries or individual clones in one step into other vectors containing new affinity tags or different promoters is very efficient [Invitrogen, Carlsbad, CA, USA (Echo cloning Gateway system)] [33,34]. This valuable method can be adapted by inserting the repeat sequences required for the recombination into the expression cassette of the ideal vector. High-throughput transformation is performed either with chemically competent cells in 96-well microtiter plates (Invitrogen, Carlsbad, CA, USA, Stratagene, La Jolla, CA, USA; Novagen, Madison, WI, USA) or by a 96-channel electroporator (BTX, San Diego, CA, USA).

## C.  System Components

### 1.  Promoter

One of the most critical factors defining an expression vector is the promoter that directs the level of gene expression. For bacterial vectors, a broad range of promoters with different efficiencies is available. Parameters influencing the usefulness of certain promoters are the repression of expression, the expression level, and the cost for the inducer. Generally, constitutively active promoters should be avoided. This may limit the number of clones by the expression of toxic gene products immediately after the transformation that may be damaging to the cells.

The first parameter, repression, is crucial for the expression of heterologous, potentially toxic, proteins in bacteria. Tightly controlled promoters like the tetracycline (TET) [35] or the arabinose (araBAD) [36] promoters allow the unaffected growth for the host bacteria under noninducing conditions. Thus, a sufficient density of cells is present when the expression is switched on by the addition of the corresponding inducer anhydrotetracyclin or arabinose. Promoters not as tightly regulated as these will show a basal level of expression, which, in some cases, may lead to cell death even before full induction.

The second parameter, the level of expression, is more critical for the subsequent purification. High levels of expression will potentially lead to the aggregation of the recombinant proteins in the form of inclusion bodies. These insoluble particles can be enriched and separated form the crude lysate but require additional purification steps, including solubilization and refolding. In some cases, refolding or denatured purification has been performed successfully directly on affinity columns [37]. This procedure is only possible with a small selection of tags, like the His tag, because the affinity matrix is also exposed to the denaturing solution and potentially destroyed.

The third parameter, which should not be neglected, particularly in a high-throughput setting where large numbers of proteins are produced and thus large volumes of culture medium are consumed, is the cost for the inducer. In this context, a broad range exists, reaching from inexpensive chemicals like the sugar arabinose for the araBAD promoter, to more expensive reagents like IPTG for induction of the bacterial *lac*I promoter. A final point to consider is a potential multifunctionality in different hosts. One example is the TET promoter that is quite efficient in mammalian and bacterial cells. The functionality may be expanded further by use of the highly efficient T7 promoter, which has been successfully employed in bacterial cells [38]. It has also gained additional significance by its utilization for in vitro transcription and translation that will be an important expansion of the current expression repertoire [39].

## 2.   Affinity Tag

In order to select a suitable affinity or epitope tag, a number of parameters must be taken into consideration [40]. A distinction can be made between relatively short peptide tags, (His-Tag [41], Strep-Tag [42], S-Tag [43] FLAG-Tag [44]), which, in general, do not interfere with the structure of the recombinant protein, and larger proteins, such as glutathione-*S*-transferase (GST) [45], cellulose-binding domain (CBD) [46], dihydrofolate reductase (DHFR) (Qiagen, Hilden, Germany), maltose-binding protein (MBP) [47], and thioredoxin (THX) [48] that often make up more than 50% of the fusion protein. Larger proteins usually stabilize their heterologous fusion proteins but might disturb the functional or structural analysis of the protein. They can be removed easily by insertion of a protease cleavage site at the transition between the tag and the heterologous protein. Large tags tend to interfere with selection procedures by preferring smaller cDNA inserts. A critical factor, which is valid for short and long affinity tags, is the potential presence of similar or identical tag sequences within the host cell. For example the purification procedure for His-Tags via immobilized metal-ion-affinity chromatography (IMAC) often results in the specific enrichment of additional proteins having metal-binding properties similar to the peptide tag [49]. When choosing an epitope tag, it must also be taken into consideration if purification under denaturing conditions will be performed. Denaturing will affect all proteins and unfold them, therefore, larger fusion proteins and tags requiring the binding to a protein ligand at the affinity matrix cannot be used for this approach. A further requirement for rapid, high-throughput purification approaches is the short contact time, or the fast on-rate of the tag toward its affinity matrix, which is dependent on the affinity constant of the tag. Ideally, the affinity matrix tolerates high flow rates and high pressure in order to minimize the processing time. It is also important that the matrix is easily regenerated by fast cleaning in place (CIP) procedures for preparation of the next sample within a short time interval and to reduce costs by reusing it [50].

## 3.   Host Strain

As mentioned previously, the choice of the host system is also critical. Prokaryotic expression hosts will never transfer all necessary posttranslational modifications which are present in typical eukaryotic proteins [51]. A circumvention of this problem might be the use of yeast [52], baculovirus [53], or mammalian cells. However, all these expression systems have a lower productivity when compared to bacterial expression. In cases where modifications can be neglected, bacteria will be the system of choice [54]. Recently, there have been several improvements adapting bacterial cells to the specific requirements of mammalian proteins. Some strains like BL21-CodonPlus-RIL (Stratagene, La Jolla, CA, USA) also contain the tRNA genes for codons rarely used in bacteria but which are very frequent in mammals. Other strains have reduced levels of proteases in order to minimize the risk of protein degradation. In some cases, secretion into the culture supernatant or the periplasmatic space is a useful alternative for proteins with specific requirements like the proper formation of disulfide bridges [55].

## 4.   Practical Examples

In order to identify the most suitable promoter composition, the expression cassette (Fig. 3) containing a bacterial protein and an N-terminal Strep-tag was equipped with a set of four different promoters, the $LacI$, T7, araBAD, and TET promoter. At optical density $OD_{600nm} = 0.4$, the constructs were induced for 4 h with 1 m$M$ IPTG for the $LacI$ and T7 promoter and 0.02% arabinose for the araBAD promoter and 0.2 $\mu$g/ml anhydroteracycline (AHT) for the TET promoter. Thereafter, the cells were spun down, boiled in sodium dodecyl salfate (SDS) sample buffer and analyzed for protein production and yield by SDS-PAGE polyacrylamide gel electrophoresis. Western blot analysis was performed with horseradish peroxidase-labelled Streptactin. No significant difference in expression levels could be detected between all four samples. In addition, no accumulation of protein in the insoluble fraction was observed (data not shown). Based on these results, the standard vector was equipped with a TET promoter, minimizing the formation of inclusion bodies and allowing a tight regulation of expression that makes this promoter very suitable for library approaches where the number of potentially toxic proteins is unknown.

## III. HIGH-THROUGHPUT PURIFICATION

The complete work flow of purification consists of several steps, starting with the growth of the individual clones, followed by the induction and expression of the heterologous proteins, lysis of the host cells, preclearing of the crude extract, and, finally, the purification and separation of the recombinant protein from the cellular proteins of the host organism. All of these individual steps must be optimized in order to minimize the number of manual interferences.

## A. Expression

The high-throughput expression of recombinant proteins is a central goal of post-genomic functional analysis of putative ORFs. Until now, only small proteomes have been structurally investigated [56,57]. Expression can also serve as tool for localization studies [58] and large-scale antibody production [59,60]. Recently, approaches for high-throughput expression in a broad range of different organisms has been summarized [61].

### 1. Growth

The ultimate goal of a purification strategy is generally to achieve as much and as pure a protein as possible. Therefore, a culture volume must be chosen which is sufficient to accomplish this task. The largest commercially available 96-well microtiter plates (MTPs) have a volume of 2.2 mL. Plates containing only 24 wells can have up to 8 mL capacity. Recently, a high-throughput expression approach in 2.2-mL deep-well MTPs using baculovirus-infected insect cells has been described [62]. When using bacterial cells, a critical factor is aeration. The transfer of sufficient oxygen can be achieved by several different technical solutions. The first and most straightforward approach is the use of a high-frequency but low-amplitude shaker, which is available from different sources (e.g., Zymark, Hopkinton, MA, USA). Another commercial solution is the combination of a high-speed shaker with a gastight hood that is constantly aerated with oxygen (GeneMachines, San Carlos, CA, USA). A third possibility is the mixing of the individual wells with miniaturized magnetic stirring bars (Variomag, Daytona Beach, FL, USA). An extremely high mixing capability and aeration can be achieved by using 96-well MTPs with a filter bottom (Whatman, Maidstone, UK). These plates can be aerated be inserting them into frames that allow a vertical gas flow, resulting in small gas bubbles in the individual wells, transferring oxygen and mixing the contents of the well.

## 2. Induction

Host cells should be in a logarithmical growth phase at the time expression is induced. Generally, this can be achieved by a 10-fold dilution of an overnight starting culture into fresh rich medium like Luria Bertanii (LB), Yeast tryptone extract (2xYT), or Teriffic broth (TB). However, this procedure is quite impractical for high-throughput approaches. For this, inoculating 1 mL medium in deep-well MTPs with cells transferred from the primary library by replicator pins creates the starting culture. The cells are grown overnight at 30°C under optimal aeration conditions. The following day, the cells are diluted 1:1 by the addition of fresh medium containing a double-concentrated inducer. This procedure reduces the number of pipetting steps by combining the dilution step with the induction. Starting at this time point, the cells are incubated for 4 h at 37°C with optimal oxygen supply for maximal protein production.

## 3. Lysis

If using an expression system that is not capable of secreting the recombinant protein to the culture medium, the cells need to be lysed in order to gain access to the expressed protein. Several procedures ranging from chemical and physical methods to genetic modifications can be utilized to lyse bacterial cells. Recently, a coupled lysis and purification technique has been described [63].

(a) *Physical Lysis.* Mostly microbial cells are disrupted by physical methods. One widely used procedure is treatment with ultrasound. The ultrasound energy is transferred by a number of different devices into the cell suspension. One type of instrument uses one or more sticks, which are inserted into the solution (Sonics, Newtown, CT, USA). However, cells can also be disrupted by incubating a complete MTP into an ultrasound water bath, transferring the sound waves through water (Misonix, Farmingdale, NY, USA). A method more amenable for high-throughput is the repeated incubation of cells in very low ($-80°C$) and high ($37°C$) temperatures. By this approach, water crystals are generated within the cells, destroying the cell wall and membranes. Other physical methods like high-pressure lysis in a French press or methods which are preferred for eukaryotic cells, such as the disruption with fast-moving blades (Ultra turrax, IKA, Germany) or Potter–Elvehjem tissue homogenizers, are not suitable for high-throughput applications.

(b)  *Chemical Lysis.*   Chemical lysis also has a long tradition for use in protein purification. The central method for the preparation of plasmid DNA relies on the lysis of cells by detergents such as SDS. Nonionic detergents such as Tween-20 or Nonidet NP40 allow a much more gentle dissolution of cell membranes [64]. Currently, some specific mixtures are available which are optimized for the lysis of bacterial cells (B-Per, Pierce, Rockford, IL, USA; Novzyme, Novagen, Madison, WI, USA). Other methods utilize the bacteriolytic properties of natural substances like lysozyme [65] or polymyxin [66]. The cell wall can be destabilized by the incubation of microbial cells in solutions containing $Ca^{2+}$ chelating compounds like EDTA or EGTA. Most of these chemicals are compatible with each other; therefore, optimal success can be achieved by combining some of these substances in a highly potent lysis mixture, leading to a gentle disruption by simultaneously targeting several critical parameters.

(c)  *Genetic Lysis.*   Microbial cells can also be disrupted by the activity of genetically regulated mechanisms. The expression of lysozyme controlled by a helper plasmid pLysS/E usually serves a tighter control of the T7 polymerase in some expression systems (pET vectors, Novagen, Madison, WI, USA) [67]. However, the intracellular presence of lysozyme clearly improves the lysis results when treating the cells with freeze–thawing cycles. Another protein, the bacterial release protein (BRP), creates pores in the cell wall and membrane of bacterial cells. BRP is also encoded on a helper plasmid and can be cotransformed into almost any *E. coli* strain [68]. The productivity of such a dual-vector expression system can be improved by regulating the lysis factor with a second independent promoter that induces the lysis only after accomplishing the expression.

(d)  *Practical Examples.*   It has proved useful to optimize throughput and minimize manual procedures required for individual steps to combine different methods in the lysis protocol. After growth and induction according to Section III.A.2, the cell suspension is cooled to 4°C and the cells are sedimented in the MTPs by centrifugation. The supernatant is removed by aspiration with a 96-channel manifold or by flipping the plate. The wet pellets are then resuspended with a 1-mL lysis mixture containing 2 mg/mL lysozyme, 1 m$M$ EDTA, and 0.1% Tween-20 in 0.1 $M$ Tris-HCl pH 7.5. In some cases, volatile buffers such as ammonium carbonate are recommended, particularly if dilute eluates have to be concentrated by evaporation. Certain additives such as Benzonase, diluted 1/10,000 to degrade the viscous nucleic acids, and a selection of different protease inhibitors, capable of preventing any form of proteolysis, are required to facilitate the subsequent purification. If using the Strep-Tag as the affinity tag, it is recommended to add 2 μg/mL avidin to mask endogenous biotinylated proteins which might contaminate the purified recombinant protein or block the affinity matrix [42].

## B. Purification

Since the introduction of the commonly used 96-well MTP, a broad range of instruments has been developed for liquid handling in this format. The pumping of different buffered solutions is the central process for liquid chromatography (LC), which is the most frequently used method for purifying individual proteins. The combination of these two entities, 96-well plates and pipetting robots, leads to a streamlined process that allows the purification of large numbers of proteins in short time intervals. Generally, the purification can be addressed in a sequential single channel or a parallel multichannel mode. The purification method ideally suited for recombinant proteins is affinity chromatography. This procedure is performed by liquid-chromatographic devices, via filtration or by other small-scale methods using magnetic beads and coated microtiter plates. A critical parameter for all of these approaches is the processing rate, dominating the achievable throughput. Because all purification methods described here consist of several steps, each individual procedure must be thoroughly optimized.

## 1. Automated Single-Channel LC

In the past, most chromatographic separations were performed with single-channel LC systems consisting of one column, one or more detectors and pumps plus a fraction collector. The first quantum leap was the introduction of high-pressure and high-flow systems (HPLC), allowing extremely sensitive and fast separations. Recently, some more improvements have been achieved by integrating sample injection and fraction collection within additional robotic units (BioCad Vision; Applied Biosystems, Foster City, CA, USA). This automation and the availability of modern chromatographic matrices lead to a dramatic increase in the processing rate of individual samples [69]. Advanced affinity matrices such as POROS (Applied Biosystems, Foster City, CA, USA) tolerate dynamic flow rates of up to 5000 cm/h. This matrix is available for the typical affinity purifications of His-Tags via metal chelate chromatography, antibody purification by protein-G matrices, and, finally, Strep-Tag-purification with immobilized streptactin. The ideal matrix must also be suited for CIP processes and regeneration, contributing significantly to a high turnover rate [70]. The typical steps, which are performed automatically, comprise the sample injection, a washing step to remove unbound or unspecific material, the elution, and finally, the cleaning and regeneration of the column. The buffer volume required in all of these steps is strictly dependent on the size of the column. Therefore, the miniaturization of the column is of highest priority, particularly if only small amounts of sample are injected, which is the case for the purification of proteins derived from expression libraries. The smallest commercial available column suited for high-pressure applications has dimensions of 2 × 30 mm (Vici Jour, Onsala, Sweden) containing a volume of ~100 μL. The number of injections tolerated by a single column is limited by the porosity of the frits closing both ends of the column. Too fine pores are blocked after only a few runs, whereas frits with 5-μm pores, which are still small enough to retain the chromatographic beads more than 10 μm in diameter, easily tolerate more than 100 injections by only a slight pressure build up. One module, which is always present in the typical HPLC configuration, is the ultraviolet (UV) detector. Modern HPLC systems can be programmed to a peak collection mode, allowing the collection of only those fractions containing recombinant protein, which is indicated by generating an absorption signal with a defined slope above a certain threshold. This feature reduces the number of samples to be analyzed after the run.

## 2. Multichannel Pipetting Robot

Single-channel systems are useful for less than 1000 samples only. In order to increase the throughput, the number of the operating channels has to be increased. Typical liquid-handling robots operate with 4–96 channels. However, only a small fraction of these automatic devices are suited for affinity chromatography. Most of those models have two major limitations. None is capable of pumping and transferring liquids from the back of the channels, and no pump can tolerate the high pressure and flow rates required for high-performance chromatography. Recently, robots have been designed specifically for solid-phase extractions (SPE), a technique which is related to affinity chromatography (EPR labautomation, Witerswil, Switzerland; CyBio, Jena, Germany). Both robots resemble the properties of a traditional HPLC with 96 individual minicolumns being addressed simultaneously in parallel with the exclusion of any type of on-line detector unit. Therefore, the quality of the purification cannot be monitored directly and no peak collection is possible. In order to circumvent this disadvantage, the method, which is applied on the multichannel system, has to be carefully tested and optimized, particularly with regard to the elution conditions on a classical HPLC. The absence of a detector also prevents a direct quality control and a rough quantitation of the eluates. Sample injection and fraction collection are performed by transporting individual MTPs via a rotating table or a conveyor belt under the central pumping station. As in the case with HPLCs, different buffer reservoirs can be switched in line by an automatic, electronic, or pneumatic valve. The columns used in these devices are either modified pipetting tips (the Zip-Tip™ of Millipore is a miniaturized example for that approach) or commercially available SPE cartridges, which can be filled with any specific chromatographic matrix such as POROS. As in the case with single-channel systems, the throughput is extremely increased if fast CIP and regeneration is possible without removing the set of columns.

## 3. Filtration Devices

Purification is not only possible by applying positive pressure onto a column, the driving force can also be introduced by a vacuum. Vacuum-purification devices using filter plates or individual columns filled with a chromatographic matrix are quite common for the large-scale preparation of plasmids (Qiagen, Hilden, Germany). An apparatus suited for that application can easily be modified to perform protein purifications. Some commercial solutions are available for His-Tags, using a metal-chelate matrix. His-Tag vacuum purification plates have successfully been used in combination with robots, pipetting sample, and buffers via 4/8 channels (Qiagen, Hilden, Germany). A clear disadvantage, which also holds true for all other multichannel applications, are clogged individual channels. A particular problem for vacuum filtration is the tendency of foaming which is enhanced if detergents are used for lysis. Filtration devices are inexpensive straightforward tools for the purification of up to several hundred samples, particularly if CIP and the reuse of the cartridges can be neglected.

## 4. Magnetic Beads

For a number of ligands (antibodies, His-tagged proteins, biotin), beads with a paramagnetic core exist. These beads can be used to automate the purification [71,72]. For that purpose, a magnetic manifold retracts the beads during washing and elution steps. There exist a number of variations suitable to work with different types of MTPs (Qiagen, Hilden, Germany; Milteny, Bergisch Gladbach, Germany; Promega, Madison, WI, USA; Pyrosequencing, Uppsala, Sweden). Recently, there have been attempts to automate this procedure, which is quite difficult because electronic magnets retain too much of the magnetic force employed. One solution is to move the MTPs via a robot between a magnetic and a nonmagnetic position. A major drawback resulting from the chemistry of the beads is increased nonspecific binding of proteins. Furthermore, there is no proper CIP for these kind of bead available, thus contributing to higher costs of this approach. For now, magnetic beads are not ideally suited for high-throughput applications.

## 5. Coated MTPs

For some small-scale approaches, MTPs with coated surfaces can be used [42]. Here, the affinity ligand is directly immobilized to the individual wells of the MTP. The purification of the crude lysate is performed by liquid-handling robots or simple plate-washing devices. This approach provides immobilized recombinant protein that can then be directly analyzed in the MTP. Alternatively, the protein is eluted from the wells and transferred to other applications. This is mostly impossible because the resulting samples are too dilute. Major drawbacks of this method are the limitation of the available binding surface, nonspecific binding toward the plastic, the short shelf life of coated plates, the impossible reuse, and high costs. All of these disadvantages prohibit the use of coated MTPs for high-throughput purification applications.

## 6. Practical Example

Five clones from an expression library were selected by blue/white screening and directionally cloned ORFs, containing an amino-terminal Strep-Tag and a carboxyl-terminal fusion with the *lacZ* alpha peptide, were grown overnight in 1 mL LB and induced as described in Section III.A.2. After lysing the cells according to Section III.A.3.d by the addition of 1 mL Lysis buffer containing 2 mg/mL lysozyme, 2 μg/mL avidin, 20 μL Benzonase, and a protease inhibitor cocktail (Complete; Roche, Basel, Switzerland) and repeated freeze–thawing, the MTPs containing the resuspended crude lysate were inserted into a HPLC robot (BioCad Vision, Applied Biosystems). The purification was performed according to the buffer conditions as described elsewhere [42] but with an optimized program allowing a processing rate of ~2 min/sample. The progress and the result of the individual runs is demonstrated by comparing the UV signals recorded by the UV detector (Fig. 4). The purity of the eluted fraction was assessed by using a highly sophisticated capillary electrophoresis (CE) device (Bioanalyzer 2100; Agilent, USA) (Fig. 5). All collected peak fractions show greater than 90% purity and the yield of the individual proteins ranges from 5 to 30 μg/mL of culture.

## IV. ARRAY-BASED ANALYSIS

The first attempts to simultaneously analyze a large number of different samples led to the development of MTPs. By diminishing the volume of the individual cavities in order to reduce the sample consumption, the number of accessible wells could be increased from the 96 positions in the first prototypes to 1536 wells, which are currently used in ultrahigh-throughput screens [73]. However the miniaturization was anticipated much earlier in multianalyte assays on solid supports [74]. Recently, a number of approaches and assays involving protein arrays have been described. Some of these methods are improvements [16,31,59] of earlier developments regarding the filter-based screening of expression libraries with antibodies [75]. A variation of this technique is called the serological analysis of autologous tumor antigens by expression cloning (SEREX) [76]. Generally, all array applications can be summarized under the term "interaction profiling." The interacting ligands can be from the class of immunological molecules, nucleic acids, small compounds, inhibitors, or substrates.

Until now, there has been a series of publications describing the development and application of antibody arrays either in the context of miniaturized ELISA [77,78] or for antigen detection by screening of expression libraries [79–83]. A very straightforward approach to employ antibody arrays is expression profiling, which could have a great potential in complementing RNA profiling and 2D-

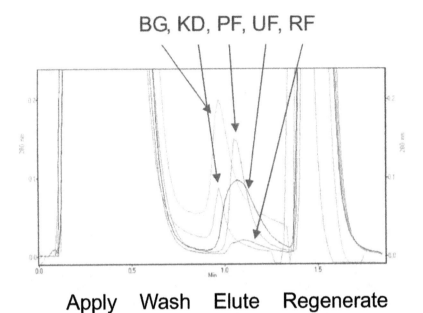

# BG, KD, PF, UF, RF

## Apply    Wash    Elute    Regenerate

**Figure 4** High-throughput chromatography. Five different proteins of microbial origin (BG, KD, PF, UF and RF) were expressed and purified according to Section III.B.6. The overlaid traces for the individual chromatograms clearly indicate a uniform behavior of all different proteins. All are eluting in an identical time interval between 0.8 and 1.2 min. The four steps comprising the application of sample, washing to remove unspecific bound protein, the elution of the pure protein, and the regeneration of the column are completed within 2 min.

GE approaches [84–90]. Arrays for global biochemical analysis are much more difficult to create because of the nonuniform nature of proteins and the lack of large numbers of purified proteins. Until now there have only been a small number of successful attempts to address these requirements [10–12,91–93]. However, the functional analysis of large numbers of proteins can only be undertaken in a parallel with simultaneous screening for activities and ligand binding.

## A.  Membrane-Based Biochemical Analysis

The biochemical analysis of a large number of proteins requires immobilized and functional protein. Individual, separate cavities or wells should be avoided because all bound proteins must be simultaneously exposed to the ligand-containing solutions to achieve maximum throughput and to minimize handling times. The

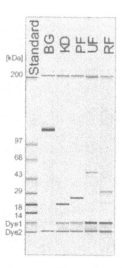

**Figure 5** Electrophoretic analysis of chromatographic fractions. The peak fractions of the eluted proteins BG, KD, PF, UF, and RF are analyzed by CE on the Bioanalyzer 2100 (Agilent, USA). The whole procedure was performed according to the manufacturer's specifications. All proteins are more than 90% pure with regard to contaminating proteins.

most direct approach is the attachment of proteins to an adsorptive surface as, for example, a nitrocellulose (NC) or poly(vinylidene difluoride) (PVDF) membrane, thus resembling a traditional dot blot. This results in a relatively stable noncovalent binding of the proteins and does not require any chemical modifications. This approach has been employed successfully for a multitude of different applications [12]. The procedure can be modified by using a membrane coated with affinity ligands (SAM Membrane; Promega, Madison WI, USA), which leads to an oriented immobilization of the proteins with an improved access to active sites but with a decreased binding capacity. Modern membranes are modified by the insertion of a fiber network to obtain a better mechanical strength. One disadvantage is the presence of capillary effects limiting the spotting density by spreading the protein dots. However, the three-dimensional structure of the membrane stabilizes the proteins and increases the local concentration. Stability problems can also be avoided by embedding proteins in a gel matrix [91] or by the coating of protein arrays with agarose [94]. Other solid substrates like silicon wafers, glass slides, plastics, or metal have to be functionalized to achieve proper binding of the proteins but have, nevertheless, also been employed successfully [92]. The immobilization of proteins onto membranes is a very straightforward and cost-efficient approach for array generation.

Recently, a number of efforts have been undertaken to investigate processes suited for the simultaneous screening of soluble expression of recombinant proteins in parallel [95,96]. This is an important prerequisite for many techniques like antibody generation, structure determination, production of therapeutic proteins, and the biochemical analysis of proteins on arrays. Currently, the most advanced example of global analysis of protein activities is the cloning, expression, purification, and screening of 5800 yeast ORFs for their ability to interact with other proteins or small molecules [97]. Another approach described recently is focused on the production of purified proteins for the generation and selection of antibodies using a dual-tag bacterial expression system optimized for immunization and immobilization [98]. Many other efforts directed to the application of automated protein expression and purification have also been reviewed lately [99]. In this context, a study aiming at the direct comparison of different expression systems which could be useful for high-throughput applications testing a selection of 336 randomly selected cDNAs must be mentioned [100].

## B.  Practical Example

In order to test the retained functionality of enzymes after immobilization onto membranes, the purified enzymes β-GAL, ALP, and POD were serially diluted 1 : 1 from a starting concentration of 1 mg/mL (2 mg/mL for GAL) in phosphate-buffered saline (PBS) and transferred to a NC membrane (Biotrace NT; Pall Gelman, Ann Arbor MI, USA) by 10 repetitive cyles of contact printing using a 250-μm pin head, thus transferring less than 1 nL per spot in total. The $8 \times 8$ pattern resulted in a density of ~250 dots/cm$^2$. The quality of the spots was assessed by staining with PonceauS. The protein microspots were probed by incubating the filters in an appropriate buffer for 30 min at ambient temperature in the presence of the following substrates:

   3-Amino-9-ethylcarbazole (AEC)
   4-Chloro-1-naphtol (4C1N)
   6-Chloro-3-indolyl-β-D-galactopyranoside (6C3IG)
   5-Bromo-6-chloro-3-indolyl-β-D-galactopyranoside (5B6C3IG)
   5-Bromo-4-chloro-3-indolyl phosphate/nitro blue tetrazolium (BCIP/NBT)
   Fast Red/Naphtol AS-TR phosphate

The filters were washed in double-distilled H$_2$O to remove excess substrates, air-dried, and analyzed with a standard office scanner at a resolution of 1200 dpi (Fig. 6). The PonceauS staining clearly detects protein spots down to a starting concentration of 125 μg/mL. The most sensitive enzymatic substrate is the combination of BCIP/NBT for ALP, which gives clear signals down to less than 8 μg/mL. All other substrates can be ranked descending from AEC (62 μg/mL), 4C1N

(125 μg/mL), and 6C3IG, 5B6C3IG, and FastRed (250 μg/mL). Improved sensitivity can be achieved by use of chemiluminescent or fluorogenic substrates.

## C. Array Analysis

There is a broad range of detection technologies available to analyze arrays. Generally, it should be distinguished between label-free and labeled detection. Most of the detection methods rely on electromagentic waves, such as fluorescence, which also includes fluorescence resonance energy transfer (FRET), and fluores-

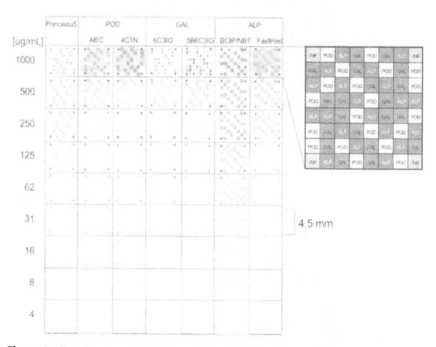

**Figure 6**  Functional analysis of immobilized proteins. The purified enzymes β-GAL, ALP, and POD were serially diluted 1 : 1 from a starting concentration of 1 mg/mL (2 mg/ml for β-GAL) in phosphate-buffered saline (PBS). The proteins were then transferred to a NC membrane (Biotrace NT; Pall Gelman Ann Arbor MI, USA) by 10 repetitive cyles of contact printing using a 250-μm pin tool, thus transferring less than 1 nL per spot in total. The $8 \times 8$ pattern results in a density of ~250 dots/cm$^2$. The quality of the spots was assessed by staining with PonceauS. The protein microspots were probed by incubating the filters in an appropriate buffer for 30 min at ambient temperature in the presence of the following substrates: AEC, 4C1N, 6C3IG, 5B6C3IG, BCIP/NBT, Fast Red/Naphtol AS-TR phosphate. After washing and air-drying, the image was taken with a standard office scanner at a resolution of 1200 dpi.

cence polarization (FP), luminescence, colorimetry, radioactivity, SPR, interference, and refraction. Other methods include mass spectrometry or electrochemical conductivity changes. A low-cost and straightforward approach that can be conducted in most laboratories is the scanning of colorimetric-stained filters. This is suitable to all assays that use secondary antibodies or other ligands coupled to the typical immunochemistry enzymes or those ligands that allow the construction of sandwich layers. In some cases, the activity of enzymes can be assessed directly by specifically labeled substrates, such as radioactive ATP for kinases. The use of fluorescent labels is not ideal in combination with filters because of the high-intrinsic-background fluorescence of those membranes. The image from array experiments can either be evaluated by a human expert or analyzed automatically (Biochipexplorer; GPC-Biotech, Martinsried, Germany).

## V. FUTURE

Based on genomic data, there will be increased efforts in the near future to massively clone and purify large numbers of proteins mainly for structural studies and for the generation of antibodies. A large number of these proteins will also be available to functional studies on arrays or for interaction and expression-profiling experiments. Protein arrays will be optimized with regard to stability of the immobilized proteins, the detection limit, and miniaturization. Based on these developments, there will be an enormous impact of protein arrays and the access to many purified proteins with know structures, ligands, and specific antibodies on the progress of diagnostics and therapeutics.

## ACKNOWLEDGMENTS

I would like to thank N. Gollmitzer, P. Sebastian, and R. Schadowski for excellent technical assistance. I greatly appreciate the preparation of protein arrays by O. Herde and the generous gift of expression clones by C. Degenhart. I also thank D. Bancroft and N. Kley for extensive discussions and P. Rayner, T. Holtschke, and H. Wirsing for critical review of the manuscript.

## REFERENCES

1. Doolittle, R.F. Microbial genomes opened up. Nature. 1998, 392(6674), 339–342.
2. Fraser, C.M.; Eisen, J.; Fleischmann, R.D.; Ketchum, K.A.; Peterson, S. Comparative genomics and understanding of microbial biology. Emerg Infect Dis. 2000, 6(5), 505–512.
3. Drews, J.; Ryser, S. The role of innovation in drug development. Nature Biotechnol. 1997, 15(13), 1318–1319.

4.  Short, J.M. Recombinant approaches for accessing biodiversity. Nature Biotechnol. 1997, 15(13), 1322–1323.

5.  Kitamura, T.; Onishi, M.; Kinoshita, S.; Shibuya, A.; Miyajima, A.; Nolan, G.P. Efficient screening of retroviral cDNA expression libraries. Proc Natl Acad Sci USA. 1995, 92(20), 9146–9150.

6.  Dalboge, H.; Lange, L. Using molecular techniques to identify new microbial bio-catalysts. Trends Biotechnol. 1998, 16(6), 265–272.

7.  Olsen, M.J.; Stephens, D.; Griffiths, D.; Daugherty, P.; Georgiou, G.; Iverson, B.L. Function-based isolation of novel enzymes from a large library. Nature Biotechnol. 2000, 18(10), 1071–1074.

8.  Pedersen, H.; Holder, S.; Sutherlin, D.P.; Schwitter, U.; King, D.S.; Schultz, P.G. A method for directed evolution and functional cloning of enzymes. Proc Natl Acad Sci USA. 1998, 95(18), 10,523–10,528.

9.  Stevens, R.C. Design of high-throughput methods of protein production for structural biology. Struct Fold Des. 2000, 8(9), R177–R185.

10. Martzen, M.R.; McCraith, S.M.; Spinelli, S.L.; Torres, F.M.; Fields, S.; Grayhack, E.J.; Phizicky, E.M. A biochemical genomics approach for identifying genes by the activity of their products. Science. 1999, 286(5442), 1153–1155.

11. Carlson, M. The awesome power of yeast biochemical genomics. Trends Genet. 2000, 16(2), 49–51.

12. Ge, H. UPA, a universal protein array system for quantitative detection of pro-tein–protein, protein–DNA, protein–RNA and protein–ligand interactions. Nucleic Acids Res. 2000, 28(2), e3.

13. Daugelat, S.; Jacobs, W.R. The *Mycobacterium tuberculosis* recA intein can be used in an ORFTRAP to select for open reading frames. Protein Sci. 1999, 8(3), 644–653.

14. Davis, C.A.; Benzer, S. Generation of cDNA expression libraries enriched for in-frame sequences. Proc Natl Acad Sci USA. 1997, 94(6), 2128–2132.

15. Hawke, N.A.; Strong, S.J.; Haire, R.N.; Litman, G.W. Vector for positive selection of in-frame genetic sequences. BioTechniques. 1997, 23(4), 619–621.

16. Bussow, K.; Cahill, D.; Nietfeld, W.; Bancroft, D.; Scherzinger, E.; Lehrach, H.; Walter, G. A method for global protein expression and antibody screening on high-density filters of an arrayed cDNA library. Nucleic Acids Res. 1998, 26(21), 5007–5008.

17. Davey, H.M.; Kell, D.B. Flow cytometry and cell sorting of heterogeneous micro-bial populations: The importance of single-cell analyses. Microbiol Rev. 1996, 60(4), 641–696.

18. Uber, D.C.; Jaklevic, J.M.; Theil, E.H.; Lishanskaya, A.; McNeely, M.R. Application of robotics and image processing to automated colony picking and arraying. BioTechniques. 1991, 11(5), 642–647.

19. Ullmann, A.; Jacob, F.; Monod, J. Characterization by in vitro complementation of a peptide corresponding to an operator-proximal segment of the beta-galactosidase structural gene of *Escherichia coli*. J Mol Biol. 1967, 24(2), 339–343.

20. Gee, K.R.; Sun, W.C.; Bhalgat, M.K.; Upson, R.H.; Klaubert, D.H.; Latham, K.A.; Haugland, R.P. Fluorogenic substrates based on fluorinated umbelliferones for con-

tinuous assays of phosphatases and beta-galactosidases. Anal Biochem. 1999, 273(1), 41–48.

21. Gerdes, H.H.; Kaether, C. Green fluorescent protein: applications in cell biology. FEBS Lett. 1996, 389(1), 44–47.
22. Lee, P.; Hruby, D.E. Detection of recombinant protein based on reporter enzyme activity: Chloramphenicol acetyltransferase. Methods Mol Biol. 1997, 63, 31–40.
23. Welsh, S.; Kay, S.A. Reporter gene expression for monitoring gene transfer. Curr Opin Biotechnol. 1997, 8(5), 617–622.
24. Suter-Crazzolara, C.; Klemm, M.; Reiss, B. Reporter genes. Methods Cell Biol. 1995, 50, 425–438.
25. Gray, M.R.; Colot, H.V.; Guarente, L.; Rosbash, M. Open reading frame cloning: Identification, cloning, and expression of open reading frame DNA. Proc Natl Acad Sci USA. 1982, 79(21), 6598–6602.
26. Shapira, S.K.; Chou, J.; Richaud, F.V.; Casadaban, M.J. New versatile plasmid vectors for expression of hybrid proteins coded by a cloned gene fused to lacZgene sequences encoding an enzymatically active carboxy-terminal portion of beta-galactosidase. Gene. 1983, 25(1), 71–82.
27. Weinstock, G.M. Use of open reading frame expression vectors. Methods Enzymol. 1987, 154, 156–163.
28. Diederich, L.; Roth, A.; Messer, W. A versatile plasmid vector system for the regulated expression of genes in *Escherichia coli*. BioTechniques. 1994, 16(5), 916–923.
29. Hengen, P.N. Reducing background colonies with positive selection vectors. Trends Biochem Sci. 1997, 22(3), 105–106.
30. Van Reeth, T.; Dreze, P.L.; Szpirer, J.; Szpirer, C.; Gabant, P. Positive selection vectors to generate fused genes for the expression of his-tagged proteins. BioTechniques. 1998, 25(5), 898–904.
31. Lueking, A.; Holz, C.; Gotthold, C.; Lehrach, H.; Cahill, D. A system for dual protein expression in *Pichia pastoris* and *Escherichia coli*. Protein Express Purif. 2000, 20(3), 372–378.
32. Schuster, M.; Wasserbauer, E.; Einhauer, A.; Ortner, C.; Jungbauer, A.; Hammerschmid, F.; Werner, G. Protein expression strategies for identification of novel target proteins. J Biomol Screen. 2000, 5(2), 89–97.
33. Liu, Q.; Li, M.Z.; Leibham, D.; Cortez, D.; Elledge, S.J. The univector plasmid-fusion system, a method for rapid construction of recombinant DNA without restriction enzymes. Curr Biol. 1998, 8(24), 1300–1309.
34. Hartley, J.L.; Temple, G.F.; Brasch, M.A. DNA cloning using in vitro site-specific recombination. Genome Res. 2000, 10(11), 1788–1795.
35. Skerra, A. Use of the tetracycline promoter for the tightly regulated production of a murine antibody fragment in *Escherichia coli*. Gene. 1994, 151(1–2), 131–135.
36. Guzman, L.M.; Belin, D.; Carson, M.J.; Beckwith, J. Tight regulation, modulation, and high-level expression by vectors containing the arabinose PBAD promoter. J Bacteriol. 1995, 177(14), 4121–4130.
37. Zouhar, J.; Nanak, E.; Brzobohaty, B. Expression, single-step purification, and matrix-assisted refolding of a maize cytokinin glucoside-specific beta-glucosidase. Protein Express Purif. 1999, 17(1), 153–162.

38. Studier, F.W.; Rosenberg, A.H.; Dunn, J.J.; Dubendorff, J.W. Use of T7 RNA polymerase to direct expression of cloned genes. Methods Enzymol. 1990, 185, 60–89.

39. Alimov, A.P.; Khmelnitsky, A.Y.u.; Simonenko, P.N.; Spirin, A.S.; Chetverin, A.B. Cell-free synthesis and affinity isolation of proteins on a nanomole scale. Biotechniques. 2000, 28(2), 338–344.

40. Nilsson, J.; Stahl, S.; Lundeberg, J.; Uhlen, M.; Nygren, P.A. Affinity fusion strategies for detection, purification, and immobilization of recombinant proteins. Protein Express Purif. 1997, 11(1), 1–16.

41. Hochuli, E.; Dobeli, H.; Schacher, A. New metal chelate adsorbent selective for proteins and peptides containing neighbouring histidine residues. J Chromatogr. 1987, 411, 177–184.

42. Skerra, A.; Schmidt, T.G. Use of the Strep-Tag and streptavidin for detection and purification of recombinant proteins. Methods Enzymol. 2000, 326, 271–304.

43. Raines, R.T.; McCormick, M.; Van Oosbree, T.R.; Mierendorf, R.C. The S.Tag fusion system for protein purification. Methods Enzymol. 2000, 326, 362–376.

44. Hopp, T.H.; Prickett, K.S.; Price, V.L.; Libby, R.T.; March, C.J.; Cerretti, D.P.; Urdal, D.L.; Conlon, P.J. A short polypeptide marker sequence useful for recombinant protein identification an purification. BioTechnology. 1988, 6, 1204–1210.

45. Smith, D.B.; Johnson, K.S. Single-step purification of polypeptides expressed in *Escherichia coli* as fusions with glutathione *S*-transferase. Gene. 1988, 67(1), 31–40.

46. Greenwood, J.M.; Ong, E.; Gilkes, N.R.; Warren, R.A.; Miller, R.C.; Kilburn, D.G. Cellulose-binding domains: potential for purification of complex proteins. Protein Eng. 1992, 5(4), 361–365.

47. Di Guan, C.; Li, P.; Riggs, P.D.; Inouye, H. Vectors that facilitate the expression and purification of foreign peptides in *Escherichia coli* by fusion to maltose-binding protein. Gene. 1988, 67(1), 21–30.

48. LaVallie, E.R.; DiBlasio, E.A.; Kovacic, S.; Grant, K.L.; Schendel, P.F.; McCoy, J.M. A thioredoxin gene fusion expression system that circumvents inclusion body formation in the *E. coli* cytoplasm. BioTechnology. 1993, 11(2), 187–193.

49. Zachariou, M.; Hearn, M.T. Protein selectivity in immobilized metal affinity chromatography based on the surface accessibility of aspartic and glutamic acid residues. J Protein Chem. 1995, 14(6), 419–430.

50. Girot, P.; Moroux, Y.; Duteil, X.P.; Nguyen, C.; Boschetti, E. Composite affinity sorbents and their cleaning in place. J Chromatogr. 1990, 510, 213–223.

51. Geisse, S.; Gram, H.; Kleuser, B.; Kocher, H.P. Eukaryotic expression systems: A comparison. Protein Express Purif. 1996, 8(3), 271–282.

52. Cereghino, G.P.; Cregg, J.M. Applications of yeast in biotechnology: protein production and genetic analysis. Curr Opin Biotechnol. 1999, 10(5), 422–427.

53. Albala, J.S.; Franke, K.; McConnell, I.R.; Pak, K.L.; Folta, P.A.; Rubinfeld, B.; Davies, A.H.; Lennon, G.G.; Clark, R. From genes to proteins: High-throughput expression and purification of the human proteome. J Cell Biochem. 2000, 80(2), 187–191.

54. LaVallie, E.R.; McCoy, J.M. Gene fusion expression systems in Escherichia coli. Curr Opin Biotechnol. 1995, 6(5), 501–506.

55. Pines, O.; Inouye, M. Expression and secretion of proteins in *E. coli*. Mol Biotechnol. 1999, 12(1), 25–34.

56. Christendat, D.; Yee, A.; Dharamsi, A.; Kluger, Y.; Savchenko, A.; Cort, J.R.; Booth, V.; Mackereth, C.D.; Saridakis, V.; Ekiel, I.; Kozlov, G.; Maxwell, K.L.; Wu, N.; McIntosh, L.P.; Gehring, K.; Kennedy, M.A.; Davidson, A.R.; Pai, E.F.; Gerstein, M.; Edwards, A.M.; Arrowsmith, C.H. Structural proteomics of an archaeon. Nat Struct Biol. 2000, 7(10), 903–909.

57. Edwards, A.M.; Arrowsmith, C.H.; Christendat, D.; Dharamsi, A.; Friesen, J.D.; Greenblatt, J.F.; Vedadi, M. Protein production: feeding the crystallographers and NMR spectroscopists. Nat Struct Biol. 2000, 7(Suppl), 970–972.

58. Simpson, J.C.; Wellenreuther, R.; Poustka, A.; Pepperkok, R.; Wiemann, S. Systematic subcellular localization of novel proteins identified by large-scale cDNA sequencing. EMBO Rep. 2000, 1(3), 287–292.

59. Bussow, K.; Nordhoff, E.; Lubbert, C.; Lehrach, H.; Walter, G. A human cDNA library for high-throughput protein expression screening. Genomics. 2000, 65(1), 1–8.

60. Larsson, M.; Graslund, S.; Yuan, L.; Brundell, E.; Uhlen, M.; Hoog, C.; Stahl, S. High-throughput protein expression of cDNA products as a tool in functional genomics. J Biotechnol. 2000, 80(2), 143–157.

61. Nasoff, M.; Bergseid, M.; Hoeffler, J.P.; Heyman, J.A. High-throughput expression of fusion proteins. Methods Enzymol. 2000, 328, 515–529.

62. Albala, J.S.; Humphery-Smith, I. Array-based proteomics: High-throughput expression and purification of IMAGE consortium cDNA clones. Curr Opin Mol Therap. 1999, 1(6), 680–684.

63. Schuster, M.; Wasserbauer, E.; Ortner, C.; Graumann, K.; Jungbauer, A.; Hammerschmid, F.; Werner, G. Short cut of protein purification by integration of cell-disrupture and affinity extraction. Bioseparation. 2000, 9(2), 59–67.

64. Gonzalez, C.; Lagos, R.; Monasterio, O. Recovery of soluble protein after expression in *Escherichia coli* depends on cellular disruption conditions. Microbios. 1996, 85(345), 205–212.

65. Davies, R.C.; Neuberger, A.; Wilson, B.M. The dependence of lysozyme activity on pH and ionic strength. Biochim Biophys Acta. 1969, 178(2), 294–305.

66. Greenwood, D. The activity of polymyxins against dense populations of *Escherichia coli*. J Gen Microbiol. 1975, 91(1), 110–118.

67. Zhang, X.; Studier, F.W. Mechanism of inhibition of bacteriophage T7 RNA polymerase by T7 lysozyme. J Mol Biol. 1997, 269(1), 10–27.

68. van der Wal, F.J.; ten Hagen-Jongman, C.M.; Oudega, B.; Luirink, J. Optimization of bacteriocin-release-protein-induced protein release by *Escherichia coli*: Extracellular production of the periplasmic molecular chaperone FaeE. Appl Microbiol Biotechnol. 1995, 44(3–4), 459–465.

69. Afeyan, N.B.; Gordon, N.F.; Mazsaroff, I.; Varady, L.; Fulton, S.P.; Yang, Y.B.; Regnier, F.E. Flow-through particles for the high-performance liquid chromatographic separation of biomolecules: perfusion chromatography. J Chromatogr. 1990, 519(1), 1–29.

70. Fulton, S.P.; Shahidi, A.J.; Gordon, N.F.; Afeyan, N.B. Large-scale processing & high-throughput perfusion chromatography. BioTechnology. 1992, 10(6), 635–639.

71. Khng, H.P.; Cunliffe, D.; Davies, S.; Turner, N.A.; Vulfson, E.N. The synthesis of sub-micron magnetic particles and their use for preparative purification of proteins. Biotechnol Bioeng. 1998, 60(4), 419–424.

72. Abudiab, T.; Beitle, R.R. Preparation of magnetic immobilized metal affinity separation media and its use in the isolation of proteins. J Chromatogr A. 1998, 795(2), 211–217.

73. Dunn, D.; Orlowski, M.; McCoy, P.; Gastgeb, F.; Appell, K.; Ozgur, L.; Webb, M.; Burbaum, J. Ultra-high throughput screen of two-million-member combinatorial compound collection in a miniaturized, 1536-well assay format. J Biomol Screen. 2000, 5(3), 177–188.

74. Ekins, R.; Chu, F.; Biggart, E. Multispot, multianalyte, immunoassay. Ann Biol Clin (Paris). 1990, 48(9), 655–666.

75. Weinberger, C.; Hollenberg, S.M.; Ong, E.S.; Harmon, J.M.; Brower, S.T.; Cidlowski, J.; Thompson, E.B.; Rosenfeld, M.G.; Evans, R.M. Identification of human glucocorticoid receptor complementary DNA clones by epitope selection. Science. 1985, 228(4700), 740–742.

76. Tureci, O.; Sahin, U.; Pfreundschuh, M. Serological analysis of human tumor antigens: Molecular definition and implications. Mol Med Today. 1997, 3(8), 342–349.

77. Mendoza, L.G.; McQuary, P.; Mongan, A.; Gangadharan, R.; Brignac, S.; Eggers, M. High-throughput microarray-based enzyme-linked immunosorbent assay (ELISA). BioTechniques. 1999, 27(4), 778–788.

78. Joos, T.O.; Schrenk, M.; Hopfl, P.; Kroger, K.; Chowdhury, U.; Stoll, D.; Schorner, D.; Durr, M.; Herick, K.; Rupp, S.; Sohn, K.; Hammerle, H. A microarray enzyme-linked immunosorbent assay for autoimmune diagnostics. Electrophoresis. 2000, 21(13), 2641–2650.

79. Bernard, A.; Michel, B.; Delamarche, E. Micromosaic immunoassays. Anal Chem. 2001, 73(1), 8–12.

80. Lueking, A.; Horn, M.; Eickhoff, H.; Bussow, K.; Lehrach, H.; Walter, G. Protein microarrays for gene expression and antibody screening. Anal Biochem. 1999, 270(1), 103–111.

81. Holt, L.J.; Bussow, K.; Walter, G.; Tomlinson, I.M. By-passing selection: Direct screening for antibody-antigen interactions using protein arrays. Nucleic Acids Res. 2000, 28(15), e72.

82. de Wildt, R.M.; Mundy, C.R.; Gorick, B.D.; Tomlinson, I.M. Antibody arrays for high-throughput screening of antibody-antigen interactions. Nature Biotechnol. 2000, 18(9), 989–994.

83. Haab, B.B.; Dunham, M.J.; O'Brown, P. Protein microarrays for highly parallel detection and quantitation of specific proteins and antibodies in complex solutions. Genome Biol. 2001, 2(2), 0004.1–0004.13.

84. Walter, G.; Bussow, K.; Cahill, D.; Lueking, A.; Lehrach, H. Protein arrays for gene expression and molecular interaction screening. Curr Opin Microbiol. 2000, 3(3), 298–302.

85. Cahill, D.J. Protein arrays: A high-throughput solution for proteomics research? In *Proteomics: A Trends Guide* Blackstock W, Mann M, eds; 2000, 49–53.

86. Borrebaeck, C.A. Antibodies in diagnostics—From immunoassays to protein chips. Immunol Today. 2000, 21(8), 379–382.

87. Holt, L.J.; Enever, C.; de Wildt, R.M.; Tomlinson, I.M. The use of recombinant antibodies in proteomics. Curr Opin Biotechnol. 2000, 11(5), 445–449.

88. Tomlinson, I.M.; Holt, L.J. Protein profiling comes of age. Genome Biol. 2001, 2(2), 1004.1–1004.3.

89. Jenkins, R.E.; Pennington, S.R. Arrays for protein expression profiling: Towards a viable alternative to two dimensional gel-electrophoresis?. Proteomics. 2001, 1, 13–29.

90. Haab, B.B. Advances in protein microarray technology for protein expression and interaction profiling. Curr Opin Drug Discov Devel. 2001, 4(1), 116–123.

91. Arenkov, P.; Kukhtin, A.; Gemmell, A.; Voloshchuk, S.; Chupeeva, V.; Mirzabekov, A. Protein microchips: Use for immunoassay and enzymatic reactions. Anal Biochem. 2000, 278(2), 123–131.

92. MacBeath, G.; Schreiber, S.L. Printing proteins as microarrays for high-throughput function determination. Science. 2000, 289(5485), 1760–1763.

93. Zhu, H.; Klemic, J.F.; Chang, S.; Bertone, P.; Casamayor, A.; Klemic, K.G.; Smith, D.; Gerstein, M.; Reed, M.A.; Snyder, M. Analysis of yeast protein kinases using protein chips. Nature Genet. 2000, 26(3), 283–289.

94. Afanassiev, V.; Hanemann, V.; Wolfl, S. Preparation of DNA and protein micro arrays on glass slides coated with an agarose film. Nucleic Acids Res. 2000, 28(12), e66.

95. Knaust, R.K.; Nordlund, P. Screening for soluble expression of recombinant proteins in a 96-well format. Anal Biochem. 2001, 297(1), 79–85.

96. Hammarstrom, M.; Hellgren, N.; van Den Berg, S.; Berglund, H.; Hard, T. Rapid screening for improved solubility of small human proteins produced as fusion proteins in *Escherichia coli*. Protein Sci. 2002, 11(2), 313–321.

97. Zhu, H.; Bilgin, M.; Bangham, R.; Hall, D.; Casamayor, A.; Bertone, P.; Lan, N.; Jansen, R.; Bidlingmaier, S.; Houfek, T.; Mitchell, T.; Miller, P.; Dean, R.A.; Gerstein, M.; Snyder, M. Global analysis of protein activities using proteome chips. Science. 2001, 293(5537), 2101–2105.

98. Graslund, S.; Falk, R.; Brundell, E.; Hoog, C.; Stahl, S. A high-stringency proteomics concept aimed for generation of antibodies specific for cDNA-encoded proteins. Biotechnol Appl Biochem. 2002, 35(Pt 2), 75–82.

99. Lesley, S.A. High-throughput proteomics: protein expression and purification in the postgenomic world. Protein Express Purif. 2001, 22(2), 159–164.

100. Braun, P.; Hu, Y.; Shen, B.; Halleck, A.; Koundinya, M.; Harlow, E.; LaBaer, J. Proteome-scale purification of human proteins from bacteria. J Proc Natl Acad Sci USA. 2002, 99(5), 2654–2659.

# 7

# Miniaturized Protein Production for Proteomics

**MICHELE GILBERT, TODD C. EDWARDS, CHRISTA PRANGE, MIKE MALFATTI, IAN R. MCCONNELL, and JOANNA S. ALBALA**
*Lawrence Livermore National Laboratory*
*Livermore, California, U.S.A.*

## I. INTRODUCTION

The development of high-throughput methods for gene discovery has paved the way for the design of new strategies for genome-scale protein analysis. We have produced a baculovirus-based system for the expression and purification of large numbers of proteins encoded by cDNA clones from the Integrated Molecular Analysis of Genomes and Their Expression (I.M.A.G.E.) Consortium collection. The system is partially automated and allows high-throughput protein expression for the analysis of the human proteome. Application of this protein expression system will greatly advance the functional and structural analysis of novel genes identified by the human and other genome projects.

## II. RECOMBINANT PROTEIN EXPRESSION

### A. Beyond Bacteria

Traditionally, *Escherichia coli* has been the workhorse of recombinant protein production in both academic and commercial labs; however, there are some inherent disadvantages to using a prokaryotic system for eukaryotic protein production. Some of the difficulties common to *E. coli*-based recombinant expression are the production of insoluble inclusion bodies requiring significant resolubilization efforts [1], misfolding of proteins, and/or lack of posttranslational modifications [2]. Inclusion body formation may be reduced by the use of a heterologous protein partner as well as provide a means for affinity purification [2]. Furthermore, to avoid improperly folded protein that often causes inclusion body formation, many labs utilize techniques such as chaperones and foldases [3] or factorial screening methods [4]. However, for high-throughput expression systems, successful refolding of proteins is unlikely to be automatable because of the distinctive nature of individual proteins.

A critical disadvantage when using prokaryotic systems is the lack of glycosylation and phosphorylation of the resultant recombinant proteins. Many proteins require these posttranslational modifications to fold correctly as well as to function properly [2,5,6]. Enzymatic changes necessary to provide these posttranslational modifications are possible after purification, but these processes are costly and can result in a large variation in the modifications [6].

## B. Eukaryotic Expression Systems

In order to overcome many of the limitations arising from prokaryotic expression for heterologous protein production, several eukaryotic systems have been developed to utilize yeast, insect, or mammalian cells for host expression. These eukaryotic systems all have the capability to properly phosphorylate and glycosylate recombinant proteins to varying degrees [5,7]. In addition, dual-use methods for recombinant expression in prokaryotic and eukaryotic systems have been devised to improve recombinant protein production [8]. Several studies have demonstrated high-throughput eukaryotic protein production in bacteria [9,10] and there is also a precedent for high-throughput expression of eukaryotic proteins in eukaryotic systems [11]. For example, the Holz et al. [11] used yeast for heterologous protein production, 96 clones from a human fetal brain cDNA expression library were randomly selected, cloned by polymerase chain reaction (PCR), and subjected to sequence analysis and protein expression. The results showed that 58 out of 96 clones were in-frame with the selection tag and the resulting recombinant proteins were detected using high-density protein arrays with an epitope-specific antibody before being produced in large quantities for functional analysis.

In pioneering work, Zhu et al. independently cloned all of the open reading frames from the genome of *Saccharomyces cerevisiae* into a yeast expression system resulting in GST-fusion proteins derived from 5800 yeast cDNAs [12]. They purified proteins in a nonautomated fashion in a 96-well format and spotted the protein onto glass slides for functional analysis. Although recombinant expression in yeast has been applied successfully to high-throughput paradigms, yeast cannot provide all of the requirements for heterologous protein production such as correct recognition of signal peptide sequences [2]. In addition to yeast, a 96-well protocol for production of human proteins in mammalian cells has been devised [13]; however, recombinant expression in mammalian cells typically results in reduced expression as compared to other eukaryotic expression systems. Because no single expression system will satisfactorily produce every protein introduced, use of multiple eukaryotic expression systems will be needed to produce more complex human proteins with extensive posttranslational modifications.

## C. Baculoviral Expression

Use of baculovirus for eukaryotic protein production overcomes many of the limitations of other eukaryotic protein expression systems. Production of more traditionally difficult proteins, such as membrane-bound proteins, has been successful using baculovirus, because insect cells are able to utilize the native signal sequence of the protein to be produced [7]. Baculoviruses have been engineered to generate a range of foreign proteins in large quantites, and although the expression system is eukaryotic, the viruses are harmless [14]. Proteins generated from recombinant baculovirus can be easily expressed in milligram quantities, and insect cells grow at ambient temperature without the need for carbon dioxide [15]. Furthermore, insect cells provide most posttranslational modifications that may be critical for the biological integrity of recombinant proteins including phosphorylation, N- and O-linked glycosylation, acylation, disulfide cross-linking, oligomeric assembly, and subcellular targeting [16]. In contrast to the more commonly used bacterial expression systems, recombinant baculoviral expression generally produces soluble proteins without the need for induction or specific temperature conditions.

## III. COMPONENTS OF A BACULOVIRUS-BASED PROTEIN PRODUCTION SYSTEM

The baculovirus system lends itself well to high-throughput protein production because the recombinant baculovirus is a stable, reusable resource. Traditionally, homologous recombination introduces foreign genes into the 130-kilobase viral genome [14]. Creation of recombinant virus begins by insertion of the cDNA of interest into a plasmid transfer vector containing a suitable viral promoter, termination signal, and flanking viral DNA sequences. Typically, the polyhedrin promoter drives protein expression, and the flanking viral DNA regions allow for recombination at the polyhedrin locus of the virus. The virus tolerates replacement of the polyhedrin gene with the cDNA because the polyhedrin gene is not essential for viral replication or production [16]. Thus, the completed virus expresses the recombinant protein to the high levels of the native gene.

Standard molecular biology protocols such as PCR amplification of cDNA, cloning of cDNA into a plasmid vector, and vector purification are routine for baculovirus production and are scalable to create recombinant baculovirus stocks. Genome centers have established many high-throughput strategies for the generation of DNA substrates for sequence analysis. These strategies can be adapted for parallelized protein production in baculovirus and can be easily miniaturized and automated.

## A.  Transfer Vectors

The first element that is critical for baculovirus-based protein production is the transfer vector. The following components are standard in most commercially available transfer vectors: a multiple-cloning site, an affinity purification tag, a polyhedrin or p10 promoter site upstream of the multiple-cloning site, and antibiotic resistance genes for screening of bacteria and insect cells. Similar to traditional bacterial vectors, baculoviral transfer vectors can incorporate multiple affinity tags or fusion protein constructs for purification. For high-throughput protein production, modification of standard multiple cloning sites to include rare restriction cutters may be beneficial to allow insertion of a wide range of cDNAs into the transfer vector. Rare-cutter restriction enzymes are less likely to cleave the cDNAs and, therefore, allow cloning of complete cDNA sequences.

## B.  Genomic DNA

The second element required for the production of a viable recombinant baculovirus is the genomic baculoviral DNA. Baculoviral DNA is commercially available from several sources and has been engineered to be devoid of proteolytic and degradative enzymes for robust recombinant baculoviral production [17]. The most important factor for efficient high-throughput protein production is the reduction of background, nonrecombinant baculoviral infection (i.e., wild-type baculovirus). Therefore, completely linearized baculoviral DNA is critical for successful recombination to occur [18].

## C.  Insect Cell Culture

Insect cells grow at 27°C under standard atmospheric conditions, and there are several cell lines available that can be maintained as adherent or suspension cultures [2]. Cultures must be handled under sterile conditions and require increased consideration in high-throughput processing. Typical suspension cultures are maintained in shaker or spinner flasks containing 50 mL to 5 L of culture media. Avoiding excess shearing upon mixing while providing the proper oxygenation of the cultures in a miniaturized format is a key issue when culturing small volumes of insect cells. Transfected insect cells are much more fragile than transformed bacterial cells, so mixing and aerating require additional care [13]. These issues are common to large-scale protein production in baculovirus using bioreactors, and many of the findings that improve cell viability, cell growth, and protein production by these means can also be applied to the miniaturized format. For example, because Pluronic F-68 decreases insect cell death and increases protein production in large cultures, it would be advantageous to include Pluronic in the medium of a miniaturized insect cell culture system [19].

## D.  Purification Components

The process of protein purification is similar for prokaryotic and eukaryotic expression systems. The primary difference lies in the preparation of cell lysates. In insect cells infected with recombinant baculovirus, proteins can be purified once the cell lysates are cleared of insoluble material. Protein production in *E. coli* frequently requires additional steps to refold proteins from insoluble inclusion bodies [20]. In either case, once soluble material is prepared, standard purification protocols apply. Most purification schemes rely on affinity purification using recombinant epitope tags or fusion proteins incorporated into the amino or carboxy terminus of the recombinant proteins.

Batch purification methods that do not require column chromatography in order to separate out components are particularly well suited for a 96-well, high-throughput format. Some affinity immobilization examples include polyhistidine interacting with a metal-chelating matrix, the chitin-binding domain interacting with chitin, and antibodies interacting with specific antigen sequences. Moreover, established 96-well plate vacuum filtration protocols for DNA plasmid purification have been modified for protein purification [21]. Some drawbacks of a vacuum filtration-based purification are potential clogging of the filter pores and excessive nonspecific binding of proteins to the filter. In this type of miniaturized system, the binding matrix may be difficult to recover, so the cost may be higher for this strategy over larger-scale systems that reuse the binding matrix. Improved stability and uniformity of affinity chromatography support matrices via dehydration (i.e., SwellGel 20) will allow long-term storage of multiwell purification plates. Rehydrating premade purification plates would overcome the batch-to-batch variation associated with pouring suspended matrix each time a new plate is needed.

Covalent attachment of affinity components to high-surface-area filter plates could also be used for high-throughput purification. Covalently attaching the affinity molecules without affecting the porosity of the filter substrate would rely on careful chemistry. This approach, although it has the possible benefit of regeneration of the filter plates, would require a high enough active surface area to ensure adequate purification of the desired proteins.

## IV.  FROM GENES TO PROTEINS

Our approach to high-throughput protein production is a PCR-based strategy, whereby unique I.M.A.G.E. cDNA clones have been used to create an array of recombinant baculoviruses in a 96-well format. All of the steps in this process, from PCR to protein production, are performed in 96-well plates and thus are amenable to automation. Each recombinant protein is engineered to incorporate an epitope tag at the amino-terminal end to allow for immunoaffinity purification.

Recombinant proteins expressed and purified using our system can be used for functional, structural and biochemical analysis.

## A.  The I.M.A.G.E. Collection

The goals of the I.M.A.G.E. Consortium are to array, sequence, map, and distribute a collection of cDNAs representing all human genes into the public domain [22], expanding upon the foundation of the Human Genome Project through the use of publicly available, arrayed cDNA libraries. The collection consists of primarily human and mouse cDNA clones; however, it also includes cDNAs from zebrafish, *Fugu*, rat, *Xenopus*, and primate species. Using high-speed robotics, the Consortium has arrayed over 5.5 million individual cDNAs into 384-well plates and has delivered replica plates to both sequencing centers and distributors worldwide. The I.M.A.G.E. Consortium now represents the largest public cDNA collection, and in collaboration with the National Institutes of Health Mammalian Gene Collection, it is creating a collection of full-length cDNA clones representing every human and mouse gene. All available information associated with an individual clone, such as its unique identification number, the library and tissue from which it was derived, and sequence data is cataloged and tracked in a database. All data are publicly available at http://image.llnl.gov.

The I.M.A.G.E. Consortium also provides several Web-based tools for access to and analysis of the data. The IMAGEne program clusters all human I.M.A.G.E. ESTs (expressed sequenced tags) to each other and to known human genes, selects a representative cDNA clone for each cluster based on full-length data, and displays sequence alignments. Recently, mouse ESTs and known genes have also been added to the IMAGEne analysis. The I.M.A.G.E. Consortium also provides biologically relevant clone sets of interest to the research community. These high-quality rearrays representing species- or tissue-specific sets of cDNAs serve as templates for both transcript profiling experiments (e.g., microarrays) and high-throughput protein production efforts targeted at specific biological pathways.

## B.  Strategy for High-Throughput Production of Recombinant Proteins

Thus, our initial strategy for high-throughput protein expression and purification began with the selection of cDNAs from the I.M.A.G.E. collection. Robots rearrayed clones from 384-well master plates into 96-well plates. We produced recombinant baculoviruses with a PCR-based method that eliminated the traditional subcloning of cDNAs into transfer vectors and plaque purification steps. To test transfection efficiency and proof-of-principle of this concept, we cloned a cDNA encoding the $\beta$-glucuronidase gene into an I.M.A.G.E. vector, creating a *Gus* construct, and transformed the vector into *E. coli*. We inoculated media in every well of a 96-well plate with the bacteria containing the *Gus* construct. Then, we amplified the cDNAs by PCR and purified the PCR products in a 96-well format. We produced the recombinant *Gus* virus by combining these PCR products with baculoviral DNA, infected insect cells, and measured protein production by β-glucuronidase activity using a colorimetric assay. Of these, 88% of recombinant *Gus* virus expressed protein from the infected insect cells, thus demonstrating the potential for efficient production of recombinant virus in a miniaturized, high-throughput format [23].

In order to test our strategy for automated protein production, we designed a "protoplate," a prototype 96-well plate, containing a variety of I.M.A.G.E. cDNAs. The protoplate comprised 8 controls and 88 cDNAs that conformed to criteria including length of the open reading frame, brevity of the 5′ untranslated region, and absence of upstream stop codons. Most protoplate cDNAs were full-length clones representing genes of known function. We selected partial sequences if the full-length cDNA failed to conform to the designated criteria or was not available for a given gene. In addition, some partial cDNAs were chosen to represent protein functional domains [23].

We amplified the selected cDNAs by high-fidelity PCR, purified the PCR products, and transfected these cDNAs with linearized baculoviral DNA into insect cells in a 96-well plate. Next, we amplified the recombinant baculoviruses, infected fresh insect cells, and harvested protein to assay for expression. Western blot analysis using an antibody against the N-terminal epitope tag demonstrated that 34 of 81 wells yielded soluble, recombinant protein.

## C. Second-Generation Strategy for High-Throughput Production of Recombinant Proteins

Our second-generation, high-throughput expression system followed a similar strategy, but it includes a custom transfer vector and additional automation. Robots set up the molecular biological reactions and a linked database tracks results throughout the process. A 1.5-ml, 96-well insect cell culture system allowed for the expression of small amounts, 2–10 μg, of epitope-tagged proteins. Use of immunoaffinity chromatography techniques in a 96-well format resulted in purified protein from the insect cell lysates.

To begin this process, selected cDNAs were amplified by high-fidelity PCR directly on a dilution of the I.M.A.G.E. bacterial cultures. PCR primers were designed to add rare restriction enzyme sites to flank cloned cDNA into our custom transfer vector. PCR products were purified using a Qiagen 96-well PCR purification kit, digested with the appropriate restriction enzymes, and the PCR products are repurified. For the creation of recombinant baculoviruses, a custom transfer vector was designed by adding an immunoaffinity tag and appropriate restriction enzyme sites into the multiple-cloning site of the Clontech pBacPAK9 transfer vector. PCR products representing each of the selected cDNAs were ligated in a miniaturized, 96-well format, after which each sample was transformed into *E. coli* and individually plated. Colonies for each cDNA clone were cultured overnight and screened to determine if the cloned PCR products were of the expected size. Recently, we automated the PCR preparation using the Genesis RSP 150 (Tecan) and Hydra 96 (Robbins Scientific) liquid handling robots. These instruments allow for faster setup of full plates and lower the incidence of errors that can occur due to inaccurate pipetting. Development of automated processes is critical for high-throughput protein production.

The recombinant baculoviral transfer vectors containing the appropriate cDNAs were transfected with linearized baculoviral DNA into adherent insect cells in a 96-well format (Fig. 1). Baculovirus production was allowed to continue for 4 days posttransfection, after which time the virus was amplified onto fresh insect cells in a new 96-well plate where viral amplification continued for 4 more days. This amplification step was repeated two to four more times to adequately increase the viral titer for protein production.

## D.  Miniaturized Protein Production

The final viral amplification was performed in a 2 ml, 96-well plate on a Carousel Magnetic Levitation Stirrer (V & P Scientific, San Diego, CA). Sterile metal

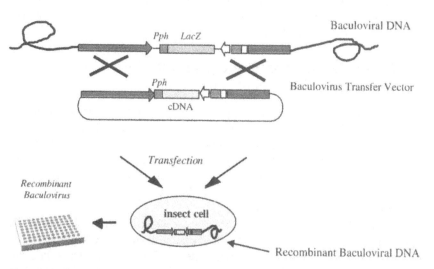

**Figure 1**   Schematic of miniaturized baculoviral production. The schematic diagram illustrates the placement of an I.M.A.G.E. cDNA into the linearized baculoviral genome. Both the cDNA and baculoviral DNA are cotransfected into insect cells in a well of a 96-well plate where homologous recombination places the cDNA in front of the polyhedrin promoter (Pph) replacing β-galactosidase gene (*LacZ*).

balls were placed within each well of the deep-well plate and the levitation Stirrer passed the plate through two horizontal magnetic fields (Fig. 2). As the steel balls rose and fell within the wells in response to the magnetic fields, the cultures were mixed and aerated. We performed several experiments to examine the effects of well volume on cell density and cell viability. We optimized the cells at a starting concentration of $5 \times 10^5$ cells/mL with a 1.5-mL well volume and demonstrated greater than 95% viability and over two population doublings in 48 h. In each well of the deep-well plate, 1.5 mL of cells were grown in suspension at a density of $1.5 \times 10^5$ cells/mL, varying concentrations of Pluronic F-68 in the media, and different size metal balls for stirring. We optimized our insect cell culture system to use a 2.38-mm steel ball, 1% Pluronic F-68 and virus at 5–10% (v/v), and approximately $10^5$–$10^7$ plaque-forming units/mL, (Fig. 3). The cells are then incubated for 4 days on the Levitation Stirrer at 27°C, after which virus is harvested by centrifugation. The same approach was applied for protein production except that the cells were incubated for 48 h, standard conditions for protein expression from baculoviral-infected insect cells. The Levitation Stirrer is a good high-throughput alternative to shaker flasks because it is designed for aerating cultures in a miniaturized format and can process multiple plates simultaneously.

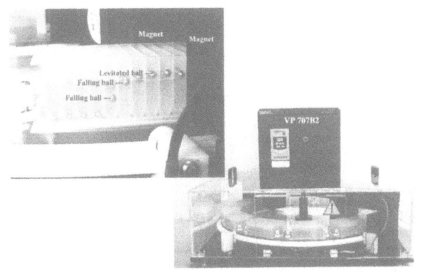

**Figure 2**   Carousel Magnetic Levitation Stirrer. The Carousel Magnetic Levitation Stirrer can simultaneously process 12 deep-well microtiter plates. Each well of the microtiter plate contains a steel ball that rises and falls as the plate passes the internal magnets, mixing and aerating the cultures, which are maintained in a temperature-controlled environment. (Courtesy of V & P Scientific, Inc., San Diego, CA.)

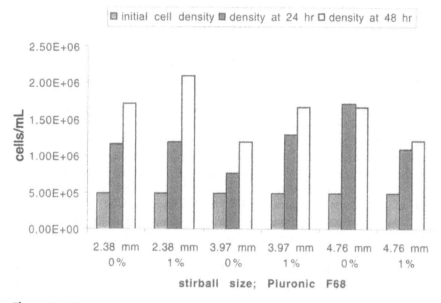

**Figure 3**  Miniaturized insect cell culture. Analysis of viability and insect cell growth using the Carousel Magnetic Levitation Stirrer shows successful doubling of *Sf9* insect cells in 48 h. Optimal conditions for growth were demonstrated with the inclusion of 1.0% Pluronic F-68 and 2.38-mm steel balls.

## E.  Protein Purification and Analysis

After growth, the cell pellets were lysed in each well of the deep-well plate. The plates were centrifuged and the supernatant containing soluble protein was transferred to a 96-well filter plate. The soluble fraction was clarified from any remaining insoluble particulates by vacuum filtration onto a second 96-well filter plate containing the epitope-tag immunoaffinity matrix in each well. The filtered lysate was allowed to interact with the immunoaffinity matrix, after which time the supernatant was pulled through the wells under low vacuum. The columns were washed, the recombinant proteins eluted by peptide competition, and the fractions collected using a vacuum manifold into a 96-well collection plate. For example, analysis of eluants from a 96-well recombinant β-glucuronidase (Gus)-only plate demonstrated that approximately 70% of the proteins (68/96 wells) were recovered after vacuum purification (Fig. 4). Soluble and insoluble protein fractions were analyzed by Western blot analysis with a primary antibody against the epitope tag. Detection was performed by enhanced chemiluminescence and analysis of protein purity was performed using standard gel electrophoresis

**Figure 4** Multiwell purification. Each well of a 96-well plate was infected with recombinant β-glucuronidase (Gus) virus at 10% (v/v), and cells were harvested after 48 h for recombinant protein production. Proteins were purified over immunoaffinity columns in a 96-well format into a collection plate. Presence of Gus protein was analysis by enzymatic hydrolysis of X-gluc (1.25 mg/mL), which is evidenced by a blue colormetric product in the well. Absorbance spectroscopy at 630 nm showed that 70% of the proteins were recovered after purification. (See the color plate.)

followed by staining with Coomassie blue. Purified recombinant proteins were maintained in an addressable format.

Preliminary results from the analysis of 16 cDNAs demonstrated that our PCR amplification and purification processes gave greater than 95% successful production of PCR products at a concentration of 1–2 μg of DNA per reaction. PCR purification using a 96-well format is almost completely accurate, resulting in very little loss of product. Ligation of PCR products in a high-throughput format, however, is problematic, as the resulting transformation efficiency shows about 55% of the clones successfully ligated, probably due to inefficient processing of the ligase. A second bottleneck for miniaturization of baculoviral production is found in the transfection of individual clones without subsequent plaque purification. These results varied widely from experiment to experiment, ranging from 50% to 100% successful transfection from a total of five experiments. Figure 5 shows a 100% transfection efficiency over several amplifications of a 96-well plate containing a single Gus construct transfected into 11 columns of the 96-well plate, with the last column containing controls (Fig. 5A–5C). The efficiency of the miniaturized transfection is evident by the second amplification of the recombinant baculoviruses (Fig. 5B). Overall, 65% of our initial 16 cDNAs expressed protein; 43% of those were soluble. These results are similar to those of our initial process that had no subcloning steps and demonstrated 42% soluble protein expression from 81 clones examined. In order to increase our overall efficiency, an improved transfer vector needs to be designed to more readily accept cDNAs. Future changes to our process may include substituting homologous recombination cloning methods for traditional subcloning through restriction enzyme digestion and DNA ligation. Throughout the process, each cDNA, virus, and protein remains in the same location within the original array to facilitate data tracking in our custom database.

## V. SUMMARY

Our strategy provides the basis for high-throughput protein expression and purification on a genomic scale. It has been designed for automation, which facilitates

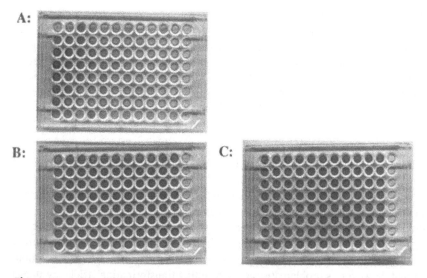

**Figure 5** A 96-well transfection. X-gluc was added to each well of a 96-well plate at 1.25 mg/mL. Enzymatic hydrolysis of this substrate by Gus results in the formation of a blue colorimetric product allowing for visual determination of the presence of the Gus protein in the well. This plate demonstrates that the r*Gus* virus and Gus protein product were produced to an efficiency of nearly 100% (excluding controls, column 12, where no baculovirus or plasmid DNA was added to the transfectant mixture). (A) Three days posttransfection; (B) first-round viral amplification, 4 days postinfection; (C) second-round viral amplification, 4 days postinfection.

the high-throughput protein production that is essential to fully investigate the human proteome. This system also provides the ability to select, array, express, and reexpress proteins from the largest public cDNA collection. This baculoviral expression paradigm could lead to several future applications for functional and structural studies [e.g., high-throughput structural analysis by nuclear magnetic resource (NMR) and x-ray crystallography, small-molecule-binding screens, and high-throughput enzymatic assays]. The versatility of our protein production system is likely to provide a powerful tool for the advancement of proteomic strategies.

## ACKNOWLEDGMENTS

This work was performed under the auspices of the U.S. Department of Energy by the University of California, Lawrence Livermore National Laboratory under Contract No. W-7405-Eng-48. The authors would like to thank Dr. Robin Clark, Dr. Anthony Davies, Dr. Ken Franke, and Ms. Bonnee Rubinfeld for their contributions to the work described in this chapter.

## REFERENCES

1. Christendat, D.; Yee, A.; Dharamsi, A.; Kluger, Y.; Gerstein, M.; Arrowsmith, C.H.; Edwards, A.M. Structural proteomics: prospects for high throughput sample preparation. Prog Biophys Mol Biol. 2000, 73, 339–345.
2. Morton, C.L.; Potter, P.M. Comparison of *Escherichia coliSaccharomyces cerevisiaePichia pastorisSpodoptera frugiperda*, and COS7 cells for recombinant gene expression. Application to a rabbit liver carboxylesterase. Mol Biotechnol. 2000, 16, 193–202.
3. Clark, E.D.B. Refolding of recombinant proteins. Curr Opin Biotechnol. 1998, 9, 157–163.
4. Armstrong, N.; de Lencastre, A.; Gouaux, E. A new protein folding screen: application to the ligand binding domains of a glutamate and kainate receptor and to lysozyme and carbonic anhydrase. Protein Sci. 1999, 8, 1475–1483.
5. Li, P.; Gao, X.G.; Arellano, R.O.; Renugopalakrishnan, V. Glycosylated and phosphorylated proteins—expression in yeast and oocytes of *Xenopus*: Prospects and challenges—relevance to expression of thermostable proteins. Protein Express Purif. 2001, 22, 369–380.
6. Jono, S.; Peinado, C.; Giachelli, C.M. Phosphorylation of osteopontin is required for inhibition of vascular smooth muscle cell calcification. J Biol Chem. 2000, 275, 20,197–20,203.
7. Coleman, T.A.; Parmelee, D.; Thotakura, N.R.; Nguyen, N.; Burgin, M.; Gentz, S.; Gentz, R. Production and purification of novel secreted human proteins. Gene. 1997, 190, 163–171.

8. Lueking, A.; Holz, C.; Gotthold, C.; Lehrach, H.; Cahill, D. A system for dual protein expression in *Pichia pastoris* and *Escherichia coli*. Protein Express Purif. 2000, 20, 372–378.

9. Lesley, S.A. High-throughput proteomics: protein expression and purification in the postgenomic world. Protein Express Purif. 2001, 22, 159–164.

10. Braun, P.; Hu, Y.; Shen, B.; Halleck, A.; Koundinya, M.; Harlow, E.; LaBaer, J. Proteome-scale purification of human proteins from bacteria. Proc Natl Acad Sci USA. 2002, 99, 2654–2659.

11. Holz, C.; Lueking, A.; Bovekamp, L.; Gutjahr, C.; Bolotina, N.; Lehrach, H.; Cahill, D.J. A human cDNA expression library in yeast enriched for open reading frames. Genome Res. 2001, 11, 1730–1735.

12. Zhu, H.; Bilgin, M.; Bangham, R.; Hall, D.; Casamayor, A.; Bertone, P.; Lan, N.; Jansen, R.; Bidlingmaier, S.; Houfek, T. Global analysis of protein activities using proteome chips. Science. 2001, 293, 2101–2105.

13. Nasoff, M.; Bergseid, M.; Hoeffler, J.P.; Heyman, J.A. High-throughput expression of fusion proteins. Methods Enzymol. 2000, 328, 515–529.

14. Summers MaS, G.E. A manual of methods for baculovirus vectors and insect cell culture procedures. Tex Agric Stn Bull. 1987, B1555, 1–56.

15. Kidd, I.M.; Emery, V.C. The use of baculoviruses as expression vectors. Appl Biochem Biotechnol. 1993, 42, 137–159.

16. Davies, A.H. Current methods for manipulating baculoviruses. Biotechnology (NY). 1994, 12, 47–50.

17. Gruenwald SaC, R. Baculovirus Expression Vector System: Procedures and Methods Manual. 5th ed.; PharMingen: San Diego: 0, 1998.

18. Kitts, P.A. Production of recombinant baculoviruses using linearized viral DNA. Methods Mol Biol. 1995, 39, 129–142.

19. Larsson, M.; Graslund, S.; Yuan, L.; Brundell, E.; Uhlen, M.; Hoog, C.; Stahl, S. High-throughput protein expression of cDNA products as a tool in functional genomics. J Biotechnol. 2000, 80, 143–157.

20. Lilie, H.; Schwarz, E.; Rudolph, R. Advances in refolding of proteins produced in *E. coli.*. Curr Opin Biotechnol. 1998, 9, 497–501.

21. Walter, G.; Bussow, K.; Cahill, D.; Lueking, A.; Lehrach, H. Protein arrays for gene expression and molecular interaction screening. Curr Opin Microbiol. 2000, 3, 298–302.

22. Lennon, G.; Auffray, C.; Polymeropoulos, M.; Soares, M.B. The I.M.A.G.E. Consortium: an integrated molecular analysis of genomes and their expression. Genomics. 1996, 33, 151–152.

23. Albala, J.S.; Franke, K.; McConnell, I.R.; Pak, K.L.; Folta, P.A.; Rubinfeld, B.; Davies, A.H.; Lennon, G.G.; Clark, R. From genes to proteins: high-throughput expression and purification of the human proteome. J Cell Biochem. 2000, 80, 187–191.

# 8

# Protein Profiling: Proteomes and Subproteomes

ERIC T. FUNG and ENRIQUE A. DALMASSO
*Ciphergen Biosystems, Inc.*
*Fremont, California, U.S.A.*

## I. INTRODUCTION

Although the value of genomics is indisputable, the need to study protein expression directly has become obvious. First, the only way to determine that an open reading frame encodes a protein is to identify that protein in a biological sample. Moreover, the paradigm of "one gene, one protein" has long been dismissed, with the discovery of alternative splicing, mRNA editing, and posttranslational modification. As there is a lack of correlation between transcription profiles and cellular protein levels, proteomics measures the proteins—those molecules directly involved in the cellular processes (additional commentary on these points can be found in Refs. 1–3). Fulfillment of the promise of proteomics depends on the power of the assays used to probe the proteome. Analysis of the proteome is a far more daunting task than analysis of the genome, as proteins are more complex than DNA and no protein amplification technique exists that is analogous to polymerase chain reaction (PCR). To successfully mine the cellular proteome, researchers must utilize multiple strategies while remaining cognizant of the respective strengths and weaknesses of each approach and, ultimately, with the intent to integrate the data obtained from these strategies into the underlying biology of the cellular system under investigation. Although critics of genomic and proteomic approaches to biological investigation decry the absence of hypothesis testing inherent to these methods, we maintain that these strategies have the potential to provide valuable information not readily attained using traditional "one gene/protein at a time" approaches. However, successful use of these strategies requires careful experimental design and the development of new and innovative approaches for high-throughput protein analysis.

The goal of this chapter is to examine protein profiling from a practical standpoint by providing the reader with background on the various techniques used for protein profiling, insights into the strengths and weaknesses of these techniques, as well as to how to counteract shortcomings. Here, we define protein profiling as a means to visualize as many proteins as possible from samples representing different cellular states (e.g., healthy versus disease, treated versus untreated). This chapter will not catalog the wealth of research resulting from protein profiling, as these can be found in other reviews (including preclinical drug development [4], cancer [5–7], microbiology [8–10], biomarker discovery [11], and neuropsychiatry [12]). In addition, this chapter will not address other protein profiling strategies, including protein biochip arrays, that mimic their DNA counterparts (high-density arrays that contain proteins of known identity at predetermined locations), protein–protein interaction assays, or strategies to determine protein structure, as those topics are covered elsewhere in this volume.

## II. TWO-DIMENSIONAL GEL ELECTROPHORESIS

Two-dimensional (2D) gel electrophoresis is by far the most established proteomic technique, having been used for over 20 years. The basic approach has not changed dramatically in that time. Proteins are separated in the first dimension on the basis of charge using isoelectric focusing, typically on an immobilized pH gradient gel strip which is then placed lengthwise into a sample well of a gel electrophoresis unit (for a review, see Ref. 13). Proteins are then separated in the second dimension by mass using sodium dodecyl sulfate–polyacrylamide gel electrophoresis (SDS–PAGE). Thus, in the vertical axis, larger proteins are above smaller proteins, and in the horizontal axis, proteins with a lower p$I$ are to the left of proteins with a higher p$I$. The resolved proteins can be visualized using a variety of methods, which include autoradiography, immunoblotting, Coomassie blue staining, silver staining, and fluorescence. Proteins of interest are identified by excision of the spots representing them from the gel, followed by an in-gel tryptic digest of the proteins. Tryptic peptide fingerprints can be matched against a database of theoretical tryptic digests. Alternatively, tandem mass spectrometry can be performed to generate peptide fragments that can be used to search against a database such as the MS-Tag website (accessible at http://prospector.ucsf.edu). Because 2D gel electrophoresis has been so widely used, its strengths and weaknesses have been well documented. As an initial approach to protein profiling, 2D gel electrophoresis is powerful; generally, over 1500 protein spots can be visualized on a single gel. The technique is sufficiently widespread that several commercial vendors now provide the gels and reagents for 2D gel electrophoresis and concerted efforts have been made to resolve some of the drawbacks of the technology, promising future improvements.

Perhaps the greatest shortcoming of 2D gel electrophoresis, and to a certain extent, any proteomics technique, is dynamic range, which is extremely limited by comparison to the range of concentration of proteins present in vivo. By one estimate, 10% of the proteins in a cell contribute to 90% of the mass of the proteome [14]. Corthals et al. note that if $10^6$ cells (approximately 20 µg of yeast extract) are loaded onto a 2D gel, only the most abundant proteins will be visualized (those present at 100,000 copies per cell or greater). Furthermore, if $10^9$ cells are loaded onto a 2D gel, only 8 ng of a 50-kDa protein would be present at 1000 per copies per cell, which is close to the detection limit of Coomassie staining [15]. Gygi et al. studied this problem further by examining a specific mass and p$I$ range of a 2D gel that had been loaded with 500 µg of yeast extract [16]. In a 4-cm$^2$ area, the authors identified the proteins present in 50 spots, demonstrating that not a single protein with a low (<0.1) Codon Adaptation Index (which roughly parallels the expected level of expression) was found, even though more than half of the yeast genes had such values. Two solutions become apparent; one must either use a more sensitive detection method or load more protein. Silver staining, which has a detection limit of approximately 0.1 ng of protein, is a problematic solution, as it does not stain protein consistently and reproducibility is poor. Silver staining has a limited dynamic range and can interfere with protein identification. The advent of new stains may resolve some of these problems in the near future. The other alternative, to increase the sample load, may result in spots representing the more abundant proteins that can comigrate and fuse, obscuring each other from detection.

There are other limitations related to the visualized proteome using 2D gel electrophoresis, including the inability to detect proteins with an extreme p$I$, proteins that are less than 15 kDa, or proteins that are hydrophobic [17]. Several strategies exist to overcome these challenges. The first, and most obvious, is to fractionate the sample prior to loading onto the 2D gel (for examples, see Refs. 18–21). Nonspecific prefractionation such as ion-exchange chromatography can be quite powerful in that fractions are generated that contain a subset of the proteome that can then be applied to separate 2D gels. Not only does such a procedure increase the number of proteins that can be visualized, but it also increases the success rate of protein identification by peptide mass fingerprinting, presumably because comigration of proteins is reduced [22]. Even the removal of albumin from serum or plasma using Cibacron blue resins can be helpful, although care must be made not to discard proteins that are associated with albumin. A second strategy is to use narrow-pH-gradient-range gels (<2 pH units wide). For example, a single pH 3–10 gradient-range gel resolved 755 protein spots from a yeast extract, but combined results from several narrow-pH-gradient-range gels (pH 3.5–4.5, 4–5, 4.5–5.5, 5–6, 5.5–6.7, and 6–9) showed that a total of distinct 2286 spots (not including the overlapping spots) could be visualized [23]. These "ultrazoom" gels can be loaded with up to 10 mg of sample, which can increase the number of proteins visualized while maintaining adequate resolution, although in many cases, particularly for clinical samples, such a sample load size is impractical [24].

In addition to the problem of dynamic range, 2D gel electrophoresis requires sample volumes that are often relatively large (e.g., 500 $\mu$g of protein was used in the experiment by Gygi et al. described earlier) and reproducibility is quite variable. In a study of 49 matched silver-stained 2D gels, Voss and Haberl showed that the overall pairwise matching success was 89% and less than 10% of the spots could be matched in 40 of the 49 gels [25]. Moreover, 25% of the spots differed in intensity by more than twofold. An additional source of variability is the position of spots, which varies from gel to gel, generating problems for automated spot-matching algorithms. For these reasons, truly high-throughput 2D electrophoresis remains untenable.

## III. PROTEINCHIP® PROTEOMICS

ProteinChip technology, designed, manufactured, and marketed by Ciphergen Biosystems (Fremont, CA), is a novel platform that uses a technique called surface-enhanced laser desorption/ionization (SELDI) to visualize the proteome [26–28]. Like 2D gel electrophoresis, the power of ProteinChip technology lies in the orthogonal separation of proteins. The first dimension of separation is performed on ProteinChip arrays and is based on the biochemical properties of the proteins in the sample but is not limited to the p$I$ of the proteins. ProteinChip arrays have surfaces that possess various chromatographic characteristics such as anion exchange, cation exchange, reverse phase, normal phase, or metal affinity. Proteins are bound to these surfaces under one set of conditions and can be washed under more stringent conditions, as needed, to enhance protein binding to the array. Wash fractions are either discarded or can be applied to additional ProteinChip arrays. Thus samples can be placed on ProteinChip arrays either in parallel or in series, with each ProteinChip array providing complementary protein profiles. The process has been termed ''retentate chromatography'' because what will ultimately be visualized are the proteins that are retained on the ProteinChip arrays. The second dimension of separation is based on the mass of the protein that is analyzed using a ProteinChip reader, which is a time-of-flight mass spectrometer.

The advantages of the ProteinChip over 2D gel electrophoresis include that protein binding to the arrays is not limited by p$I$ because separation of the cellular protein is based on the inherent chromatographic property of the chip. In addition, the number of proteins binding to the arrays can be manipulated by the binding and washing conditions applied based on the biophysical characteristics of the proteins applied to ProteinChips. Visualization of small proteins is enhanced by use of a time-of-flight mass spectrometer. ProteinChip proteomics has an additional advantage over to 2D gel analysis in that the technology requires a much reduced sample size. In studies using clinically derived samples, where there is typically a limited amount of sample available and prefractionation is prohibited, the desired end point of these studies is simply to compare protein profiles. A typical ProteinChip profiling experiment requires 10 μg of cell lysates per spot (there are eight spots per ProteinChip), which is within the sample collection size of most clinical samples. In particular, ProteinChip proteomics has found a niche when used in conjunction with laser capture microdissection (LCM), a technique in which specific cell populations can be selected for analysis by genomic or proteomic techniques (reviewed in Refs. 29–31). Successful ProteinChip profiling of LCM-derived samples can be performed on fewer than 100 cells, although 2000 cells are used typically, and has been a successful technique to profile a variety of cancers [6,32,33].

Conversely, ProteinChip proteomics does not visualize large proteins as well as 2D gels, and as with 2D gels, dynamic range sensitivity of the proteome is also limiting with ProteinChip technology. Consequently, similar to 2D gels, it is recommended that complex samples be prefractionated either nonspecifically (e.g., using gel filtration or ion-exchange chromatography) or specifically (by affinity chromatography or for specific subproteomes) prior to binding the samples onto the ProteinChip arrays (see Fig. 1).

Protein identification using ProteinChip proteomics requires some sample enrichment and purification, which can be performed either on the chip or in spin columns, followed by tryptic digestion of the sample (either on-chip or in-gel, if the purified protein is run on a 1D gel). As with 2D gel identification, the tryptic peptide mass fingerprint can be matched against theoretical digests of proteins present in databases or the tryptic peptides fragmented by tandem mass spectrometry and the fragments searched against a database (for an example, see Ref. 34). Because protein identification on the ProteinChip platform is somewhat more laborious than it is using the 2D gel approach, most researchers using the ProteinChip system have opted to perform validation studies prior to entering the identification phase. Unlike 2D gel studies, ProteinChip studies can be easily scaled up to include many samples, so the typical profiling experiment begins with a discovery phase in which experimental conditions are optimized (such as chip type and wash buffers) to identify candidate biomarkers. Then, a larger-scale validation study is performed to confirm that the candidate biomarkers are, indeed, worthy of further study.

It is noteworthy that in many cases actual identification of the candidate biomarkers is not necessary. A multimarker approach to biomarker discovery is particular relevant when using proteomics technologies, in which patterns of markers (as in SELDI spectra) can form the basis of diagnostic decisions, often with much greater sensitivity and specificity than a single marker might provide. For example, analysis of the receiver operator characteristic (ROC) curve for

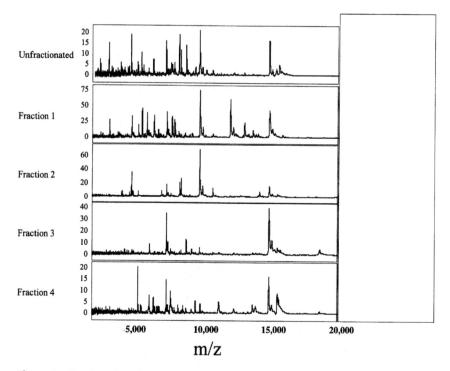

**Figure 1**  Fractionation of liver lysates increases protein visualization. Mouse liver lysate was applied to a weak-cation-exchange (WCX) ProteinChip® array. The retained proteins were visualized using a ProteinChip reader. The top spectrum shows the unfractionated lysate; the bottom four spectra show the results of fractionation of the lysates on an anion-exchange column (Q Sepharose). The fractions are obtained by a stepwise pH elution gradient.

serum prostate-specific antigen (PSA) showed that no single value for PSA con-centration can be used to provide both high sensitivity and specificity scores (defined as greater than 80%) [35]. This can be contrasted against a study per-formed in using the ProteinChip platform for a biomarker discovery project study-ing transitional cell carcinoma (TCC) of the bladder [36]. This study examined cells isolated from the urine of patients with TCC, benign bladder disease (e.g. inflammatory), or healthy controls and found five candidate markers, none with sensitivity level greater than 70%. By combining three markers, sensitivity of 83% could be achieved while maintaining 67% specificity. Multimarker panels for diagnosis are likely to become more common with the identification of additional markers allowing for additional testing for high sensitivity and specificity. Finally, the ProteinChip array format provides a platform for the development of a quantitative multiplex immunoassay [33]. Antibodies to specific analytes can be covalently coupled to specialized ProteinChip arrays, which are then used to survey the abundance of these analytes in clinical specimens.

# IV. OTHER PROTEIN PROFILING TECHNIQUES

The combination of liquid chromatography (LC) or capillary electrophoresis (CE) with mass spectrometry (e.g., LC-MS/MS and CE coupled with Fourier transform ion-cyclotron resonance mass spectrometry) promises to provide valuable insight into the composition of the proteome [37]. A complex mixture can be digested by an endoprotease and the resultant peptides separated by liquid chromatography. Eluted peptides can then be sequenced by tandem mass spectrometry [38–43]. This method is particularly useful if the flow rate is decreased to 20 nL/min [44], and other modifications have been implemented to increase sensitivity [45–47]. In one of the most ambitious implementations of LC-MS/MS, 1484 proteins from the yeast proteome were separated and identified [48], including proteins that are not typically visualized on 2D gels such as membrane proteins and proteins with a high p*I*. However, these techniques are low throughput and, thus far, provide little quantitative information and are, therefore, limited for comparative profiling.

In fact, no proteomics profiling technique is as yet rigorously quantitative. For example, silver staining of 2D gels is notoriously uneven and nonlinear. Although novel stains such as SYPRO Ruby (Molecular Probes) promise to be linear over a larger dynamic range (stated to be 2–2000 ng), their use is hampered by the need to minimize quenching as well as to develop accessible detection methods [49]. Additionally, mass spectrometry is notoriously nonquantitative. This is, in part, because the ionization properties of any given protein or peptide is unique and, therefore, the most accurate standard for quantitation of a given peptide or protein is itself. One technique that has been used to address this problem is stable isotope dilution, in which one sample is grown in one isotopic condition and a second sample is grown in a different isotopic condition [50]. The ratio of the two isotopes seen by mass spectrometry reflects the relative abundance of the protein in the two samples. For example, one sample of yeast was grown in natural nitrogen isotopic distribution (99.6% $^{14}$N and 0.4% $^{15}$N) and a second sample of yeast was grown in a media containing predominantly $^{15}$N (>96%). The samples were pooled, proteins separated by reverse phase–high-performance liquid chromatographs (RP-HPLC) followed by 1D SDS-PAGE, and selected proteins digested by trypsin and analyzed by MALDI-MS (matrix-assisted laser desorption/ionization-mass spectrometry) [51]. Peptides derived from proteins of cells grown in the heavy-isotope condition differed by one mass unit for each incorporated $^{15}$N. Although this approach can be used in cells grown in culture, it obviously cannot be used on samples derived from other sources such as tissue. Isotope-coded affinity tag (ICAT) extends the idea of using the difference in isotope mass to a greater number of applications [52,53]. In this protocol, a tag that targets cysteinyl residues on the protein is used. One sample is labeled with a "light" tag and the other is labeled with a "heavy" tag, which substitutes deuterium for hydrogen at eight sites along the carbon backbone. The labeled protein mixtures are then combined and digested, and the labeled fragments are affinity purified using a tag (an avidin–biotin interaction, in this case). The purified peptides can then be further separated by liquid chromatography and then identified by tandem mass spectrometry. The ratio of the heavy to light tag forms the basis of quantitation. The broad concept of this method can also be extended to in vivo labeling of cells grown under conditions of different isotopic distribution (e.g., $^{15}N$) [54]. One drawback that can seriously limit the versatility of these techniques is that although it can be useful for pairwise comparisons, extension to the study of multiple samples or multiple sample groups is not possible. Additionally, the study of posttranslational modifications is restricted to fragments containing both the isotope tag as well as the posttranslational modification.

## V.  PROFILING OF SUBPROTEOMES

There are several strategies for the study of subproteomes that have become of great interest for several reasons. First, as has been discussed, limitations in dynamic range and sensitivity using traditional proteomics techniques makes scanning the entire proteome in a single experiment impractical. Second, specific classes of proteins can be studied as a unit, thereby allowing more a direct comparison of members. Third, a specific cellular process can be studied in greater detail. Complex cellular states can be simplified into subproteomes that can be more easily compared both at the technical level (within the resolving power of the techniques used) and at the computational level. By studying several subproteomes simultaneously from a sample, one can "reconstruct" a single cohesive image of the entire proteome for the given sample. The following discussion will survey the subproteomes of modified proteins (e.g., phosphorylated proteins), protein families (e.g., enzymes), and subcellular organelles.

One example of a subproteome is that composed of proteins containing covalent modifications such as phosphorylation or glycosylation. Because protein modification has long been known to regulate protein function, several technologies to probe these subproteomes have been developed. Two-dimensional gel electrophoresis combined with immunoblotting has been used to examine phosphoproteins; however, the low abundance of phosphoproteins within the cell as well as the limited availability of antiphosphoserine and antiphosphothreonine antibodies has hampered the utility of this approach. The ProteinChip system can be used to examine the phosphoproteome, either by immobilizing antiphosphoamino acid antibodies onto a preactivated ProteinChip array or, in some cases, by using an immobilized metal-affinity chromatography (IMAC) ProteinChip array. However, these approaches suffer from some of the same disadvantages as with the 2D gel approach. Zhu et al. recently described the construction of a biochip consisting of kinase substrates covalently attached to the walls of individual wells in a 96-well plate [55]. Protein kinases were expressed by recombinant strategies, applied to the chip, and kinase assays were performed in the presence of radioactive ATP. A phosphorimager was used to detect phosphorylated substrates. Although the powers of this technique are obvious, it is limited by the need to produce each kinase, which is not always possible. Moreover, the kinase reactions were performed in vitro, which can have both positive (kinase activity in vitro that would not occur in vivo) and negative (absence kinase activity in vitro that would occur in vivo) artifacts and are limited by the difficulty in studying upregulation or downregulation of kinase activity by perturbation of signaling cascades.

Two methods to enrich for phosphorylated proteins using chemical methods have been described. One takes advantage of the alkaline-labile property of the phosphate moiety and replaces the endogenous phosphate with a biotinylated functional group that can be used to enrich for these modified proteins [56]. In this scheme, proteins are first oxidized to protect cysteines, then incubated in the presence of lithium hydroxide, neutralized, followed by incubation in the presence of a biotinylating agent. After dialysis to remove the unused biotinylating agent, biotinylated proteins can either be digested with trypsin and the biotinylated peptides captured on avidin beads, or the entire protein can be captured on avidin beads, eluted, and digested with trypsin. The tryptic peptides can then be sequenced by MALDI-MS/MS or SELDI-MS/MS with subsequent identification of the phosphorylation site. A second chemical method to purify phosphorylated proteins is to modify phosphopeptides with free sulfhydryls, which can be subsequently trapped on a solid phase consisting of iodacetic acid-linked glass beads [57]. Acid elution regenerates the phosphopeptides, which are then analyzed and sequenced by tandem mass spectrometry. While the proof-of-concept experiments are compelling, additional studies are needed to demonstrate the quantitative nature of this technique, particularly when studying in vivo systems. Moreover, these complex series of chemical steps can also modify nonphosphorylated proteins, sometimes in unpredictable ways, and therefore obviate the study of those proteins.

Glycosylated proteins present a different set of challenges for proteomics. Glycoproteins typically exhibit microheterogeneity, with the presence of variable numbers and, often, types of glycan moiety attached to several sites on the protein. Additionally, many glycosylated proteins are large and contain hydrophobic domains, characteristics that have limited their appearance in 2D gels. For example, enrichment for glycosylated proteins in mitochondria by chromatography on ConA lectin resin resulted in the majority of the proteins remaining unresolved on the 2D gel, and the gel contained numerous horizontal and vertical streaks [58]. Examination of the glycoproteome is much easier in reducing, denaturing 1D gels combined with MALDI-MS. Mapping of glycosylation sites can be performed by incubating the isolated glycoprotein with a site-specific protease to generate peptides and then a portion of the sample can be treated with glycosidases. Shifts in mass upon deglycosylation can then be visualized by mass spectrometry (either MALDI or SELDI). One potential technical limitation to this approach is that glycopeptides generally produce weaker signals than do their nonglycosylated counterparts and so there remains the risk of failure to detect the glycopeptides. Using a series of glycosidases, structural determination of glycan moieties can be performed. A more thorough analysis of the glycoproteome can be found in Ref. 59 and some applications can be found in Refs. 60 and 61.

Enzymes provide another class of proteins whose expression levels are often below the threshold of the dynamic range of traditional proteomics techniques. One of the first methods to scan the proteome for a class of enzymes was developed by Liu et al. and termed "activity-based protein profiling" [62]. The basic requirement of this strategy is the development of a moiety that binds irreversibly to a conserved domain in the family of enzymes (most obviously, the active site) and that can be derivatized with a group that can be used to monitor capture of the enzyme. Liu et al. created a biotinylated long-chain fluorophosphonate that binds and irreversibly inhibits catalytically active serine hydrolases but not cysteine, aspartyl, metallohydrolases, or catalytically inactive serine hydrolases. Detection of the enzymes can be performed by Western blot using an avidin probe, and if enough is purified in this manner, the enzymes can be identified using mass spectrometry. Another technique to isolate enzymes with a specific substrate is to immobilize the protein substrate on beads and bind the capture extract (e.g., cell lysate) to the beads. After washing, an endoprotease is added to generate peptide fragments that can be identified by mass spectrometry. Such a technique was used to purify a β-secretase-like activity and to identify the protease as cathepsin D [63]. For these assays to be successful, the substrate must be immobilized in such a manner that following the enzymatic reaction, both the enzyme and substrate remain bound to each other and attached to the beads. Additionally, proteins without catalytic activity could bind nonspecifically to the beads, so any molecules identified in this manner have to be characterized for in vitro and in vivo enzymatic activity.

The ProteinChip platform has been used to study the cleavage of β-amyloid by BACE, a protein present in plaques of patients with Alzheimer's disease. The various members of the secretase family (BACE1, BACE2, γ-secretase) cleave β-amyloid with differing site specificities, and multiple species of processed β-amyloid can be generated. In patients with Alzheimer's disease, there is a preponderance of specific amyloidogenic forms of β-amyloid (particularly, the 42-residue form). By coupling an antibody to a preactivated surface ProteinChip array, these various forms of β-amyloid can be captured from a complex sample and analyzed in the ProteinChip reader. The various forms of β-amyloid can be identified based on their mass, and such an experiment can also reveal the relative quantities of each isoform form of β-amyloid in the mixture [64–67]. This type of information would not be obtained using a standard enzyme-linked immunoabsorbant assay (ELISA).

Subcellular fractionation is another way to generate subproteomes that can provide valuable information as well as eliminate limitations in dynamic range and proteome visualization discussed earlier. By analyzing organelles separately, one can enrich for a subset of proteins involved in a specific biological process. The purified organelle preparation can then be subjected to 1D electrophoresis, 2D electrophoresis, ProteinChip proteomics, or any of the other above-outlined techniques (reviewed in Ref. 68]. It should be noted that, just as in the analysis of the entire proteome, further fractionation of the organelle preparations can lead to more complete visualization of their respective protein content.

A pair of studies analyzing the Golgi subproteome highlights the utility of multiple fractionation procedures to maximize the number of proteins that can be visualized. In an initial study of isolated stacked Golgi from rat liver, a 300–400-fold enrichment of Golgi markers was achieved leading to the identification of 174 proteins, most of them known to be of high abundance [69]. The addition of a Triton X-114 for increased solubilization and separation followed by anion-exchange chromatography of two of the Triton fractions resulted in the identification of additional proteins of lower abundance. Even so, the authors point out that further fractionation protocols will be necessary to completely visualize the Golgi proteome [58]. Another example of the utility of fractionating organelle preparations comes from a study of the mitochondrial subproteome, which is of interest because of the role of the mitochondria in biosynthetic, metabolic, and cell-death pathways [70,71]. Whereas profiling of mitochondria as a whole sample can result in visualization of part of the proteome (300–500 spots stained by Coomassie blue, 1500 stained by silver), additional information can be obtained with a few simple prefractionation steps [71]. For example, calcium-binding proteins could be enriched by using IMAC spin columns charged with calcium, leading to the detection of 819 proteins, whereas enrichment of hydrophobic proteins on phenyl Sepharose resin led to the detection of 736 protein spots.

The plasma membrane subproteome has been of particular interest, for therein lies the receptors that form the basis for intercellular communication and the initiation of countless signal transduction cascades. It is further assumed that membrane receptors and transporters will be targets for therapy; this is of interest to pharmaceutical and biotechnology companies which have been formed to study these proteins and develop drugs that target them. One of the most extensive analyses of the membrane subproteome included the characterization of solubilization properties of various plasma membrane proteins such that a model could be generated [termed "additive main effects with multiplicative interaction" (AMMI)] that would allow researchers to classify proteins according to their solubility properties [72]. This type of analysis should allow for the design of experiments that specifically target specific subtypes of membrane receptors. Glycosyl phosphatidylinositol (GPI)-linked proteins are one subclass of membrane proteins that participate in signal transduction as well as plasma membrane organization (reviewed in Ref. 73). Not surprisingly, these proteins are difficult to resolve when the entire membrane fraction is applied to a 2D gel. However, when GPI-anchored proteins are enriched by separation on a sucrose gradient following detergent extraction, the GPI-linked proteins are more readily detected [74]. Analysis of this class of proteins by 2D gel electrophoresis is limited by the presence of the lipid group, and only after cleavage of this group by phosphatidylinositol-specific phospholipase C can the proteins be clearly resolved, and for definitive identification, deglycosylation by $N$-glycosidase F treatment is necessary.

## VI.  CLOSING COMMENTS

Proteomics is an evolving multidisciplinary endeavor, requiring knowledge of the biology of the system being studied, biochemistry for appropriate use of separation techniques, chemistry for advancing protein visualization technologies, mass spectrometry for protein visualization and identification, and statistics and bioinformatics for the analysis and interpretation of data. Although 2D gel electrophoresis has been the mainstay of proteomics in the past, novel technologies such as ProteinChip proteomics have emerged. Combined use of these technologies to visualize the many cellular subproteomes individually and simultaneously will consequently provide a cohesive understanding of the cellular proteome. Finally, these data must be integrated with genomic information, the biological context of the system, and comparisons across species for a comprehensive analysis of an integrated cellular proteome.

## REFERENCES

1. Gygi, S.P.; Rochon, Y.; Franza, B.R.; Aebersold, R. Mol Cell Biol. 1999, 19, 1720–1730.
2. Anderson, L.; Seilhamer, J. Electrophor. 1997, 18, 533–537.
3. Pandey, A.; Mann, M. Nature. 2000, 405, 837–846.
4. Steiner, S.; Witzmann, F.A. Electrophoresis. 2000, 21, 2099–2104.
5. Jain, K.K. Pharmacogenomics. 2000, 1, 385–393.

6.  Eggeling, F.; Davies, H.; Lomas, L.; Fiedler, W.; Junker, K.; Claussen, U.; Ernst, G. BioTechniques. 2000, 29, 1066–1070.
7.  Alaiya, A.A.; Franzen, B.; Auer, G.; Linder, S. Electrophoresis. 2000, 21, 1210–1217.
8.  Washburn, M.P.; Yates, J.R. Curr Opin Microbiol. 2000, 3, 292–297.
9.  O'Connor, C.D.; Adams, P.; Alefounder, P.; Farris, M.; Kinsella, N.; Li, Y.; Payot, S.; Skipp, P. Electrophoresis. 2000, 21, 1178–1186.
10. Cash, P. Electrophoresis. 2000, 21, 1187–1201.
11. Fung, E.T.; GL, J.r.; Dalmasso, E.A. Curr Opin Mol Ther. 2000, 2, 643–650.
12. Rohlff, C. Electrophoresis. 2000, 21, 1227–1234.
13. Gorg, A.; Obermaier, C.; Boguth, G.; Harder, A.; Scheibe, B.; Wildgruber, R.; Weiss, W. Electrophoresis. 2000, 21, 1037–1053.
14. Miklos, G.L.; Maleszka, R. Electrophoresis. 2001, 22, 169–178.
15. Corthals, G.L.; Wasinger, V.C.; Hochstrasser, D.F.; Sanchez, J.C. Electrophoresis. 2000, 21, 1104–1115.
16. Gygi, S.P.; Corthals, G.L.; Zhang, Y.; Rochon, Y.; Aebersold, R. Proc Natl Acad Sci USA. 2000, 97, 9390–9395.
17. Harry, J.L.; Wilkins, M.R.; Herbert, B.R.; Packer, N.H.; Gooley, A.A.; Williams, K.L. Electrophoresis. 2000, 21, 1071–1081.
18. Fountoulakis, M.; Langen, H.; Evers, S.; Gray, C.; Takacs, B. Electrophoresis. 1997, 18, 1193–1202.
19. Fountoulakis, M.; Langen, H.; Gray, C.; Takacs, B. J Chromatogr A. 1998, 806, 279–291.
20. Fountoulakis, M.; Takacs, M.F.; Berndt, P.; Langen, H.; Takacs, B. Electrophoresis. 1999, 20, 2181–2195.
21. Fountoulakis, M.; Takacs, M.F.; Takacs, B. J Chromatogr A. 1999, 833, 157–168.
22. Butt, A.; Davison, M.D.; Smith, G.J.; Young, J.A.; Gaskell, S.J.; Oliver, S.G.; Beynon, R.J. Proteomics. 2001, 1, 42–53.
23. Wildgruber, R.; Harder, A.; Obermaier, C.; Boguth, G.; Weiss, W.; Fey, S.J.; Larsen, P.M.; Gorg, A. Electrophoresis. 2000, 21, 2610–2616.
24. Hoving, S.; Voshol, H.; van Oostrum, J. Electrophoresis. 2000, 21, 2617–2621.
25. Voss, T.; Haberl, P. Electrophoresis. 2000, 21, 3345–3350.
26. Weinberger, S.R.; Morris, T.S.; Pawlak, M. Pharmacogenomics. 2000, 1, 395–416.
27. Fung, E.T.; Thulasiraman, V.V.; Weinberger, S.R.; Dalmasso, E.A. Curr Opin Biotechnol. 2001, 12, 65–69.
28. Merchant, M.; Weinberger, S.R. Electrophoresis. 2000, 21, 1164–1177.
29. Gillespie, J.W.; Ahram, M.; Best, C.J.; Swalwell, J.I.; Krizman, D.B.; Petricoin, E.F.; Liotta, L.A.; Emmert-Buck, M.R. Cancer J. 2001, 7, 32–39.
30. Banks, R.E.; Dunn, M.J.; Forbes, M.A.; Stanley, A.; Pappin, D.; Naven, T.; Gough, M.; Harnden, P.; Selby, P.J. Electrophoresis. 1999, 20, 689–700.
31. Bichsel, V.E.; Liotta, L.A.; Petricoin, E.F. Cancer J. 2001, 7, 69–78.
32. Paweletz, C.P.; Gillespie, J.W.; Ornstein, D.K.; Simone, N.L.; Brown, M.R.; Cole, K.A.; Wang, Q.-.H.; Huang, J.; Hu, N.; Yip, T.-.T. Drug Dev Res. 2000, 49, 34–42.
33. Wright, G.L.; Cazares, L.H.; Leung, S.-.M.; Nasim, S.; Adam, B.-.L.; Yip, T.-.T.; Schellhammer, P.F.; Gong, L.; Vlahou, A. Prostate Cancer Prostatic Dis. 2000, 2, 264–276.

34. Thulasiraman, V.; McCutchen-Maloney, S.L.; Motin, V.L.; Garcia, E. Biotechniques. 2001, 30, 428–432.
35. Catalona, W.J.; Partin, A.W.; Slawin, K.M.; Brawer, M.K.; Flanigan, R.C.; Patel, A.; Richie, J.P.; deKernion, J.B.; Walsh, P.C.; Scardino, P.T. JAMA. 1998, 279, 1542–1547.
36. Vlahou, A.; Schellhammer, P.F.; Mendrinos, S.; Patel, K.; Kondylis, F.I.; Gong, L.; Nasim, S.; GL, J.r. Am J Pathol. 2001, 158, 1491–1502.
37. Chalmers, M.J.; Gaskell, S.J. Curr Opin Biotechnol. 2000, 11, 384–390.
38. Link, A.J.; Eng, J.; Schieltz, D.M.; Carmack, E.; Mize, G.J.; Morris, D.R.; Garvik, B.M.; Yates, J.R. Nature Biotechnol. 1999, 17, 676–682.
39. Martin, S.E.; Shabanowitz, J.; Hunt, D.F.; Marto, J.A. Anal Chem. 2000, 72, 4266–4274.
40. Ducret, A.; Van Oostveen, I.; Eng, J.K.; Yates, J.R.; Aebersold, R. Protein Sci. 1998, 7, 706–719.
41. Dongre, A.R.; Eng, J.K.; Yates, J.R. Trends Biotechnol. 1997, 15, 418–425.
42. Haynes, P.A.; Yates, J.R. Yeast. 2000, 17, 81–87.
43. Shen, Y.; Zhao, R.; Belov, M.E.; Conrads, T.P.; Anderson, G.A.; Tang, K.-; Tolic, L.; Veenstra, T.D.; Lipton, M.S.; Udseth, H.R. Anal Chem. 2001, 73, 1766–1775.
44. Shabanowitz, J.; Settlage, R.E.; Marto, J.A.; Christian, R.E.; White, F.M.; Russo, P.S.; Martin, S.E.; Hunt, D.F. In Mass Spectrometry in Biology and Medicine Burlingame AL, ed; Humana Press: Totowa: 0, 2000, 163–177.
45. Opiteck, G.J.; Ramirez, S.M.; Jorgenson, J.W.; Moseley, M.A. Anal Biochem. 1998, 258, 349–361.
46. Tong, W.; Link, A.; Eng, J.K.; Yates, J.R. Anal Chem. 1999, 71, 2270–2278.
47. Spahr, C.S.; Susin, S.A.; Bures, E.J.; Robinson, J.H.; Davis, M.T.; McGinley, M.D.; Kroemer, G.; Patterson, S.D. Electrophoresis. 2000, 21, 1635–1650.
48. Washburn, M.P.; Wolters, D.; Yates, J.R. Nature Biotechnol. 2001, 19, 242–247.
49. Lopez, M.F.; Berggren, K.; Chernokalskaya, E.; Lazarev, A.; Robinson, M.; Patton, W.F. Electrophoresis. 2000, 21, 3673–3683.
50. Gygi, S.P.; Rist, B.; Aebersold, R. Curr Opin Biotechnol. 2000, 11, 396–401.
51. Oda, Y.; Huang, K.; Cross, F.R.; Cowburn, D.; Chait, B.T. PNAS. 1999, 96, 6591–6596.
52. Gygi, S.P.; Rist, B.; Gerber, S.A.; Turecek, F.; Gelb, M.H.; Aebersold, R. Nature Biotechnol. 1999, 17, 994–999.
53. Griffin, T.J.; Gygi, S.P.; Rist, B.; Aebersold, R.; Loboda, A.; Jilkine, A.; Ens, W.; Standing, K.G. Anal Chem. 2001, 73, 978–986.
54. Conrads, T.P.; Alving, K.; Veenstra, T.D.; Belov, M.E.; Anderson, G.A.; Anderson, D.J.; Lipton, M.S.; Pasa-Tolic, L.; Udseth, H.R.; Chrisler, W.B. Anal Chem. 2001, 73(9), 2132–2139.
55. Zhu, H.; Klemic, J.F.; Chang, S.; Bertone, P.; Casamayor, A.; Klemic, K.G.; Smith, D.; Gerstein, M.; Reed, M.A.; Snyder, M. Nature Genet. 2000, 26, 283–289.
56. Oda, Y.; Nagasu, T.; Chait, B.T. Nature Biotechnol. 2001, 19, 379–382.
57. Zhou, H.; Watts, J.D.; Aebersold, R. Nature Biotechnol. 2001, 19, 375–378.
58. Taylor, R.S.; Wu, C.C.; Hays, L.G.; Eng, J.K.; Yates, J.R.; Howell, K.E. Electrophoresis. 2000, 21, 3441–3459.

59. Kuster, B.; Krogh, T.N.; Mortz, E.; Harvey, D.J. Proteomics. 2001, 1, 350–361.
60. Rudd, P.M.; Colominas, C.; Royle, L.; Murphy, N.; Hart, E.; Merry, A.H.; Hebestreit, H.F.; Dwek, R.A. Electrophoresis. 2001, 22, 285–294.
61. Geng, M.; Zhang, X.; Bina, M.; Regnier, F. J Chromatogr B: Biomed Sci Appl. 2001, 752, 293–306.
62. Liu, Y.; Patricelli, M.P.; Cravatt, B.F. Proc Natl Acad Sci USA. 1999, 96, 14694–14699.
63. Gruninger-Leitch, F.; Berndt, P.; Langen, H.; Nelboeck, P.; Dobeli, H. Nature Biotechnol. 2000, 18, 66–70.
64. Cai, H.; Wang, Y.; McCarthy, D.; Wen, H.; Borchelt, D.R.; Price, D.L.; Wong, P.C. Nature Neurosci. 2001, 4, 233–234.
65. Austen, B.M.; Frears, E.R.; Davies, H. J Peptide Sci. 2000, 6, 459–469.
66. Davies, H.; Lomas, L.; Austen, B. BioTechniques. 1999, 27, 1258–1261.
67. Frears, E.R.; Stephens, D.J.; Walters, C.E.; Davies, H.; Austen, B.M. NeuroReport. 1999, 10, 1699–1705.
68. Jung, E.; Heller, M.; Sanchez, J.C.; Hochstrasser, D.F. Electrophoresis. 2000, 21, 3369–3377.
69. Taylor, R.S.; Fialka, I.; Jones, S.M.; Huber, L.A.; Howell, K.E. Electrophoresis. 1997, 18, 2601–2612.
70. Koc, E.C.; Burkhart, W.; Blackburn, K.; Moseley, A.; Koc, H.; Spremulli, L.L. J Biol Chem. 2000, 275, 32,585–32,591.
71. Lopez, M.F.; Kristal, B.S.; Chernokalskaya, E.; Lazarev, A.; Shestopalov, A.I.; Bogdanova, A.; Robinson, M. Electrophoresis. 2000, 21, 3427–3440.
72. Santoni, V.; Kieffer, S.; Desclaux, D.; Masson, F.; Rabilloud, T. Electrophoresis. 2000, 21, 3329–3344.
73. Anderson, R.G. Annu Rev Biochem. 1998, 67, 199–225.
74. Fivaz, M.; Vilbois, F.; Pasquali, C.; van der Goot, F.G. Electrophoresis. 2000, 21, 3351–3356.

# 9

# Shotgun Proteomics and Its Applications to the Yeast Proteome

**ANITA SARAF and JOHN R. YATES III**
*The Scripps Research Institute*
*La Jolla, California, U.S.A.*

## I. INTRODUCTION

The availability of completed genome sequences of several prokaryotes, viruses, and eukaryotes has transformed biomedical research in a short period of time. Several postgenomic technologies including SAGE (serial analysis of gene expression), [1] and cDNA microarrays [2,3] have been developed to globally and quantitatively measure gene expression at the mRNA level. However, no strict linear correlation between gene expression and the protein complement or "proteome" of a cell has been determined. In fact, many studies have shown quite a poor correlation between mRNA and protein expression levels [4–7]. Proteins, not genes, are responsible for most functions in the cell; thus, studying the cellular proteome can provide a picture of the protein environment at any given time.

The word "proteomics" was introduced in the mid-1990s [8] to describe the functional analysis of proteins on a large scale, including expression level, protein–protein interactions, modifications, and localization. The goal of these analyses is to obtain a more global and integrated view of cellular and disease processes, as well as regulatory networks. Proteomics can be divided into two major categories: protein expression proteomics and "cell map" proteomics [9]. Expression proteomics attempts to provide the quantitative picture of protein expression from cells or tissues. By comparing the changes in protein expression in two or more proteomes, this "differential display" proteomics enables the study of global changes in protein expression and has potential application in studying a wide range of diseases and can serve as a guide to drug design. Expression proteomics can also be used to study coregulation of protein expression by measuring expression levels of proteins under different cellular conditions and states. "Cell map" proteomics involves the determination of the subcellular location of proteins and the systematic study of protein–protein interactions in purified protein complexes and cellular organelles. It has become increasingly clear that most proteins do not work alone, but carry out physiological processes in concert with other cellular proteins. The presence of a protein in a complex indicates a direct involvement in the biological process rather than association by coregulation, and its interaction with other proteins in a multiprotein complex gives insight into protein function within the overall context of the cell. Cell map proteomics enables the study and characterization of the potential role of a target protein from a specific group of proteins in a cellular process. Such studies provide information about protein signaling, interaction between different cellular pathways, or disease mechanisms.

Mass spectrometry has emerged as a powerful analytical method for proteome analysis. Refinements to ionization methods have resulted in improved sensitivities and integration with liquid-separation techniques. These improvements, combined with tandem mass spectrometry, have created powerful techniques for peptide sequencing and protein analysis. Typically, a protein mixture is digested with a specific protease such as trypsin and then analyzed by mass spectrometry. Hunt et al. first described the basic process for peptide sequencing using a triple quadrupole instrument [10]. Peptide ions are separated first in the mass analyzer and then passed into a gas-phase collision cell, where they are activated through low-energy gas-phase collisions and fragmented. The fragment ions are then transferred to and separated in a second mass analyzer and passed on to the detector. Peptides readily fragment along their peptide backbones and the sequence of the peptides can be determined by interpreting the resulting fragmentation pattern [11,12]. Other types of mass spectrometer that are also useful for peptide sequencing include ion traps and quadrupole time-of-flight (TOF) instruments. A drawback to peptide sequencing using tandem mass spectrometry is that interpretation of the data can be time intensive. Ionization techniques such as electrospray ionization (ESI) provide a better and more robust integration of tandem mass spectrometry methods with liquid chromatography, improving the sensitivity of analysis [13,14]. Tandem mass spectrometry coupled to on-line separation devices such as high-pressure liquid chromatography (HPLC) and capillary electrophoresis (CE) has made the procedure sensitive enough to analyze peptides and proteins at low enough levels to address many interesting biological research questions. In order to obtain maximum sensitivity, efforts have focused on coupling nanoscale liquid chromatography, at submicroliter flow rates, to the highly sensitive microscale ESI source. Currently, detection limits of a few femtomoles of peptide material loaded on the liquid-chromatographic columns make this technique compatible with silver-stained, fluorescently-labeled, or faintly stained Coomassie gel bands and thus capable of detecting very low-abundance proteins and peptides. Peptide mixtures can be directly loaded on a reverse-phase liquid-chromatography column, separated on the basis of their hydrophobicity, and eluted directly in the tandem mass spectrometer. Using data-dependent acquisition (the ability of an instrument program to select specific peptides for collision-induced dissociation), tandem mass spectra can be collected with a higher efficiency than manually controlled conditions. In this manner, many more mass spectra are acquired, thereby requiring automated tools for data interpretation.

Tandem mass-spectrometry data is used to search protein and/or nucleotide databases to identify the amino acid sequence that best matches the spectrum and, from the peptide sequence, identify the protein. Information contained in each tandem mass spectrum is highly specific and unique and enables confident identification of individual proteins even from complex protein mixtures. Eng et al. used tandem mass spectrometry in conjunction with a database search algorithm, SEQUEST, to identify proteins present in mixtures [15]. This new strategy, labeled ''shotgun proteomics,'' allows routine identification of proteins from complex mixtures. McCormack et al. further validated the use of this protein identification strategy for analysis of protein mixtures obtained from different biological experiments [16]. This approach has also been successfully used to identify proteins enriched from subcellular compartments [17] and has been extended to the analyses of protein complexes [18–23], whole cell lystaes [24], and tissues [25]. Further extension of this strategy also enabled the identification and characterization of posttranslational modifications from a variety of protein sources [25]. These strategies have proved to be highly sensitive and allow for rapid and comprehensive analysis of proteins from complex mixtures.

## II. TOOLS FOR PROTEOMICS

The tools used for the analysis of a proteome still lag behind the tools used for DNA or RNA analysis. Microarray technology has been successfully used to identify and quantitate DNA and RNA molecules. However, due to the heterogeneous and complicated nature of the proteins, it is inherently more difficult to perform similar types of experiment for proteome analysis. Analytical methods used for proteome analysis should be high throughput, highly sensitive, able to resolve proteins mixtures/complexes into their individual components, capable of discriminating differentially modified proteins, and able to quantitatively display and analyze all the proteins present within a sample. Currently, proteomic research is performed in three steps: (1) separation of the protein mixtures, (2) characterization of the separated proteins/polypeptides using mass spectrometry (MS), and (3) database searching (Fig. 1).

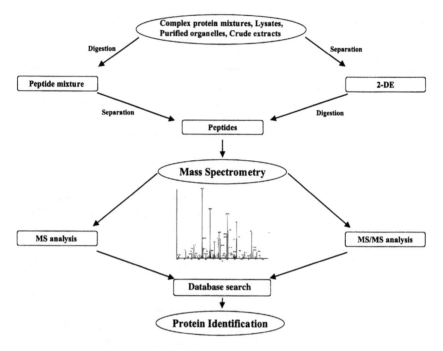

**Figure 1**   General flow scheme for proteomic analysis.


## A.   Two-Dimensional Electrophoresis

A common technique for global proteome analysis is two-dimensional polyacryl-amide gel electrophoresis (2-DE) in which the proteins are separated in one dimension by the isoelectric point (p$I$) and in the second dimension by the molecular mass. Over the years, improvements, including the introduction of immobilized pH gradients, specialized pH gradients, and use of fluorescent dyes, have greatly increased the reproducibility, resolution, and sensitivity of 2-DE [26–28]. As a result, this separation technique can resolve more than 10,000 protein spots on a single gel [29,30]. Even though a single 2-DE gel can resolve thousands of protein isoforms, each individual spot from the gel must be individually extracted, digested, and then analyzed—a time-intensive process.

Methods that have been used for identification of individual proteins from gels include immunoblotting, N-terminal sequencing [31,32], and internal peptide sequencing [33,34]. Currently, the most widely used method for identifying proteins resolved by 2-DE involves the excision of spots from gels followed by ''in-gel'' digestion of the proteins with a protease and elution of the peptides from the gel [34–36]. The eluted peptides are then analyzed by mass spectrometry (MS) or tandem mass spectrometry (MS/MS); the mass spectral data derived from the peptides are then correlated with the information contained in the databases. Over the years, the 2-DE technique has been improved both in terms of speed and automation. The 2-DE technique has also been coupled with software programs to facilitate identification and quantitation [37].

The primary application of 2-DE is protein expression profiling, in which the protein expression of any two samples can be compared (e.g., normal versus diseased state); the intensity of the spots provides quantitative information about expression. The effect on the state of a cellular system in response to a particular treatment can be directly visualized by monitoring differences in the protein spots. Despite improvements in automation, 2-DE is still labor intensive and time-consuming; each spot to be identified from the gel must be processed and analyzed individually. In addition, 2-DE is limited in several ways. Large and hydrophobic proteins (including integral membrane proteins) are difficult to get into the gel and proteins with extreme acidity (p$I$ <3) or basicity (p$I$ >11) are not well resolved [27]. Another limitation of 2-DE is its insufficient dynamic range (the range in relative abundance of proteins in total protein preparations). Several studies have shown that 2-DE analysis of whole-cell lysates results in identification of only the most abundant proteins. Even though Futcher et al. were able to resolve 1400 spots from a *Saccharomyces cerevisiae* lysate, only 148 proteins could be identified [38]. Perrot et al. identified approximately 279 proteins from a whole-cell lysate of *S. cerevisiae* [39], but, to date, no low-abundance proteins have been detected by 2-DE [6,40]. Thus, the limitations of 2-DE have resulted in the development of other analytical approaches for proteomics.

## B. Sample Fractionation and Enrichment

A single genome can give rise to an enormous number of proteomes, and the dynamic range of abundance of proteins in biological samples can be as high as $10^6$–$10^9$ [41]. Several approaches can be used to reduce the complexity of the biological sample and also enable the analysis of low-abundance proteins. These methods include subcellular fractionation and affinity protein enrichment. In addition to these, several electrophoretic/chromatographic methods can be used for protein prefractionation. The method of choice should be simple and involve minimal sample loss.

Subcellular fractionation can be used to reduce the complexity and diversity of a proteome, as only a subset of proteins from a particular cellular organelle can be selected for analysis. A common strategy for subcellular fractionation involves the disruption of cellular organization and fractionation of the resulting homogenate to separate different cellular organelles. The fractionation is generally achieved by "classical" density gradient separations in which differential density gradients are used to separate and isolate specific organelles, including nuclei [42,43], mitochondria [44,45], lysosomes [46,47], endosomes [48,49], and plasma membrane [50,51]. Analysis of the proteins from these organelles can be used to identify and characterize the biological association of the proteins and their function. Organelle purification has been proposed as a way to address some of the limitations of proteomic analysis [41,52,53].

The problem with dynamic range can also be circumvented by the use of interactive proteomics (baits). It has become increasingly evident that many proteins form multiprotein complexes to carry out their biochemical and enzymatic processes. The number of the proteins involved in such a multiprotein complex can range from two to several hundred. Identifying these protein–protein interactions can help distinguish the proteins involved in particular biological processes or pathways and thereby give additional insight into gene function and, ultimately, how cells work as a system. There are several affinity-based approaches used by biologists to purify multiprotein complexes or enrich certain components in crude protein mixtures (Fig. 2).

A common strategy for the identification of interacting proteins is affinity purification of the complex from a cell extract with an antibody to one or more of its components (immunoafinity purification). The success of this approach depends on the availability of a suitable antibody that binds specifically to the protein of interest. A related method is to engineer a DNA encoding for a particular protein, to express and incorporate an affinity tag, against which antibodies or other affinity reagents are readily available. Thus, protein can be expressed in the cells, and the complex affinity purified. A modification to this approach is the use of two tags [54], which decreases nonspecific binding and improves the specific recovery of the protein complex. An alternate approach is the use of a "bait" protein, in which the protein of interest is immobilized on a solid support. There are several ways to link the target protein to a solid support. Recombinant proteins containing affinity sequences can be generated which can be bound to immobilized resins. Cell extracts or other protein mixtures are then run over the immobilized protein, resulting in proteins with an affinity for the immobilized protein retained on the beads.

## C. Mass Spectrometry

The most significant breakthrough in the field of proteomics has been the use of mass spectrometry for protein identification and characterization. A mass spectrometer consists of three basic elements: (1) an ionization source, (2) one or more mass analyzers, and (3) a detector. The instruments are generally named after the type of the ionization source and the mass analyzers found in the instrument, and they are used to measure the mass to charge ($m/z$) ratio of the analytes. In principle, proteins can be identified by their molecular mass, but they are

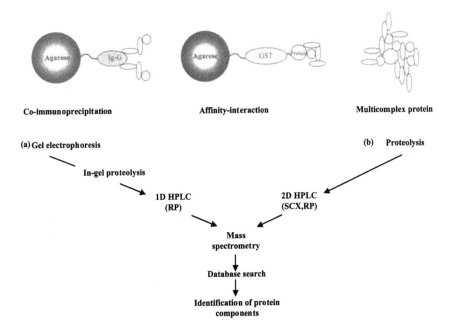

**Figure 2** Three biochemical methods to determine protein–protein interactions. The first is coimmunoprecipitation, in which an antibody is used to precipitate a specific protein along with its protein partners. The second method is protein–affinity interaction chromatography that uses a bound protein as "bait" to bind interacting proteins. The last method is purification of an intact protein complex. (a) Proteins to be analyzed are fractionated by SDS-PAGE, excised, and digested with protease. Resulting peptides are fractionated by liquid chromatography and analyzed by ESI-MS or spotted onto grids for analysis by MALDI/TOF. (b) Proteins to be analyzed by MudPIT are subjected to proteolysis directly and the resulting peptides are subjected to two-dimensional liquid chromatography and analyzed by ESI-MS/MS. Proteins are identified by searching of protein databases.

usually identified by the molecular mass of their digested peptides or through further fragmentation of their peptides through tandem mass spectrometry. There are several advantages to protein identification by analysis of peptides rather than proteins: The sensitivity is better; mass measurements are accurate; and the data obtained can be taken directly for comparison to protein sequences derived from protein- and nucleotide-sequence databases. There are two main approaches for analysis of peptides using mass spectrometry: (1) "peptide-mass mapping" and (2) tandem mass-spectrometry peptide sequencing.

The "peptide-mass mapping" concept was introduced by Henzel [55] and measures the mass of individual peptides in a mixture to create a mass spectrum.

Matrix-assisted laser-desorption ionization time-of-flight (MALDI-TOF) is commonly used for peptide-mass mapping and is frequently the first step in quick and automated protein identification. MALDI is a soft-ionization technique in which the analyte is incorporated into a chemical matrix that contains small ultraviolet (UV)-absorbing organic molecules; a laser pulse is used to irradiate the matrix in vacuum, and gas-phase ions are generated, which travel down a flight tube. A TOF instrument is one of the simplest mass analyzers in which the $m/z$ ratio of the ion is determined by the time it takes the ion to traverse the length of a flight tube. Because of its high speed, MALDI-TOF is a method of choice for rapid identification of protein spots from gels. However, as the complexity of the sample increases, the ambiguities in protein identification using MALDI-TOF also increase.

Electrospray ionization tandem mass spectrometry (ESI-MS/MS) is currently the method of choice for amino acid sequencing of peptides. In ESI, the peptides are ionized directly from the liquid phase by applying a potential difference between a capillary and the inlet of the mass spectrometer. As the flow stream exits the capillary, it sprays a fine mist of droplets. The droplets containing the peptides are desolvated; the gas-phase charged ions are desorbed from the droplets and pass from the source into the mass analyzer, where they are separated according to their $m/z$ ratios. Three types of tandem mass analyzers are commonly coupled with the ESI sources for proteomic studies and include triple quadrupole, ion trap, and quadrupole time of flight. Although these three analyzers differ in way they work, they all perform similar analyses.

## D.  Databases and Algorithms for Protein Identification

Databases allow mass spectrometry to be used for protein identification. The genome of the yeast *S. cerevisiae* [56], as well as dozens of microbes [57], has been entirely sequenced. The genome of *S. cerevisiae* was the first eukaryotic genome to be sequenced and contains approximately 6200 genes; the proteomic community has had tremendous benefit since it became available. Along with microbial genomes, numerous higher eukaryotes including *Caenorhabditis elegans*, *Drosphilia melanogaster*, mouse, human, and pufferfish among others are either completed or approaching completion [58–62]. Three types of databases, namely protein, expressed sequence tag, and genome databases, can be searched against mass spectrometric data. Protein sequence databases can be searched both using mass fingerprint and tandem mass-spectrometric data. The success of database searching depends on (1) the quality of the data obtained from the mass-spectrometric analyses, (2) the quality of the database searched, and (3) the software used to search the databases.

An important development in biological mass spectrometry has been the development of algorithms for analysis of tandem mass-spectral data. The first algorithm defined to identify proteins by matching tandem mass spectrometry data to database sequences is SEQUEST, which was introduced from this laboratory by Yates and Eng in 1995 [15,63,64]. Numerous algorithms have been developed to search MS and MS/MS data (reviewed in Ref. 65), but due to the space limitation, we will not discuss them here. The SEQUEST algorithm uses both the molecular mass of the peptide and its subsequent fragmentation pattern. For those peptides within the database whose mass is consistent with the mass of the experimental peptide, two scoring criteria are used in the sequence analysis. The preliminary score for each amino acid sequence is calculated by matching the predicted ions, the continuity of the fragment ions in the sequence, and the length of the amino acid sequence; cross-correlation scores are generated by comparing theoretical mass spectra for the top 500 scoring amino acid sequences in the preliminary score to the observed tandem mass spectrum. A typical LC/MS/MS analysis using data-dependent acquisition yields thousands of tandem mass spectra, so data analysis can be time-consuming. To overcome this problem, SEQUEST searches are run on multiprocess computers connected in a cluster.

When using tandem MS with nominal mass resolution, it is difficult for the software to uniquely determine the charge states of multiply-charged peptides, thereby increasing the weight of the tandem mass spectra to be searched. This occurs because a molecule is calculated for both $+2$ and $+3$ charge states for each tandem MS. Our new code for charge state determination, 2 to 3, determines the charge states of the multiply charged parent peptides and eliminates spectra of low quality in a way that does not affect protein identification results [66]. The program 2 to 3 also identifies and filters out spectra containing a prominent 98-Da neutral loss specific for phosphorylated peptides. A modified version of the SEQUEST (SEQUEST-PHOS) that considers the unique MS/MS fragmentation patterns of phosphopeptides is used to identify and characterize the phosphorylation sites on phosphopeptides.

SEQUEST will return top-matching peptides for each observed tandem spectrum, but it requires additional software to assemble the information at protein level. Recently, a software program called DTASelect was introduced from our lab to assemble and evaluate shotgun proteomics data [67]. The program applies multiple layers of filtering to SEQUEST search results and produces several analytical reports in a variety of formats. ''Contrast'' is an accompanying tool of DTASelect and can be used for differential analysis of different samples or criteria sets.

## E.  Multidimensional Protein Identification Technology

Promising new developments that bypass the gel-based 2-DE approach for proteome analyses have emerged. Using the ''shotgun approach,'' identification of proteins from complex protein mixtures is made possible by combining on-line high-resolution, high-pressure microcapillary liquid chromatography with tandem mass spectrometry to characterize each component in a mixture of many. Automated data acquisition adds to the "walk-away" ability of this technique. There are several advantages of this shotgun approach over the 2-DE approach. In the shotgun approach, peptides are produced before chromatography and are independent of the molecular mass (Mr), isoelectric point (pI), and size and hydrophobicity of the proteins. To perform protein mixture analysis, proteolytic digestion is performed in solution; protein mixtures are denatured in urea, reduced and alkylated, and then digested with a protease, such as trypsin, to produce peptide fragments that are suitable for MS analysis. The resulting peptide mixture is then separated using HPLC. Gatlin et al. developed a microcolumn microelectrospray interface that eliminates the use of frits and junctions and enables the achievement of low flow rates at the tip of the HPLC column [68]. The peptide mixture is loaded on to a microcapillary column packed with C18 reverse phase resin and the peptides resolved by applying a reverse-phase gradient. Moderately complex protein mixtures have been successfully analyzed using this strategy [16–20].

Identification of proteins from more complex protein mixtures presents a particular challenge for separations. These mixtures produce vastly more complicated peptide mixtures, which require better separation techniques to allow acquisition of mass-spectral data on much greater number of peptides. In 1999, Link et al. developed an approach, called MudPIT (Multidimensional Protein Identification Technology), which allowed for the comprehensive component identification in complex mixtures [18] (Fig. 3). MudPIT uses a biphasic microcapillary column with strong cation-exchange (SCX) resin upstream of RP resin. Peptide mixtures at low pH are loaded onto the SCX phase and discrete fractions of the peptides are eluted off the SCX phase onto the RP using salt pulses. Peptides from the RP are then resolved using an organic gradient and analyzed by MS/MS. This approach uses the independent physical properties of charge and hydrophobicity of peptides to resolve complex protein mixtures before mass spectrometry [18]. Multiple salt pulses with increased salt concentration are applied to displace additional peptide fractions from the SCX onto the RP [69,70]. Depending on the complexity of the sample, the salt pulses are introduced in an iterative manner, typically involving 3–18 steps. This on-line fractionation of the complex protein mixtures using two-dimensional fractionation helps to increase the dynamic range of the whole analysis and enables one to collect MS/MS data on an increased number of peptides.

## III. MUDPIT AND ITS APPLICATIONS TO YEAST PROTEOMICS

### A. Proteome Analysis

The classical proteomic approach for separating whole-cell lysates is based on 2-DE and has a limited application due to the drawbacks discussed earlier. *S.*

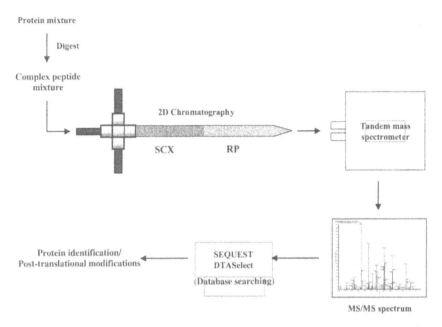

Protein mixture

Complex peptide mixture

2D Chromatography

SCX          RP

Tandem mass spectrometer

Protein identification/
Post-translational modifications

SEQUEST
DTASelect
(Database searching)

MS/MS spectrum

**Figure 3** Proteomic analysis by multidimensional liquid chromatography. A complex mixture of proteins is digested and the resulting peptides are loaded directly onto a capillary column packed with strong cation-exchange resin (SCX) and a reverse-phase resin (RP). Peptides are sequentially displaced from the SCX phase onto the RP, where they are resolved and eluted directly into the mass spectrometer. The resulting tandem mass spectra are searched through a protein database using SEQUEST. Software is used to assign peptides to their respective proteins and thus identify the proteins present.

*cerevisiae* proteins have generally been analyzed by 2-DE [38,39,55,71,72] and the greatest number of proteins identified previously in a single study was 279 [39]. Previous studies that incorporated non-gel-based multidimensional separations with tandem mass spectrometry were not able to identify more than 200 proteins from a single sample [18]. An impressive application of the MudPIT was the direct analysis of whole-cell lysate of *S. cerevisiae* [24]. This lysate was fractionated into three fractions: soluble fraction, lightly washed insoluble fraction, and heavily washed insoluble fraction to reduce the complexity of the sample. All three fractions were separately analyzed using MudPIT. A total of 1484 proteins were identified; to date, the highest number of proteins was identified in *S. cerevisiae*. Since the number of the genes in *S. cerevisiae* genome is approximately 6000, the number of the proteins identified using the MudPIT approach does not likely represent the whole proteome of *S. cerevisiae*, but it still represents a large-scale and global view of the *S. cerevisiae* proteome. The identified set of proteins included integral membrane proteins, low-abundance proteins, transcription factors, proteins with p$I$ >10 and proteins with higher molecular weight. As mentioned previously, these proteins are rarely seen using 2-D-E-based techniques. Although low-molecular-weight proteins (<10 kDa) and proteins with low p$I$'s (<4.3) were also identified, this methodology seemed to be slightly biased against this class of proteins, perhaps because acidic proteins produce a lower number of tryptic peptides. Interpretation of the data using the MIPS catalogs showed a good representation of every major functional category and protein class. Over 80% of the proteins in the yeast genome have Codon Adaptation Index (CAI) [73] values between 0 and 0.2, implying that 80% of the proteins in yeast are low-abundance proteins. Previous studies in yeast have identified few proteins with CAIs < 0.2 [6,38,39]; this study identified 791 proteins that had a CAI < 0.2, indicating that MudPIT is sensitive enough to identify low-abundance proteins.

## B.  Protein Complexes

Identification of the subcellular localization of a protein and of other proteins with which a particular protein interacts will help in understanding the function of that protein within a cell [74]. Several biochemical strategies have been adapted for use with a MudPIT approach to determine protein–protein interactions in cellular systems [22,23,75,76]. In order to study protein complexes and their function, the proteins from a particular cell organelle are enriched, as mentioned previously, and analyzed by one- or two-dimensional LC coupled to MS/MS.

Link et al. used this approach to identify approximately 80 proteins from periplasmic space of *E. coli* at a time when only 80–90% of the *E. coli* genome was sequenced [17]. The technology was further improved by the introduction of the biphasic LC column that enabled the identification of a total of 75 out of the total 78 proteins present in the ribosomal complex of *S. cerevisiae* [18]. Comparative analysis of the same fraction using 2-DE coupled to mass spectrometry enabled the identification of only 63 proteins from the ribosomal complex. MudPIT analysis of the affinity-purified 26S proteosome enabled the identification of every subunit of this large multiprotein complex; furthermore, it enabled the identification of an additional set of proteins that potentially interact with the proteosome [19,20]. The first comprehensive analysis of the spliceosome complex from *S. pombe* and *S. cerevisiae* using MudPIT identified at least 26 proteins, including some novel proteins that were not been previously identified and also showed that the two yeast complexes were nearly identical in composition [21]. MudPIT analysis of the affinity-purified Dam1p protein complex from budding yeast, which is required for both spindle integrity and kinetochore function, identified a total of seven proteins, four of which have not been previously reported [22].

## C.  Protein Modification

Posttranslational modification of proteins play a key role in protein regulatory mechanisms, and characterization of protein modifications is important for understanding protein function. Although genomic technologies provide significant information about gene structure and protein sequences, they provide no means to determine the posttranslational modifications on a protein. At least 200 different posttranslational modifications have been described [77]. Most of these modifications are believed to play an important role in cellular processes, but due to the time- and labor-intensive nature of the methods to characterize such modifications, only a few have been extensively studied. One well-studied modification is phosphorylation, which has been shown to play key roles in signal transduction and the regulation of cellular processes [78–80]. Many of the traditional techniques used to map protein modifications over the past 50 years could only determine whether a protein is modified, but could not easily map the exact site(s) where the protein was modified. Mass spectrometry is emerging as one of the most powerful tools for the analysis of protein modifications because virtually any type of protein modification can be identified. Several different strategies have been used to study protein modifications. Again, most of these are capable of targeting only a particular modification. For example, phosphopeptides can be enriched by immobilized metal-affinity chromatography (IMAC) [81,82] prior to analysis by MS. The disadvantage is that iron or gallium used in the IMAC column may bind nonphosphorylated residues such as glutamic acid or aspartic acid, as they also carry a negative charge. This particular problem was recently circumvented to some extent by preparing methyl esters of the peptides before passing them through the IMAC column [83].

A key to characterize posttranslational modification is to get high sequence coverage (peptides covering the entire length of the protein) for the proteins of interest. We achieved this by combining high-resolution MudPIT with proteolytic digestion of proteome with enzymes of different selectivities. This approach not only measures phosphopeptides but also peptides containing other modifications that produce a mass shift in the peptide mass that can be detected by MS/MS analysis. Overlapping peptides are generated which increase the sequence coverage for the proteins, reduce the ambiguity in mapping modifications, and increase the chance of obtaining a peptide with a good quality MS/MS spectrum. One application involved the characterization of the posttranslational modifications associated with the cell-cycle-regulating cyclin-dependent kinase in *S. pombe*–Cdc2p. It exists as a multiprotein complex and the phosphorylation sites in Cdc2p are well characterized [84]. Cdc2p was affinity purified using a TAP tag [85], subjected to multiproteolytic digestion using trypsin, subtilisin, and elastase, and analyzed by MudPIT. Two previously reported phosphorylation sites, Y15 and T167 in Cdc2p, were detected with multiple overlapping peptides.

Interestingly, we were able to identify novel phosphorylation sites in two of the cyclin partners of Cdc2p, Cdc13p, and Cig1p (Table 1). In addition to phosphorylation, we were able to find multiple methylation and oxidation sites within Cdc2p (Table 1). An important aspect of this study was that modifications were detected in TAP-purified protein complexes making it possible to combine the identification of the proteins in the complex with the analysis of their posttranslational modifications in a single experiment. MudPIT analysis of kinetochore complex in budding yeast identified 18 phosphorylation sites [23].

## IV. QUANTITATION

A variety of approaches are underway to develop methods for quantitative proteomics [40,86–92]. Quantitation involves the measurement of the relative abundance of the proteins in a system in two or more different states. Stable isotope methods have been recently introduced to measure the relative changes in the two states of the proteome. The method involves the incorporation of a stable isotope in one of the two states to be compared that shifts the mass of the peptides by a significant amount. The proteins from each state are labeled as "heavy" or "light." Different approaches to incorporate this mass difference into proteins/peptides have been established. This can be incorporated at the protein synthesis stage by growing micro-organisms in one state in normal medium and in another state in medium containing $^{15}N$ instead of $^{14}N$ (Fig. 4). This early labeling is possible when one can control the growth of an organism and this method has been used for several microbes, including S. cerevisiae [86]. As an alternative, Gygi et al. introduced an isotopic tag on cysteines residues after cell lysis to achieve quantitation and used a non-gel-based approach for identifying the peptides [40] (Fig. 4). However, this method is limited to proteins containing cysteine residues. Washburn et al. used the MudPIT approach coupled with $^{15}N$ labeling

**Table 1** Protein Modifications of S. *pombe* Cdc2–TAP Purified Complex Identified Using Multiproteolytic Digestion in Conjunction with Analysis by MudPIT

| Protein | % Coverage[a] | Phosphorylation | Oxidation | Methylation |
|---------|---------------|-----------------|-----------|-------------|
| Cdc2 | 88.2 | Y15, T167 | M89, M253, M273 | R249. K251, R252, K257 |
| Cdc13 | 91.7 | S176, S180, or S183 | M129, M225, M447 | |
| Cig1 | 57.8 | S108 | M159 | |

[a]Percentage of amino acid sequence coverage obtained for proteins using DTASelect filter cutoff values chosen for this study [25].

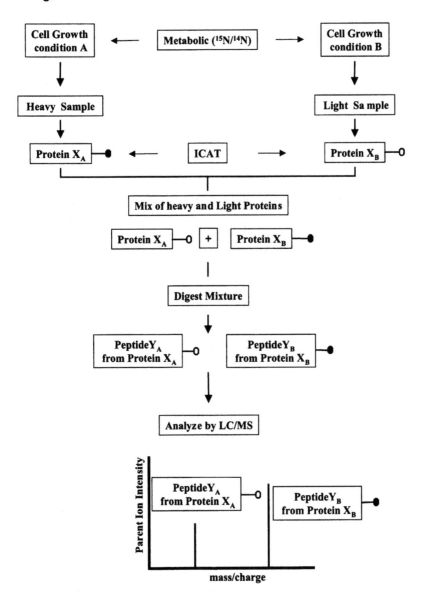

**Figure 4**   Strategies for quantifying proteomic expression. Cells are grown under two different conditions to alter protein expression. Metabolic labeling introduces a label during the early growth of an organism and is the earliest point of introduction of heavy and light labels. The proteins from the cell state A and cell state B are harvested, denatured, reduced, and can be labeled with light or heavy ICAT reagents. The proteins from cell state A and B are mixed, digested, and analyzed by MS. The peak ratios of identical peptides from each growth condition yield relative quantitative information.

to quantitate and validate the dynamic range of *S. cerevisiae* proteome [69]. The dynamic range of the system between the two labeled states was at least 10:1 and more than 800 unique proteins for each state were identified, demonstrating the potential of metabolic labeling coupled to MudPIT as a tool for quantitative proteomic analysis.

## V. CONCLUSIONS

The rapidly progressing field of proteomics is playing a key role in unraveling gene function directly at protein level and furthering our understanding of the mechanisms of biological processes. High-throughput and sensitive proteomics technologies such as MudPIT are proving useful for the characterization of proteins from whole-cell lysates and multiprotein complexes. Future technological improvements may allow for the characterization of the entire proteome. Strides are being made in the analysis of posttranslational modifications, which can now be studied in a high-throughput manner. Quantitation of large number of proteins in a short span of time is still a challenge to the field of proteomics, but with the improvements in technologies, quantitation of proteins in total cell lysates will be possible. Finally, overall improvements in technologies and methodologies will allow proteomics to contribute to biological research as much as gene expression studies.

## ACKNOWLEDGMENTS

The authors thank Claire Delahaunty and Hayes McDonald for critical reading and comments of this chapter. A.S. is supported by NIH RO1 EY13288-03. J.R.Y. is supported by Office of Naval Research N00014-01-2-003, Office of Naval Research N00014-00-1-0421, NIH RR11823, and NIH RO1 EY13288.

## REFERENCES

1. Velculescu, V.E.; Zhang, L.; Vogelstein, B.; Kinzler, K.W. Serial analysis of gene expression. Science. 1995, 270, 484–487.
2. Schena, M.; Shalon, D.; Davis, R.W.; Brown, P.O. Quantitative monitoring of gene expression patterns with a complementary DNA microarray. Science. 1995, 270, 467–470.
3. Shalon, D.; Smith, S.J.; Brown, P.O. A DNA microarray system for analyzing complex DNA samples using two-color fluorescent probe hybridization. Genome Res. 1996, 6, 639–645.
4. Abbott, A. A post-genomic challenge: learning to read patterns of protein synthesis. Nature. 1999, 402, 715–720.

5. Anderson, L.; Seilhamer, J. A comparison of selected mRNA and protein abundances in human liver. Electrophoresis. 1997, 18, 533–537.

6. Gygi, S.P.; Rochon, Y.; Franza, B.R.; Aebersold, R. Correlation between protein and mRNA abundance in yeast. Mol. Cell. Biol. 1999, 19, 1720–1730.

7. Ideker, T., et al. Integrated genomic and proteomic analyses of a systematically perturbed metabolic network. Science. 2001, 292, 929–934.

8. Wasinger, V.C., et al. Progress with gene-product mapping of the Mollicutes: *Myco-plasma genitalium*. Electrophoresis. 1995, 16, 1090–1094.

9. Blackstock, W.P.; Weir, M.P. Proteomics: quantitative and physical mapping of cellular proteins. Trends Biotechnol. 1999, 17, 121–127.

10. Hunt, D.F.; Yates, J.R.; Shabanowitz, J.; Winston, S.; Hauer, C.R. Protein sequencing by tandem mass spectrometry. Proc. Natl. Acad. Sci. USA. 1986, 83, 6233–6237.

11. Biemann, K.; Scoble, H.A. Characterization by tandem mass spectrometry of structural modifications in proteins. Science. 1987, 237, 992–998.

12. Hunt, D.F., et al. Tandem quadrupole Fourier-transform mass spectrometry of oligopeptides and small proteins. Proc. Natl. Acad. Sci. USA. 1987, 84, 620–623.

13. Covey, T.R.; Huang, E.C.; Henion, J.D. Structural characterization of protein tryptic peptides via liquid chromatography/mass spectrometry and collision-induced dissociation of their doubly charged molecular ions. Anal. Chem. 1991, 63, 1193–1200.

14. Griffin, P.R.; Coffman, J.A.; Yates, J.R. Structural studies of proteins by capillary HPLC electrospray tandem mass spectrometry. Int. J. Mass Spectrom. Ion Proc. 1991, 111, 131–149.

15. Eng, J.K.; McCormack, A.L.; Yates, J.R. An approach to correlate tandem mass spectral data of peptides with amino acid sequences in a protein database. J. Am. Soc. Mass Spectrom. 1994, 5, 976–989.

16. McCormack, A.L., et al. Direct analysis and identification of proteins in mixtures by LC/MS/MS and database searching at the low-femtomole level. Anal. Chem. 1997, 69, 767–776.

17. Link, A.J.; Hays, L.G.; Carmack, E.B.; Yates, J.R. Identifying the major proteome components of Haemophilus influenzae type-strain NCTC 8143. Electrophoresis. 1997, 18, 1314–1334.

18. Link, A.J., et al. Direct analysis of protein complexes using mass spectrometry. Nature Biotechnol. 1999, 17, 676–682.

19. Verma, R., et al. Proteasomal proteomics: identification of nucleotide-sensitive proteasome-interacting proteins by mass spectrometric analysis of affinity-purified proteasomes. Mol. Biol. Cell. 2000, 11, 3425–3439.

20. Verma, R.; McDonald, H.; Yates, J.R.; Deshaies, R.J. Selective degradation of ubiquitinated Sic1 by purified 26S proteasome yields active S phase cyclin-Cdk. Mol. Cell. 2001, 8, 439–448.

21. Ohi, M.D., et al. Proteomics analysis reveals stable multiprotein complexes in both fission and budding yeasts containing Myb-related Cdc5p/Cef1p, novel pre-mRNA splicing factors, and snRNAs. Mol. Cell. Biol. 2002, 22, 2011–2024.

22. Cheeseman, I.M., et al. Implication of a novel multiprotein Dam1p complex in outer kinetochore function. J. Cell Biol. 2001, 155, 1137–1145.

23. Cheeseman, I.M.; Anderson, S.; Jwa, M.; Green, E.; Kang, J-S.; Yates, J.R.; Chan, C.S.; Drubin, D.G.; Barnes, G. Phospho-regulation of kinetochore–microtubule attachments by the aurora kinase Ipl1p. Cell. 2002, 111, 1–20.

24. Washburn, M.P.; Wolters, D.; Yates, J.R. Large-scale analysis of the yeast proteome by multidimensional protein identification technology. Nature Biotechnol. 2001, 19, 242–247.

25. MacCoss, M.J., et al. Shotgun identification of protein modifications from protein complexes and lens tissue. Proc. Natl. Acad. Sci. USA. 2002, 99, 7900–7905.

26. Bjellqvist, B.; Pasquali, C.; Ravier, F.; Sanchez, J.C.; Hochstrasser, D. A. nonlinear wide-range immobilized pH gradient for two-dimensional electrophoresis and its definition in a relevant pH scale. Electrophoresis. 1993, 14, 1357–1365.

27. Gorg, A., et al. The current state of two-dimensional electrophoresis with immobilized pH gradients. Electrophoresis. 2000, 21, 1037–1053.

28. Patton, W.F. A thousand points of light: the application of fluorescence detection technologies to two-dimensional gel electrophoresis and proteomics. Electrophoresis. 2000, 21, 1123–1144.

29. Klose, J. Genotypes and phenotypes. Electrophoresis. 1999, 20, 643–652.

30. Klose, J., et al. Genetic analysis of the mouse brain proteome. Nature Genet. 2002, 30, 385–393.

31. Aebersold, R.H.; Teplow, D.B.; Hood, L.E.; Kent, S.B. Electroblotting onto activated glass. High efficiency preparation of proteins from analytical sodium dodecyl sulfate–polyacrylamide gels for direct sequence analysis. J. Biol. Chem. 1986, 261, 4229–4238.

32. Matsudaira, P. Sequence from picomole quantities of proteins electroblotted onto polyvinylidene difluoride membranes. J. Biol. Chem. 1987, 262, 10,035–10,038.

33. Aebersold, R.H.; Leavitt, J.; Saavedra, R.A.; Hood, L.E.; Kent, S.B. Internal amino acid sequence analysis of proteins separated by one- or two-dimensional gel electrophoresis after in situ protease digestion on nitrocellulose. Proc. Natl. Acad. Sci. USA. 1987, 84, 6970–6974.

34. Rosenfeld, J.; Capdevielle, J.; Guillemot, J.C.; Ferrara, P. In-gel digestion of proteins for internal sequence analysis after one- or two-dimensional gel electrophoresis. Anal. Biochem. 1992, 203, 173–179.

35. Shevchenko, A.; Wilm, M.; Vorm, O.; Mann, M. Mass spectrometric sequencing of proteins silver-stained polyacrylamide gels. Anal. Chem. 1996, 68, 850–858.

36. Wilm, M.; Mann, M. Analytical properties of the nanoelectrospray ion source. Anal. Chem. 1996, 68, 1–8.

37. Yan, J.X.; Sanchez, J.C.; Tonella, L.; Williams, K.L.; Hochstrasser, D.F. Studies of quantitative analysis of protein expression in Saccharomyces cerevisiae. Electrophoresis. 1999, 20, 738–742.

38. Futcher, B.; Latter, G.I.; Monardo, P.; McLaughlin, C.S.; Garrels, J.I. A sampling of the yeast proteome. Mol. Cell. Biol. 1999, 19, 7357–7368.

39. Perrot, M., et al. Two-dimensional gel protein database of Saccharomyces cerevisiae (update 1999). Electrophoresis. 1999, 20, 2280–2298.

40. Gygi, S.P., et al. Quantitative analysis of complex protein mixtures using isotope-coded affinity tags. Nature Biotechnol. 1999, 17, 994–999.

41. Corthals, G.L.; Wasinger, V.C.; Hochstrasser, D.F.; Sanchez, J.C. The dynamic range of protein expression: a challenge for proteomic research. Electrophoresis. 2000, 21, 1104–1115.

42. Prasad, S.C.; Dritschilo, A. High-resolution two-dimensional electrophoresis of nuclear proteins: A comparison of HeLa nuclei prepared by three different methods. Anal. Biochem. 1992, 207, 121–128.
43. Baciu, P.C.; Durham, J.P. A procedure for the extraction and high resolution two-dimensional gel electrophoresis of total nuclear phosphoproteins from isotonically purified nuclei. Electrophoresis. 1990, 11, 162–166.
44. Pavlica, R.J.; Hesler, C.B.; Lipfert, L.; Hirshfield, I.N.; Haldar, D. Two-dimensional gel electrophoretic resolution of the polypeptides of rat liver mitochondria and the outer membrane. Biochim. Biophys. Acta. 1990, 1022, 115–125.
45. Rabilloud, T., et al. Two-dimensional electrophoresis of human placental mitochondria and protein identification by mass spectrometry: Toward a human mitochondrial proteome. Electrophoresis. 1998, 19, 1006–1014.
46. Temesvari, L.; Rodriguez-Paris, J.; Bush, J.; Steck, T.L.; Cardelli, J. Characterization of lysosomal membrane proteins of *Dictyostelium discoideum*. A complex population of acidic integral membrane glycoproteins, Rab GTP-binding proteins and vacuolar ATPase subunits. J. Biol. Chem. 1994, 269, 25,719–25,727.
47. Diettrich, O.; Mills, K.; Johnson, A.W.; Hasilik, A.; Winchester, B.G. Application of magnetic chromatography to the isolation of lysosomes from fibroblasts of patients with lysosomal storage disorders. FEBS Lett. 1998, 441, 369–372.
48. Huber, L.A., et al. Endosomal fractions from viral K-ras-transformed MDCK cells reveal transformation specific changes on two-dimensional gel maps. Electrophoresis. 1996, 17, 1734–1740.
49. Pol, A.; Enrich, C. Membrane transport in rat liver endocytic pathways: Preparation, biochemical properties and functional roles of hepatic endosomes. Electrophoresis. 1997, 18, 2548–2557.
50. Ishimura, R.; Noda, K.; Hattori, N.; Shiota, K.; Ogawa, T. Analysis of rat placental plasma membrane proteins by two-dimensional gel electrophoresis. Mol. Cell. Endocrinol. 1995, 115, 149–159.
51. Rouquie, D., et al. Construction of a directory of tobacco plasma membrane proteins by combined two-dimensional gel electrophoresis and protein sequencing. Electrophoresis. 1997, 18, 654–660.
52. Cordwell, S.J.; Nouwens, A.S.; Verrills, N.M.; Basseal, D.J.; Walsh, B.J. Subproteomics based upon protein cellular location and relative solubilities in conjunction with composite two-dimensional electrophoresis gels. Electrophoresis. 2000, 21, 1094–1103.
53. Lopez, M.F. Better approaches to finding the needle in a haystack: optimizing proteome analysis through automation. Electrophoresis. 2000, 21, 1082–1093.
54. Rigaut, G., et al. A generic protein purification method for protein complex characterization and proteome exploration. Nature Biotechnol. 1999, 17, 1030–1032.
55. Henzel, W.J., et al. Identifying proteins from two-dimensional gels by molecular mass searching of peptide fragments in protein sequence databases. Proc. Natl. Acad. Sci. USA. 1993, 90, 5011–5015.
56. Goffeau, A., et al. Life with 6000 genes. Science. 1996, 274, 546–567.
57. Humphery-Smith, I.; Cordwell, S.J.; Blackstock, W.P. Proteome research: Complementarity and limitations with respect to the RNA and DNA worlds. Electrophoresis. 1997, 18, 1217–1242.

58. Consortium, T.C.E.S. Genome Sequence of the nematode *C. elegans*: A platform for investigating biology. Science. 1998, 282, 2012–2018.

59. Adams, M.D., et al. The genome sequence of *Drosophila melanogaster*. Science. 2000, 287, 2185–2195.

60. Aparicio, S., et al. Whole-genome shotgun assembly and analysis of the genome of Fugu rubripes. Science. 2002, 297, 1301–1310.

61. Gregory, S.G., et al. A physical map of the mouse genome. Nature. 2002, 418, 743–750.

62. Venter, J.C., et al. The sequence of the human genome. Science. 2001, 291, 1304–1351.

63. Yates, J.R.; Eng, J.K.; McCormack, A.L.; Schieltz, D. Method to correlate tandem mass spectra of modified peptides to amino acid sequences in the protein database. Anal. Chem. 1995, 67, 1426–1436.

64. Yates, J.R.; McCormack, A.L.; Eng, J. Mining genomes with MS. Anal. Chem. 1996, 68, 534A–540A.

65. Fenyo, D.; Beavis, R.C. Informatics and data management in proteomics. Trends Biotechnol. 2002 Dec, 20(12 Suppl), S35–8.

66. Sadygov, R.G.; Eng, J.; Durr, E.; Saraf, A.; McDonald, W.H.; MacCoss, M.J.; Yates, J.R. Code Developments to improve the efficiency of automated MS/MS spectra interpretation. J. Proteome Res. 2002, 1, 211–215.

67. Tabb, D.L.; McDonald, W.H.; Yates, J.R. DTASelect and Contrast: tool for assembling and comparing protein identification from shotgun proteomics. J. Proteome Res. 2002, 1, 21–26.

68. Gatlin, C.L.; Kleemann, G.R.; Hays, L.G.; Link, A.J.; Yates, J.R. Protein identification at the low femtomole level from silver-stained gels using a new fritless electrospray interface for liquid chromatography–microspray and nanospray mass spectrometry. Anal. Biochem. 1998, 263, 93–101.

69. Washburn, M.P.; Ulaszek, R.; Deciu, C.; Schieltz, D.M.; Yates, J.R. Analysis of quantitative proteomic data generated via multidimensional protein identification technology. Anal. Chem. 2002, 74, 1650–1657.

70. Wolters, D.A.; Washburn, M.P.; Yates, J.R. An automated multidimensional protein identification technology for shotgun proteomics. Anal. Chem. 2001, 73, 5683–5690.

71. Shevchenko, A., et al. Linking genome and proteome by mass spectrometry: Large-scale identification of yeast proteins from two dimensional gels. Proc. Natl. Acad. Sci. USA. 1996, 93, 14,440–14,445.

72. Garrels, J.I., et al. Proteome studies of *Saccharomyces cerevisiae*: Identification and characterization of abundant proteins. Electrophoresis. 1997, 18, 1347–1360.

73. Sharp, P.M.; Li, W.H. The Codon Adaptation Index—A measure of directional synonymous codon usage bias, and its potential applications. Nucleic Acids Res. 1987, 15, 1281–1295.

74. Pawson, T.; Scott, J.D. Signaling through scaffold, anchoring, and adaptor proteins. Science. 1997, 278, 2075–2080.

75. Verma, R., et al. Molecular Biol. Cell. 2000, 11, 3425–3439.

76. Verma, R., et al. Molecular Cell. 2001, 8, 439–448.

77. Krishna, R.G.; Wold, F. Post-translational modification of proteins. Adv. Enzymol. Related Areas Mol. Biol. 1993, 67, 265–298.

78. Hunter, T. Protein kinases and phosphatases: The yin and yang of protein phosphory-lation and signaling. Cell. 1995, 80, 225–236.

79. Krebs, E.G. The growth of research on protein phosphorylation. Trends Biochem. Sci. 1994, 19, 439.

80. Sun, H.; Tonks, N.K. The coordinated action of protein tyrosine phosphatases and kinases in cell signaling. Trends Biochem. Sci. 1994, 19, 480–485.

81. Posewitz, M.C.; Tempst, P. Immobilized gallium(III) affinity chromatography of phosphopeptides. Anal. Chem. 1999, 71, 2883–2892.

82. Nuwaysir, L.M.; Stults, S.J. Electrospray ionization mass spectrometry of phospho-peptides isolated by on-line immobilized metal ion affinity chromatography. J. Am. Soc. Mass Spectrom. 1993, 4, 662–669.

83. Ficarro, S.B., et al. Phosphoproteome analysis by mass spectrometry and its applica-tion to *Saccharomyces cerevisiae*. Nature Biotechnol. 2002, 20, 301–305.

84. Berry, L.D.; Gould, K.L. Regulation of Cdc2 activity by phosphorylation at T14/Y15. Prog. Cell Cycle Res. 1996, 2, 99–105.

85. Tasto, J.J.; Carnahan, R.H.; McDonald, W.H.; Gould, K.L. Vectors and gene target-ing modules for tandem affinity purification in *Schizosaccharomyces pombe*. Yeast. 2001, 18, 657–662.

86. Oda, Y.; Huang, K.; Cross, F.R.; Cowburn, D.; Chait, B.T. Accurate quantitation of protein expression and site-specific phosphorylation. Proc. Natl. Acad. Sci. USA. 1999, 96, 6591–6596.

87. Pasa-Tolic, L.; Anderson, G.A.; Lipton, M.S.; Peden, K.K.; Martinovic, S.; Tolic, N.; Bruce, J.E.; Smith, R.D. High-throughput proteome wide precision measurements of protein expression using mass spectrometry. J. Am. Chem. Soc. 1999, 121, 7949–7950.

88. Conrads, T.P., et al. Quantitative analysis of bacterial and mammalian proteomes using a combination of cysteine affinity tags and 15N-metabolic labeling. Anal. Chem. 2001, 73, 2132–2139.

89. Munchbach, M.; Quadroni, M.; Miotto, G.; James, P. Quantitation and facilitated de novo sequencing of proteins by isotopic N-terminal labeling of peptides with a fragmentation-directing moiety. Anal. Chem. 2000, 72, 4047–4057.

90. Yao, X.; Freas, A.; Ramirez, J.; Demirev, P.A.; Fenselau, C. Proteolytic 18O labeling for comparative proteomics: model studies with two serotypes of adenovirus. Anal. Chem. 2001, 73, 2836–2842.

91. Goodlett, D.R., et al. Differential stable isotope labeling of peptides for quantitation and de novo sequence derivation. Rapid Commun. Mass Spectrom. 2001, 15, 1214–1221.

92. Griffin, T.J., et al. Quantitative proteomic analysis using a MALDI quadrupole time-of-flight mass spectrometer. Anal. Chem. 2001, 73, 978–986.

# 10

# Forward and Reverse Proteomics: It Takes Two (or More) to Tango

**DAVID E. HILL, NICOLAS BERTIN, and MARC VIDAL**
*Dana-Farber Cancer Institute and Harvard Medical School*
*Boston, Massachusetts, U.S.A.*

With the release of two drafts of the human genome sequence [1,2] and its completion expected next year, the Human Genome Project has gradually shifted into what is often referred to as the Human Proteome Project (HPP). Among the long-term goals of HPP, it will be important to identify for humans and a few model organisms: (1) the majority of expressed proteins, (2) their individual posttranslational modifications, and (3) most of the macromolecules with which they interact; or collectively, create a comprehensive description of the corresponding proteomes. The already available genome sequences will be crucial to accomplish these tasks. Indeed, protein-encoding open reading frames (ORFs), spanning from the initiation codon to the stop codon, can be inferred using genome annotation algorithms [3,4]. This information is currently used in two ways. In 'forward proteomics,' predicted ORFs serve as guide for the identification of endogenous proteins purified from cellular extracts, whereas in 'reverse proteomics' [5], cloned ORFs are used to express proteins in heterologous and/or exogenous systems (Fig. 1). In this chapter, we will focus on some of the challenges and strategies of reverse proteomics. In particular, we will discuss the challenge of cloning (nearly) complete sets of ORFs, or 'ORFeomes,' and using such cloned ORFeomes to generate protein–protein interaction maps.

## I. INTRODUCTION

The proteome, by analogy to the genome, is the complete set of expressed proteins of an organism, including their various isoforms, generated by alternative splicing of primary transcripts and/or posttranslational modifications (reviewed in Ref. 6;

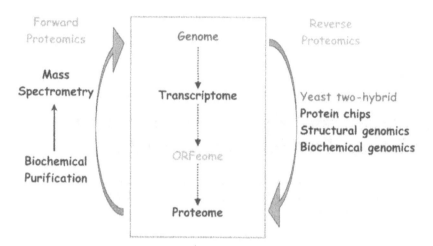

**Figure 1** Forward and reverse proteomics and the Human Proteome Project. 'Forward proteomics' uses predicted ORFs to serve as guides for the identification of endogenous proteins purified from cellular extracts. This process employs biochemical purifications and mass spectrometry to obtain information on expressed proteins and the constituents of protein complexes, whereas 'reverse proteomics' utilizes cloned ORFs to express proteins in heterologous and/or exogenous systems to perform Y2H for protein–protein interaction mapping.

see also Refs. 7 and 8). There is however, a major conceptual difference between genome and proteome. A genome can be considered finite in first approximation; that is, there are a discrete number of nucleotides to be sequenced, which should give rise after annotation to a specific number of predicted genes. On the other hand, it is more difficult to consider the proteome as a discrete and finite entity. At any given point in time, the expressed proteome will be a subset of the complete proteome. Over the duration of the cell cycle, in different cell types, in response to cellular and metabolic stresses, and throughout the life span of the organism, there will be many different proteome subsets. A major consequence of this multiplicity of proteome subsets is that a variety of approaches will be required to accurately describe a complete proteome. Additionally, protein complexes and the protein–protein interactions that form them are at the heart of most biological processes. Thus, identifying all individual proteins in any given complex and determining their specific interactions will be critical for the HPP. In addition, validation of the data arising from both forward and reverse proteomics approaches by independent methods will be an essential element of the HPP.

## II. CLONING AN ENTIRE ORFEOME

The goal of ORFeome cloning projects is the isolation of a complete set of all expressed genes of an organism of interest as cloned ORFs and their use in various functional genomic projects, including the exogenous expression of a complete proteome and the generation of a comprehensive protein–protein interaction map. To help in developing the concepts and technologies needed to undertake a human ORFeome cloning project, we decided to focus our attention on *Caenorhabditis elegans*.

Based on *in silico* analysis of its complete genome sequence, *C. elegans* is predicted to have approximately 19,000 ORFs [9]. Less than 10% of the predicted genes have been thoroughly characterized using conventional approaches and only 50% have been annotated at the level of cDNA or expressed sequence tags (ESTs). This leaves nearly 9000 predicted genes for which there is no functional information supporting their existence, expression, and structure. To address the issue of how many genes are expressed in *C. elegans* and to develop a resource for functional and comparative genomics, we embarked on obtaining all of the expressed genes as cloned ORFs using a highly representative cDNA library as a source of clones. Additionally, we elected to use recombinational cloning employing the Gateway system [10,11] (discussed below) instead of traditional restriction enzyme-based cloning in order to accomplish a genome-scale cloning project. In a preliminary report of this effort, we analyzed over 2200 ORFs, of which nearly 1100 had not been previously annotated; this analysis suggested that there are at least 17,300 protein-coding genes expressed in *C. elegans* [12]. As part of the structural annotation of the predicted genes of *C. elegans*, we were able to correct nearly 12% of all predicted exons and 27% of predicted ORFs. In specific cases, the cloning and sequencing of these mispredicted ORFs led to functional annotations based on sequence orthologies. That project has now been completed and has led to the creation of a resource of over 12,000 cloned ORFs (Reboul et al., *Nature Genetics*, 2003; in press). These ORFs are now being utilized in numerous functional genomic projects; in particular, we are using the cloned ORFs and the yeast two-hybrid system (Y2H) for proteomewide protein-interaction studies. Information on individual cloned ORFs is available through our website at worfdb.dfci.harvard.edu [12].

Genome-scale cloning projects require facile and standardized manipulations of multiple DNA molecules, preferably using robotic platforms whenever possible. However, standard molecular cloning involving restriction digestions and ligations of thousands of ORFs is not very amenable to high-throughput, robotic-based methods; in many cases, cDNAs cloned into traditional vectors cannot be readily transferred to alternative vectors for other experimental protocols. To overcome these limitations, an alternative strategy has been developed that uses in vitro recombination between a DNA of interest and a plasmid vector [10,11,13]. Subsequent generation of plasmids for expression of the gene or ORF of interest is also accomplished by an in vitro recombination event between the initial cloned DNA fragment and the expression vector of interest. Several methods using in vitro recombination have been developed and commercialized. For the initial cloning of the *C. elegans* ORFeome and for subsequent transfer of all cloned ORFs into Y2H vectors and other expression vectors, the Gateway system of recombinational cloning has been employed [10–12]. The Gateway system is based on bacteriophage λ integration into and excision from the *Escherichia coli* chromosome. A site-specific recombination event between the attP site in the phage genome and the attB site in the *E. coli* chromosome, with no net gain or loss of DNA, leads to integration and the creation of two novel flanking sites: attL and attR. An intrachromosomal recombination between the flanking attL and attR sites leads to excision of the phage genome. In the first implementation of Gateway used for ORF cloning [10], all four wild-type λ recombination sites (attB, attP, attL, and attR) were duplicated and modified into two separate artificial sites (B1 and B2 for attB, P1 and P2 for attP, L1 and L2 for attL, R1 and R2 for attR) such that B1 recombines with P1, and B2 with P2, but B1 fails to recombine with P2 and B2 with P1. Likewise, L1 recombines with R1, and L2 with R2, but L1 fails to recombine with R2 and L2 with R1. Thus, any fragment such as an ORF of interest flanked by B1 in its 5′ end and B2 in its 3′ end can be cloned *unidirectionally* into a vector (referred to as the 'donor' plasmid) containing P1 and P2. B1–ORF–B2 fragments are produced by polymerase chain reaction (PCR) using ORF-specific primers synthesized with B1 and B2 sequences at their 5′ ends, respectively, as well as a comprehensive cDNA library as template DNA (Fig. 2A).

Importantly, once ORFs or other DNA fragments are cloned as entry clones, they can then be transferred into expression vectors referred to as destination plasmids (Fig. 2b). Such plasmids contain R1 and R2 sites. Because B1 and B2 have been designed not to contain a stop codon, N-terminal and/or C-terminal fusion proteins can be generated from destination expression clones. Currently, dozens of different destination vectors are available, comprising a selection of some of the most popular prokaryotic and eukaryotic expression vectors. Most standard expression vectors are readily converted to destination vectors simply by inserting a 1.8-kb cassette that contains all of the necessary Gateway sequences. The power of the Gateway cloning system is that a collection of ORFs 'stored' in Gateway Clones can be transferred in parallel to one or more destination vectors in a simple reaction that can take place in 96-well plates.

One significant consequence of having the *C. elegans* ORFeome cloned in the Gateway system is that any ORF identified in an Y2H screen employing conventional cDNA libraries is already cloned and available for either retesting in Y2H or for other functional tests. In fact, in three separate Y2H screens involving 125 baits, over 400 interacting proteins were obtained and 80% of the ORFs

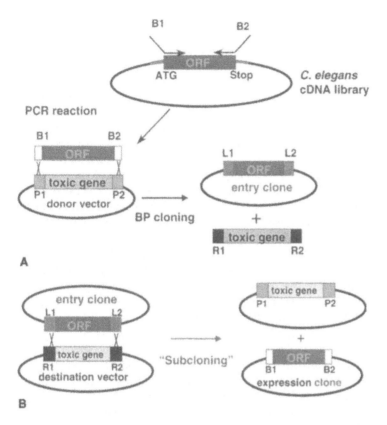

**Figure 2** (A) Creation of Gateway 'entry clones:' In vitro recombination between a PCR product and 'donor' vector leads to the creation of an entry clone. (B) Creation of Gateway 'destination expression clones:' In vitro recombination between an entry clone and a 'destination' vector leads to the creation of an expression clone. All DNA manipulations performed in vitro; no restriction–ligation reactions.

encoding those proteins were already available in our *C. elegans* ORFeome collection as full-length clones immediately available for subsequent use.

## III. FORWARD AND REVERSE PROTEOMICS SYSTEMS

As mentioned above (see Fig. 1), proteomics, defined as the field of proteome identification and analyses, can be approached from either a protein-based direction ('forward proteomics'), or from a gene-based direction ('reverse proteomics'). Forward proteomics is best exemplified by two major techniques that have come to be associated with the field of proteomics: chromatography and mass spectrometry (MS). Various chromatographic procedures, of which two-dimensional (2D) gel electrophoresis is a highly specialized form, are initially employed to achieve a high degree of purification, either as isolated proteins or as protein complexes, from whole-cell lysates. MS is then used to obtain a 'peptide signature' for each protein, thereby identifying specific proteins based on identical signatures or patterns found in peptide and protein databases. Chromatographic and MS-based analyses of proteomes are (1) capable of identifying upward of several thousand proteins concomitantly expressed in a cell [14] and (2) determining the constituents of macromolecular complexes present as a consequence of the instantaneous state of the cell or organism [15–18]. Basically, it is possible to obtain a very detailed and potentially comprehensive picture of the instantaneous components of the proteome. However, both 2D gels and liquid chromatography have a limited dynamic range (i.e., the effective concentration range over which specific proteins can be unambiguously identified). In practice, this means that proteins expressed at very low abundance levels are not detected by MS. What is also missing in the forward proteomic approach is the knowledge of the pairwise protein–protein interactions that make up a particular protein complex, how these interactions can affect the state of the cell, and the biochemical and organismal function of all proteins involved.

Reverse proteomics, on the other hand, makes use of genomic sequences and their resulting predicted ORFs, heterologous expression, and ex vivo analyses to determine protein–protein interactions and establish protein function. By determining the interacting partners for multiple proteins predicted to be involved in a particular biological function, one can assign putative function to otherwise unknown proteins. The data from protein interaction studies can be assembled into protein interaction networks, which provide hypotheses about putative protein complexes and how functional complexes potentially interact [19–24]; see also reviews in Refs. 5 and 25–27.

Recently, high-throughput protein expression systems utilizing hundreds to thousands of exogenously expressed proteins have been developed and utilized for functional assays [28–30]. One exciting technology, patterned on nucleic acid-based microarrays, is 'protein chips,' in which large numbers of proteins can be simultaneously tested in an in vitro assay for structure–function studies [31–33]. By starting with cloned genes and expressing the encoded proteins, 'reverse proteomics' nicely complements the information obtained by 'forward' proteomics approaches.

## IV.  IDENTIFYING PROTEIN–PROTEIN INTERACTIONS

One of the major tools of 'reverse proteomics' has been the (Y2H) system, originally described by Fields and Song [34,35] (Fig. 3A), and the various derivatives

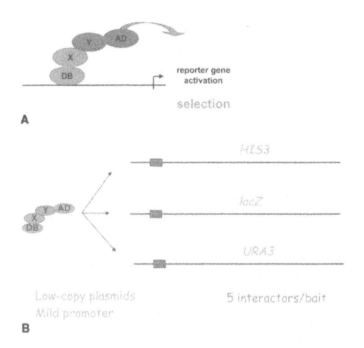

A

B

**Figure 3** (A) Prototypic Y2H system: Interactions between the bait (X) and the prey (Y) leads to reconstitution of a transcription factor, which then allows transcriptional activation of a reporter gene. (B) Minimizing spurious interactions: Employing three reporter genes with 'weak' promoters and using DB-X and AD-Y expressed from single-copy plasmid vectors reduces the overall rate of false positives in the prototypic Y2H.

that have been developed (reviewed in Ref. 36; see also Refs. 37–41). Although Y2H is the most commonly utilized system for identifying protein–protein interactions, bacterial and mammalian two-hybrid systems have also been successfully employed [42–45]. The basic readout of all varieties of two-hybrid systems is the functional reconstitution of an activity that can either confer selective growth advantage or can produce a detectable signal only when two or more macromolecules interact with each other. The prototypic Y2H employs the DNA-binding domain of the yeast Gal4 protein (DB) fused to protein X, generally referred to as the 'bait,' and the separable Gal4 domain capable of transcriptional activation (AD) fused to protein Y, termed the 'prey' [46]. In principal, any sequence-specific DB can be paired with any transcriptional activation domain (see Ref. 36 for additional examples). When DB-X and AD-Y are coexpressed in cells, a functional transcription factor is reconstituted if bait and prey can interact, which then allows transcriptional activation of a reporter gene.

## V.  Y2H SCREENING APPROACHES

Dependent on the number of baits and preys to be screened, there are several screening strategies that can be employed and a number of specific methods have been published [47–51]. Briefly, the first choice is between using a yeast mating strategy versus cotransformation of a single yeast strain. In the mating strategy, all DB-X and AD-Y are separately introduced into yeast strains of the opposite mating type. Each haploid strain carries complementary mutations such that diploid yeast cells are directly obtained by selection for prototrophic growth on the appropriate minimal growth medium. In cotransformation, DB-X and AD-Y carry distinct selectable markers and the yeast host is sequentially transformed by bait and prey plasmids. In either case, selection for Y2H target gene expression is then imposed and yeast cells capable of growing on selective media are initially presumed to express a bait–prey interacting pair.

Proteomewide Y2H screens with large numbers of baits can be performed in several ways. When a cDNA library is used as a source of preys, individual baits are screened against the entire library. This approach maximizes the number of potential interactors, but the identity of individual preys is unknown until recovered clones are sequenced. Alternatively, all pairwise combinations of DB-X and AD-Y can be tested in an array or matrix format. This approach works well with small numbers of candidate ORFs [13,19,52], but for the entire collection of 17,300 expressed *C. elegans* ORFs, this approach is not feasible with currently available automation devices. One solution is to reduce the complexity of the prey library to be tested (see Section IX) and/or to pool multiple baits for screening against pools of prey. Regardless of the screening approaches taken, proteomewide screens will require some degree of robotics simply to accommodate all of the liquid-handling steps.

## VI.  MAXIMIZING INTERACTIONS

In practical terms, two-hybrid systems typically underestimate the number of interactions for several reasons. All Y2H systems target proteins to a specific location, as in the nucleus for transcriptional activation with DB-X and AD-Y. Forced subcellular localization may preclude certain interactions from taking place, such as those involving integral membrane proteins. Alternatively, proteins that are secreted or have transmembrane domains may simply not even get into the nucleus in spite of the nuclear targeting abilities of DB and AD. Second, many higher eukaryotic interactions are based on specific posttranslational modifications that must occur prior to or coincident with a protein–protein interaction. Unless the specific enzymes responsible for such modification are also present, certain interactions may not take place. Third, any heterologous protein that is deleterious to the yeast host, such as a fusion to either the DB or AD partner, will not allow growth of the expression vector host cells and thus not be detected as an interactor by screening or by direct selection. Thus, two-hybrid systems have an inherent false-negative rate that limits the number of potential protein–protein interactions that can be obtained. Recognition of this limitation has led to the development of alternative strategies for identifying protein–protein interactions such as the use of phage display [53–55] that can complement Y2H.

To overcome the inherent limitations of the canonical DB-X/AD-Y Y2H, alternative versions have been developed that allow for interactions outside of the nucleus [39–4156] and/or incorporate posttranslational modifications of either the bait or the prey [57–60]. For example, the original Y2H system using transcription factors has been reconfigured such that the interactions take place at or near the plasma membrane. In one embodiment of the Y2H, interacting proteins reconstitute an ubiquitin-mediated cleavage of a transcription factor, which is then free to enter the nucleus and turn on target gene expression [61]. This system specifically covers those circumstances where either membrane association is essential to an interaction or where interactions in the nucleus are precluded. An alternative system requires proteins to be properly targeted to the membrane, and bait–prey interaction leads to productive Ras-based signaling cascade and stimulation of cell proliferation [37,56]. Many enzymes have very high turnover rates and corresponding rapid substrate dissociation rates. In this case, the interaction between an enzyme and its substrate is too transient to allow productive Y2H complexes and subsequent transcriptional activation. One mechanism that has been successfully employed to study enzyme–substrate complexes is to use substrate-trapping mutants in which a mutation in the active site of the enzyme allows substrate binding but precludes enzyme activity and subsequent release of the bound substrate [20,62]. The resulting DB-enzyme/substrate-AD complex is now sufficiently stable to be scored as a positive interaction in standard Y2H screens. One class of protein interactions that have been difficult to study by Y2H is those involving integral membrane proteins. Recently, two novel versions of Y2H have been developed for identifying integral membrane protein interactions [39,40]. Another class of proteins that are difficult to study by Y2H is transcriptional activators. When used as baits, the resulting DB-X will activate reporter gene expression without the need for an interacting AD-Y. One solution is to simply swap bait and prey so that individual AD-Xs, where X is a known transcriptional activator, are screened against a library of DB-Y [63]. Alternatively, a system utilizing active repression of transcription of a toxic gene can be used with transactivator proteins as baits. In this system, any prey protein binding to a DB transactivator will repress the transcription of the toxic gene [64] and allow growth of the expression vector host cells. Clearly, the fundamental paradigm for Y2H of the bait–prey interaction is not limited to reconstitution of a transcription factor but is limited only by imagination and basic cell biology. What remains to be seen with these novel varieties of the canonical Y2H is whether they can be used in high-throughput settings with a large number of distinct baits to identify novel interacting partners. Because no single Y2H method appears to be capable of detecting all possible protein interactions, expanding the universe of potential interactors through use of multiple Y2H methods will be necessary in order to obtain a proteomewide collection of protein–protein interactions.

By determining the number of positive interactions obtained in a standard DB-X/AD-Y Y2H screen using over 75 pairs of baits and preys known to interact, we recovered approximately 50% of expected bait–prey interactions (results compiled from Refs. 13, 19, and 52; see also vidal.dfci.harvard.edu/interactome for a complete list of interactions). These results establish a false-negative rate of nearly 50%, or perhaps even higher because the three studies were done with intracellular proteins that would be expected to work in DB-X/AD-Y versions of Y2H. An obvious solution to reducing the false-negative rate is to perform additional Y2H screens that use the ras-recruitment system and/or the split ubiquitin system, for example, instead of the traditional DB-X/AD-Y methodology. This, however, can be an enormous challenge when dealing with hundreds or more 'bait' proteins that need to be subcloned into the relevant vector, not to mention the need for creation of multiple libraries of 'prey' proteins. Whereas standard methods of cloning DNA preclude such an endeavor, the use of recombinational cloning systems such as the Gateway system [5,10] allows one to generate novel bait and/or prey expression plasmids and libraries in a facile high-throughput, semiautomated manner.

## VII.  MINIMIZING SPURIOUS INTERACTIONS

In addition to false-negative rates of 50% or more, early versions of Y2H were compromised by a relatively high rate of false positives that were due in part to the technical features of the system, namely reliance on a single DB-X/AD-Y reporter gene, the use of multicopy vectors, and the use of strong promoters driving the expression of both DB-X and AD-Y. However, by incorporating multiple reporter genes for transcription activation following interaction between a DB-X and AD-Y pair, by employing different DNA sequences for binding by DB, by using low-copy-number vectors, and by employing dual baits, the system achieved greater specificity [65–69]. One such scheme for using multiple promoters and selectable markers is shown in Fig. 3b. In particular, the use of selectable markers such as yeast *URA3* and *HIS3* genes also allows for additional selection based on sensitivity to 5-fluoro-orotic acid and resistance to increasing concentrations of the imidazole analog 3-aminotriazole, respectively [51]. By also incorporating *LacZ* as one of the DB-X/AD-Y target genes, it is possible to score for an unselected phenotype based on β-galactosidase activity. In addition, β-galactosidase assays can be performed with liquid cultures, which are more amenable to robotics-based screening systems [70].

Two-hybrid systems are also susceptible to other forms of false positives independent of the molecular details of the system. In particular are those false positives due to irrelevant peptides arising from AD fusion to any wrong reading frame in a cDNA of interest. This problem occurs most often with cDNA libraries and is readily apparent upon sequencing AD-Y clones. On the other hand, false positives can also arise from full-length clones in the proper reading frame encoding normally expressed proteins. These false positives can be due to 'sticky' proteins that seem able to interact with any number of different, biologically unrelated baits; for example, in all of our *C. elegans* Y2H screens, we routinely obtain the homeobox protein PAL-1. By creating AD-Y libraries comprised of comprehensive sets of defined genes, one can eliminate the more notorious of this class of 'false positives' (see Section IX). An equally problematic false positive is one that occurs by virtue of the 'prey' interacting with DB independently of the bait [71]. This can be avoided by using Gateway recombinational cloning to introduce any given bait into different DB vectors and then screen for interactors. A bona fide X–Y interaction should generally occur independent of the DB and AD used.

One can also encounter false positives in Y2H screens that are due to bona fide interactions between bait and prey, but the two proteins are never expressed in the same cellular compartment at the same time [36]. This type of false positive is initially difficult to identify, requiring independent confirmation such as transcriptional profiling or immunological studies.

All two-hybrid systems are plagued with some degree of 'self-activators,' which, in the standard DB-X/AD-Y version of Y2H, would be a DB-X capable of activating gene expression in the absence of any AD-Y. In small-scale Y2H screening, self-activators are more of a nuisance and can be easily identified and eliminated. In large-scale, genomewide screens, the presence of self-activators makes it impossible to obtain any useful results. However, self-activators can be eliminated by employing methods that selectively inhibit the growth of strains in which Y2H target gene expression is independent of AD-Y interacting with DB-X [72].

## VIII. GENOMEWIDE PROTEIN INTERACTION MAPS

The Y2H technique has been an extremely valuable technique for analyzing specific interactions between two proteins or mapping interaction domains (e.g., see Refs. 73–75). However, the real promise for Y2H is in large-scale, genomewide screens that will identify protein interactions that define networks or interaction maps of cellular function [26,27,36,76]. Already, Y2H has been used to generate protein interaction maps for biological processes, including the transcription, splicing, and protein degradation machinery [52,77–79], signal transduction pathways Ras and Rb, DNA damage response pathways [13,19], and cellular differentiation [20]. Critical to all of these studies, and numerous others as well, is the fact that Y2H is readily amenable to high-throughput, semiautomated methodologies [51,70,80–84]. When used in conjunction with a library of 'baits' encoding all expressed proteins and a comprehensive library of 'preys,' the yeast two-hybrid system is, in principle, capable of generating a proteome-wide set of protein-protein interactions.

Recently, several groups have begun proteomewide screens of protein interactions from *Helicobacter pylori, Saccharomyces cerevisiae,* and *C. elegans* [21,22,24,85]. Two separate studies with *S. cerevisiae* have led to the identification of over 3000 potential protein–protein interactions [21,24,82]. Interestingly, the two different yeast projects show limited overlap in the collection of interactions obtained and the combined number of interactors per bait was less than one. The *H. pylori* screen, on the other hand, recovered over three interactors per bait and covered approximately 45% of the predicted proteome [22]. Nevertheless, the results demonstrate that the strategy of proteomewide screens can generate large numbers of potential interactions. However, to obtain full coverage, it may be necessary to conduct multiple proteomewide screens under different conditions or perform a systematic series of smaller screens focused on known biological functions.

An alternative to the global or genomewide approach is to use as baits all proteins known to be involved in a particular cellular process such as a signaling pathway or known molecular complex. These so-called 'modular' screens are usually done in order to identify all possible protein interactions associated with the desired cellular function as first demonstrated for the RNA-splicing machinery [79]. When smaller, 'modular' screens are done with a limited number of baits, the number of interactors per bait is seen to increase [27]. For example, we observe between three and five interactors per bait in modular screens involving the ras and Rb pathways for vulval development, the DNA damage response, and the proteasome machinery in the worm [13,19,52]. However, as indicated earlier, not all known interactions are recovered even in these 'modular' screens, again suggesting that multiple, different two-hybrid methods will need to be employed to achieve a comprehensive result.

## IX.  CREATION OF GENOMEWIDE Y2H BAIT AND PREY LIBRARIES

Any Y2H project requires separately cloning genes or ORFs of interest into bait and prey expression vectors. One of the rate-limiting aspects of Y2H protein–protein interaction studies is the source of 'prey' clones fused to AD. Typically, individual baits are separately subcloned as fusions with a DNA-binding domain, and the availability of a cloned ORFeome accomplishes the need for a genomewide collection of 'baits' [12,24,81]. Prey clones, on the other hand, are generally obtained by constructing cDNA libraries in which the cDNA inserts are fused to an activation domain, and the subsequent success of a Y2H screen is directly related to the overall quality of the library. As with cDNA libraries in general, the inserts fused to AD are not necessarily full-length clones nor are they always cloned in the correct reading frame which leads to spurious peptides that can score as positives in Y2H screens (see Section VII). Regardless of the structural limitations seen in cDNA libraries, expressing an entire proteome in the context of Y2H requires only the manipulation of DNA molecules. An additional limitation of cDNA libraries is that they are composed of a nonuniform distribution of those mRNAs present at the time of RNA purification. Consequently, cDNAs from genes expressed at very low levels or various alternatively spliced isoforms may not be present in sufficient quantities unless one is willing to screen many millions of independent transformants. In Y2H screens of *C. elegans* proteins, one is required to screen routinely over 2 million yeast transformants in order to achieve a minimal coverage of all cDNAs expressed from a genome that contains less then 20,000 genes [72]. In many cases, the 'prey' clones that are recovered do not encode full-length proteins, but, rather, are truncated clones that arise during the creation of the cDNA library. Although the bait–prey interaction between DB-X and AD-truncated Y may correctly reflect the actual interaction between X and Y, there is also the prospect that the truncated Y exposes an epitope that would not otherwise interact with any 'bait' or could also be a nonsensical polypeptide translated from the wrong reading frame. Furthermore, in those cases in which preys are recovered as truncated clones, full-length clones still need to be isolated for further studies.

Various strategies have been employed to create 'better' libraries. One solution has been the use of 'normalized' libraries which attempt to minimize the range of expression levels between highly expressed genes and very low expressed genes seen in standard libraries [86,87]. However, it will be difficult to create a comprehensive library of human cDNAs that encompasses all tissues, cell types, and developmental stages. An alternative approach is to generate a library in which all of the cDNAs from all of the expressed genes are present in roughly equimolar amounts. Under these conditions, one would need to screen many fewer transformants in order to achieve adequate coverage of an entire genome. For example, a comprehensive library containing equimolar amounts of 20,000 unique ORFs would require only 200,000 yeast transformants in order to achieve 10-fold coverage of an entire genome. The genomewide Y2H screens by Uetz et al. [21] and Ito et al. [24] used 'equimolar' libraries created by cloning the 6000 yeast genes into an AD vector. For metazoans, overall genome size, higher gene numbers, complex exon/intron structure, and alternative splicing precludes using the yeast approach. Nevertheless, several different approaches have been used to create nearly 'equimolar' metazoan libraries. The availability of large numbers of arrayed human ESTs from the I.M.A.G.E. (Integrated Molecular Analysis of Genomes and their Expression) Consortium [88] has led to the creation of an AD-Y prey library of human clones for use in Y2H screens [81]. By performing gene-specific PCR on individual, known ESTs, the resulting library exhibits greater complexity and better representation of low-abundance clones than that seen in traditional libraries, although it does not appear that 'full-length' clones were preferentially chosen. A related effort with over 3000 mouse full-length cDNAs has been undertaken as well [83], and we have used the Gateway system to transfer all of our cloned *C. elegans* ORFs into a Y2H AD vector (Reboul et al., *Nature Genetics*, 2003; in press). Having collections of individually cloned ORFs or ESTs allows one to create 'equimolar' metazoan libraries with the hope that novel, additional protein–protein interactions will be obtained by Y2H. One immediate prediction for an equimolar library of full-length ORFs is that it will enhance the possibility of obtaining interacting proteins that must either be full length in order for an interaction to occur or must be present at higher concentrations than those obtained in conventional libraries. Furthermore, the use of such a library also supports the above-described notion that alternative strategies will be required to identify all possible interacting proteins, particularly in view of the relatively high false-negative rate seen with standard Y2H methods. In the absence of both a cloned ORFeome and availability of recombinational cloning, the creation of an AD library having equimolar representation and the identification of additional interacting proteins for each of the baits tested will not be possible.

## X.  PROTEIN-INTERACTION CLUSTERS

As a consequence of various improvements that have reduced the rate of false positives, Y2H has become the most widely used method for identifying protein–protein interactions. When multiple proteins involved in the same biological process, a biological 'module,' are used as baits for Y2H, the ensuing set of interacting partners and their cognate baits can be assembled into a protein interaction map (reviewed in Ref. 89). Such maps should help identify novel, unknown proteins that contribute to that process. Within these modular maps, one or more sets of interconnected protein interactions or protein clusters can be observed. In the simplest maps, clusters form closed loops of the type 'A binds B binds C binds A' (see Fig. 4 for a 'hypothetical' cluster). As the number of baits to be tested increases, however, small clusters are seen to connect to other clusters [19,20]. The resulting modular map may contain several distinct 'clusters' of proteins involved in multiple intracluster and intercluster connections. In addition, a Y2H screen of one 'module' may identify interacting proteins that are also found in a Y2H screen for distinctly different modules [19,52]. Such 'cross-cluster' interactions serve to suggest ways in which different biological processes communicate with one another (Fig. 5).

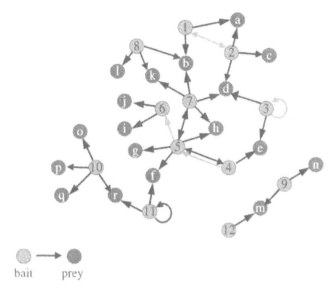

**Figure 4** Hypothetical protein-interaction network: A set of 12 baits (light gray circles) identifies (dark or light gray arrows), 18 unique prey (dark gray circles), and 2 homodimers (semi circular arrows). Light gray arrows designate interactions from a pairwise matrix Y2H, whereas dark gray arrows show interactions obtained from a library Y2H screen.

## XI. PROTEOMEWIDE MAPS

To achieve a genomewide map requires either a large number of modular Y2H screens or a single genomewide screen. In either case, highly complex maps that serve to connect multiple functions can be obtained from Y2H screens. In the various genomewide screens that have been conducted [22,24,82], interactions have been obtained between and among clusters that include many novel or otherwise unknown proteins, thereby suggesting hypothetical functions for these proteins. Additionally, proteins of known function have been found in interaction clusters that are of completely different function, suggesting potentially new functions for these 'known' proteins. However, as mentioned earlier, the various genomewide approaches have clearly not reached saturation, principally because no single Y2H method is capable of detecting all possible interacting partners for any given bait protein. Nevertheless, the availability of cloned ORFeomes coupled with robotic systems employing recombinational cloning techniques allows one to perform multiple, different Y2H screens, thereby achieving the goal of identifying all possible protein–protein interactions.

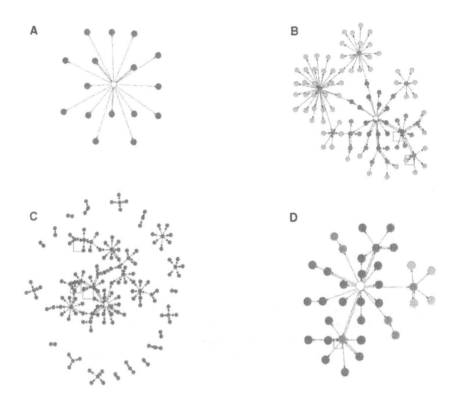

**Figure 5** Generation of a protein-interaction network. The *C. elegans* MRT-2 gene was used as bait in a Y2H screen in conjunction with other DNA damage response (DDR) genes [9]. Circles denote ORFs and the arrows between two circles indicate the bait–prey interaction. Arrows are from bait to prey, and double lines show reciprocal interactions. (A) The 16 prey (green circles) obtained with MRT-2 as bait (yellow circle); 3 of the MRT-2 prey, when used as baits interacted with MRT-2 as prey. (B) The expanded network when additional DDR genes are used as baits (purple circles) or are found as prey (orange circles) and the resulting direct (green circles) and indirect (purple circles) connections to MRT-2 (yellow circle). (C) The core DDR interactome network [9] centered on MRT-2 (yellow circle). (D) Direct connections as in (A) (dark green circles) with MRT-2 (yellow circle) as bait and connections arising from non-DDR Y2H screens (blue, orange, and purple circles). Light green circles are prey that were obtained from both MRT-2 and non-DDR Y2H screens. (See the color plate.)

## ACKNOWLEDGMENTS

We thank A.J.M. Walhout for comments on the manuscript and members of the Vidal lab for support. This work was supported by grants 5R01HG01715–02 (National Human Genome Research Institute), P01CA80111–02 and 7 R33 CA81658–02 (National Cancer Institute), and 232 (Merck Genome Research Institute) awarded to M.V.

## REFERENCES

1.  Lander, E. S.; Linton, L. M.; Birren, B.; Nusbaum, C.; Zody, M. C.; Baldwin, J.; Devon, K.; Dewar, K.; Doyle, M.; FitzHugh, W.; Funke, R.; Gage, D.; Harris, K.; Heaford, A. Initial sequencing and analysis of the human genome. Nature. 2001, 409, 860–921.
2.  Venter, J. C.; Adams, M. D.; Myers, E. W.; Li, P. W.; Mural, R. J.; Sutton, G. G.; Smith, H. O.; Yandell, M.; Evans, C. A.; Holt, R. A.; Gocayne, J. D.; Amanatides, P. The sequence of the human genome. Science. 2001, 291, 1304–1351.
3.  Durbin, R.; Thierry-Mieg, J. The ACeDB genome database. In Computational Methods in Genome Research Suhai S, ed; Plenum: New York, 1994.
4.  Stein, L.; Sternberg, P.; Durbin, R.; Thierry-Mieg, J.; Spieth, J. WormBase: Network access to the genome and biology of *Caenorhabditis elegans*. Nucleic Acids Res. 2001, 29, 82–86.
5.  Walhout, A. J.; Vidal, M. Protein interaction maps for model organisms. Nat Rev Mol Cell Biol. 2001, 2, 55–62.
6.  Harrison, P. M.; Kumar, A.; Lang, N.; Snyder, M.; Gerstein, M. A question of size: The eukaryotic proteome and the problems in defining it. Nucleic Acids Res. 2002, 30, 1083–1090.
7.  Kahn, P. Molecular biology: From genome to proteome: Looking at a Cell's Proteins. Science. 1995, 270, 369–370.
8.  Service, R. F. PROTEOMICS: High-speed biologists search for gold in proteins. Science. 2001, 294, 2074–2077.
9.  Consortium, T. C. E. S. Genome sequence of the nematode *C. elegans*: A platform for investigating biology. The *C. elegans* Sequencing Consortium. Science. 1998, 282, 2012–2018.
10. Hartley, J. L.; Temple, G. F.; Brasch, M. A. DNA cloning using in vitro site-specific recombination. Genome Res. 2000, 10, 1788–1795.
11. Walhout, A. J.; Temple, G. F.; Brasch, M. A.; Hartley, J. L.; Lorson, M. A.; van den Heuvel, S.; Vidaly, M. GATEWAY recombinational cloning: Application to the cloning of large numbers of open reading frames or ORFeomes. Methods Enzymol. 2000, 328, 575–592.
12. Reboul, J.; Vaglio, P.; Tzellas, N.; Thierry-Mieg, N.; Moore, T.; Jackson, C.; Shin-i, T.; Kohara, Y.; Thierry-Mieg, D.; Thierry-Mieg, J.; Lee, H.; Hitti, J. Open-reading-frame sequence tags (OSTs) support the existence of at least 17,300 genes in *C. elegans*. Nature Genet. 2001, 27, 332–336.

13. Walhout, A. J.; Sordella, R.; Lu, X.; Hartley, J. L.; Temple, G. F.; Brasch, M. A.; Thierry-Mieg, N.; Vidal, M. Protein interaction mapping in *C. elegans* using proteins involved in vulval development. Science. 2000, 287, 116–122.

14. Washburn, M. P.; Wolters, D.; Yates, J. R. I. Large-scale analysis of the yeast proteome by multidimensional protein identification technology. Nature Biotechnol. 2001, 19, 242–247.

15. Gavin, A. C.; Bosche, M.; Krause, R.; Grandi, P.; Marzioch, M.; Bauer, A.; Schultz, J.; Rick, J. M.; Michon, A. M.; Cruciat, C. M.; Remor, M.; Hofert, C.; Hofert, M. Functional organization of the yeast proteome by systematic analysis of protein complexes. Nature. 2002, 415, 141–147.

16. Ho, Y.; Gruhler, A.; Heilbut, A.; Bader, G. D.; Moore, L.; Adams, S. L.; Millar, A.; Taylor, P.; Bennett, K.; Boutilier, K.; Yang, L.; Wolting, C.; Donaldson, I.; Schandorff, S. Systematic identification of protein complexes in *Saccharomyces cerevisiae* by mass spectrometry. Nature. 2002, 415, 180–183.

17. Yates, J. R. Mass spectrometry. From genomics to proteomics. Trends Genet. 2000, 16, 5–8.

18. Zhou, H.; Ranish, J. A.; Watts, J. D.; Aebersold, R. Quantitative proteome analysis by solid-phase isotope tagging and mass spectrometry. Nature Biotechnol. 2002, 20, 512–515.

19. Boulton, S. J.; Gartner, A.; Reboul, J.; Vaglio, P.; Dyson, N.; Hill, D. E.; Vidal, M. Combined functional genomic maps of the C. elegans DNA damage response. Science. 2002, 295, 127–131.

20. Drees, B. L.; Sundin, B.; Brazeau, E.; Caviston, J. P.; Chen, G. C.; Guo, W.; Kozminski, K. G.; Lau, M. W.; Moskow, J. J.; Tong, A.; Schenkman, L. R.; McKenzie, A. A protein interaction map for cell polarity development. J Cell Biol. 2001, 154, 549–571.

21. Ito, T.; Chiba, T.; Ozawa, R.; Yoshida, M.; Hattori, M.; Sakaki, Y. A comprehensive two-hybrid analysis to explore the yeast protein interactome. Proc Natl Acad Sci USA. 2001, 98, 4569–4574.

22. Rain, J. C.; Selig, L.; De Reuse, H.; Battaglia, V.; Reverdy, C.; Simon, S.; Lenzen, G.; Petel, F.; Wojcik, J.; Schachter, V.; Chemama, Y.; Labigne, A.; Legrain, P. The protein–protein interaction map of *Helicobacter pylori*. Nature. 2001, 409, 211–215.

23. Stanyon, C. A; Finley, R. L. Progress and potential of *Drosophila* protein interaction maps. Pharmacogenomics. 2000, 1, 417–431.

24. Uetz, P.; Giot, L.; Cagney, G.; Mansfield, T. A.; Judson, R. S.; Knight, J. R.; Lockshon, D.; Narayan, V.; Srinivasan, M.; Pochart, P.; Qureshi-Emili, A.; Li, Y. A comprehensive analysis of protein–protein interactions in *Saccharomyces cerevisiae*. Nature. 2000, 403, 623–627.

25. Legrain, P.; Selig, L. Genome-wide protein interaction maps using two-hybrid systems. FEBS Lett. 2000, 480, 32–36.

26. Legrain, P.; Wojcik, J.; Gauthier, J. M. Protein–protein interaction maps: A lead towards cellular functions. Trends Genet. 2001, 17, 346–352.

27. Walhout, A. J.; Boulton, S. J.; Vidal, M. Yeast two-hybrid systems and protein interaction mapping projects for yeast and worm. Yeast. 2000, 17, 88–94.

28. Albala, J. S.; Franke, K.; McConnell, I. R.; Pak, K. L.; Folta, P. A.; Rubinfeld, B.; Davies, A. H.; Lennon, G. G. From genes to proteins: High-throughput expression and purification of the human proteome. J Cell Biochem. 2000, 80, 187–191.

29. Braun, P.; Hu, Y.; Shen, B.; Halleck, A.; Koundinya, M.; Harlow, E.; LaBaer, J. Proteome-scale purification of human proteins from bacteria. Proc Natl Acad Sci USA. 2002, 99, 2654–2659.

30. Knaust, R. K.; Nordlund, P. Screening for soluble expression of recombinant proteins in a 96-well format. Anal Biochem. 2001, 297, 79–85.

31. Albala, J. S. Array-based proteomics: the latest chip challenge. Expert Rev Mol Diagn. 2001, 1, 145–152.

32. MacBeath, G.; Schreiber, S. L. Printing proteins as microarrays for high-throughput function determination. Science. 2000, 289, 1760–1763.

33. Zhu, H.; Bilgin, M.; Bangham, R.; Hall, D.; Casamayor, A.; Bertone, P.; Lan, N.; Jansen, R.; Bidlingmaier, S.; Houfek, T.; Mitchell, T.; Miller, P.; Dean, R. A.; Gerstein, M.; Snyder, M. Global analysis of protein activities using proteome chips. Science. 2001, 293, 2101–2105.

34. Fields, S.; Song, O. A novel genetic system to detect protein–protein interactions. Nature. 1989, 340, 245–246.

35. Fields, S.; Sternglanz, R. The two-hybrid system: An assay for protein–protein interactions. Trends Genet. 1994, 10, 286–292.

36. Vidal, M.; Legrain, P. Yeast forward and reverse "n"-hybrid systems. Nucleic Acids Res. 1999, 27, 919–929.

37. Aronheim, A. Improved efficiency sos recruitment system: Expression of the mammalian GAP reduces isolation of Ras GTPase false positives. Nucleic Acids Res. 1997, 25, 3373–3374.

38. Aronheim, A.; Zandi, E.; Hennemann, H.; Elledge, S. J.; Karin, M. Isolation of an AP-1 repressor by a novel method for detecting protein–protein interactions. Mol Cell Biol. 1997, 17, 3094–3102.

39. Ehrhard, K. N.; Jacoby, J. J.; Fu, X. Y.; Jahn, R.; Dohlman, H. G. Use of G-protein fusions to monitor integral membrane protein-protein interactions in yeast. Nature Biotechnol. 2000, 18, 1075–1079.

40. Hubsman, M.; Yudkovsky, G.; Aronheim, A. A novel approach for the identification of protein–protein interaction with integral membrane proteins. Nucleic Acids Res. 2001, 29, E18.

41. Johnsson, N.; Varshavsky, A. Split ubiquitin as a sensor of protein interactions in vivo. Proc Natl Acad Sci USA. 1994, 91, 10,340–10,344.

42. Dang, C. V.; Barrett, J.; Villa-Garcia, M.; Resar, L. M.; Kato, G. J.; Fearon, E. R. Intracellular leucine zipper interactions suggest c-Myc hetero-oligomerization. Mol Cell Biol. 1991, 11, 954–962.

43. Hays, L. B.; Chen, Y. S.; Hu, J. C. Two-hybrid system for characterization of protein–protein interactions in E. coli. BioTechniques. 2000, 29, 288–290, 292, 294 passim.

44. Joung, J. K.; Ramm, E. I.; Pabo, C. O. A bacterial two-hybrid selection system for studying protein-DNA and protein–protein interactions. Proc Natl Acad Sci USA. 2000, 97, 7382–7387.

45. Karimova, G.; Pidoux, J.; Ullmann, A.; Ladant, D. A bacterial two-hybrid system based on a reconstituted signal transduction pathway. Proc Natl Acad Sci USA. 1998, 95, 5752–5756.

46. Chien, C. T.; Bartel, P. L.; Sternglanz, R.; Fields, S. The two-hybrid system: A method to identify and clone genes for proteins that interact with a protein of interest. Proc Natl Acad Sci USA. 1991, 88, 9578–9582.

47. Bartel, P.; Fields, S. The Yeast Two-Hybrid System; Oxford University Press: New York, 1997.

48. Bartel, P. L.; Fields, S. Analyzing protein–protein interactions using two-hybrid system. Methods Enzymol. 1995, 254, 241–263.

49. Kolonin, M. G.; Zhong, J.; Finley, R. L. Interaction mating methods in two-hybrid systems. Methods Enzymol. 2000, 328, 26–46.

50. MacDonald, P. N. Two-hybrid Systems: Methods and Protocols; Humana Press: Totowa: 0, 2001; Vol. 177.

51. Walhout, A. J.; Vidal, M. High-throughput yeast two-hybrid assays for large-scale protein interaction mapping. Methods. 2001, 24, 297–306.

52. Davy, A.; Bello, P.; Thierry-Mieg, N.; Vaglio, P.; Hitti, J.; Doucette-Stamm, L.; Thierry-Mieg, D.; Reboul, J.; Boulton, S.; Walhout, A. J.; Coux, O.; Vidal, M. A protein–protein interaction map of the *Caenorhabditis elegans* 26S proteasome. EMBO Rep. 2001, 2, 821–828.

53. Crameri, R.; Jaussi, R.; Menz, G.; Blaser, K. Display of expression products of cDNA libraries on phage surfaces. A versatile screening system for selective isolation of genes by specific gene-product/ligand interaction. Eur J Biochem. 1994, 226, 53–58.

54. Palzkill, T.; Huang, W.; Weinstock, G. M. Mapping protein–ligand interactions using whole genome phage display libraries. Gene. 1998, 221, 79–83.

55. Zozulya, S.; Lioubin, M.; Hill, R. J.; Abram, C.; Gishizky, M. L. Mapping signal transduction pathways by phage display. Nature Biotechnol. 1999, 17, 1193–1198.

56. Broder, Y. C.; Katz, S.; Aronheim, A. The ras recruitment system, a novel approach to the study of protein–protein interactions. Curr Biol. 1998, 8, 1121–1124.

57. Cao, H.; Courchesne, W. E.; Mastick, C. C. A phosphotyrosine-dependent protein interaction screen reveals a role for phosphorylation of caveolin-1 on tyrosine 14: Recruitment of C- terminal Src kinase. J Biol Chem. 2002, 277, 8771–8774.

58. Keegan, K.; Cooper, J. A. Use of the two hybrid system to detect the association of the protein–tyrosine-phosphatase, SHPTP2, with another SH2-containing protein, Grb7. Oncogene. 1996, 12, 1537–1544.

59. Lioubin, M. N.; Algate, P. A.; Tsai, S.; Carlberg, K.; Aebersold, A.; Rohrschneider, L. R. p150Ship, a signal transduction molecule with inositol polyphosphate-5-phosphatase activity. Genes Dev. 1996, 10, 1084–1095.

60. Osborne, M. A.; Dalton, S.; Kochan, J. P. The yeast tribrid system—Genetic detection of trans-phosphorylated ITAM-SH2-interactions. BioTechnology (NY). 1995, 13, 1474–1478.

61. Stagljar, I.; Korostensky, C.; Johnsson, N.; te Heesen, S. A genetic system based on split-ubiquitin for the analysis of interactions between membrane proteins in vivo. Proc Natl Acad Sci USA. 1998, 95, 5187–5192.

62. Kawachi, H.; Fujikawa, A.; Maeda, N.; Noda, M. Identification of GIT1/Cat-1 as a substrate molecule of protein tyrosine phosphatase zeta/beta by the yeast substrate-trapping system. Proc Natl Acad Sci USA. 2001, 98, 6593–6598.

63. Du, W.; Vidal, M.; Xie, J. E.; Dyson, N. RBF, a novel RB-related gene that regulates E2F activity and interacts with cyclin Er in *Drosophila*. Genes Dev. 1996, 10, 1206–1218.

64. Hirst, M.; Ho, C.; Sabourin, L.; Rudnicki, M.; Penn, L.; Sadowski, I. A two-hybrid system for transactivator bait proteins. Proc Natl Acad Sci USA. 2001, 98, 8726–8731.

65. Bartel, P.; Chien, C. T.; Sternglanz, R.; Fields, S. Elimination of false positives that arise in using the two-hybrid system. Biotechniques. 1993, 14, 920–924.

66. Chevray, P.; Nathans, D. Protein interaction cloning in yeast: Identification of mammalian proteins that react with the leucine zipper of Jun. Proc Natl Acad Sci USA. 1992, 89, 5789–5793.

67. James, P.; Halladay, J.; Craig, E. A. Genomic libraries and a host strain designed for highly efficient two- hybrid selection in yeast. Genetics. 1996, 144, 1425–1436.

68. Serebriiskii, I.; Khazak, V.; Golemis, E. A. A two-hybrid dual bait system to discriminate specificity of protein interactions. J Biol Chem. 1999, 274, 17,080–17,087.

69. Vidal, M. The reverse two-hybrid system. In The Yeast Two-Hybrid System Bartels P, Fields S, eds; Oxford University Press: New York, 1997, 109–147.

70. Serebriiskii, I. G.; Toby, G. G.; Golemis, E. A. Streamlined yeast colorimetric reporter activity assays using scanners and plate readers. BioTechniques. 2000, 29, 278–288.

71. Nordgard, O.; Dahle, O.; Andersen, T. O.; Gabrielsen, O. S. JAB1/CSN5 interacts with the GAL4 DNA binding domain: a note of caution about two-hybrid interactions. Biochimie. 2001, 83, 969–971.

72. Walhout, A. J.; Vidal, M. A genetic strategy to eliminate self-activator baits prior to high- throughput yeast two-hybrid screens. Genome Res. 1999, 9, 1128–1134.

73. Chaudhuri, B.; Hammerle, M.; Furst, P. The interaction between the catalytic A subunit of calcineurin and its autoinhibitory domain, in the yeast two-hybrid system, is disrupted by cyclosporin A and FK506. FEBS Lett. 1995, 357, 221–226.

74. Colonna, T. E.; Huynh, L.; Fambrough, D. M. Subunit interactions in the Na,K-ATPase explored with the yeast two-hybrid system. J Biol Chem. 1997, 272, 12366–12372.

75. Iwabuchi, K.; Li, B.; Bartel, P.; Fields, S. Use of the two-hybrid system to identify the domain of p53 involved in oligomerization. Oncogene. 1993, 8, 1693–1696.

76. Boulton, S. J.; Vincent, S.; Vidal, M. Use of protein-interaction maps to formulate biological questions. Curr Opin Chem Biol. 2001, 5, 57–62.

77. Cagney, G.; Uetz, P.; Fields, S. Two-hybrid analysis of the *Saccharomyces cerevisiae* 26S proteasome. Physiol Genomics. 2001, 7, 27–34.

78. Flores, A.; Briand, J. F.; Gadal, O.; Andrau, J. C.; Rubbi, L.; Van Mullem, V.; Boschiero, C.; Goussot, M.; Marck, C.; Carles, C.; Thuriaux, P.; Sentenac, A.; Werner, M. A protein–protein interaction map of yeast RNA polymerase III. Proc Natl Acad Sci USA. 1999, 96, 7815–7820.

79. Fromont-Racine, M.; Rain, J. C.; Legrain, P. Toward a functional analysis of the yeast genome through exhaustive two- hybrid screens. Nature Genet. 1997, 16, 277–282.

80. Buckholz, R. G.; Simmons, C. A.; Stuart, J. M.; Weiner, M. P. Automation of yeast two-hybrid screening. J Mol Microbiol Biotechnol. 1999, 1, 135–140.

81. Hua, S. B.; Luo, Y.; Qiu, M.; Chan, E.; Zhou, H.; Zhu, L. Construction of a modular yeast two-hybrid cDNA library from human EST clones for the human genome protein linkage map. Gene. 1998, 215, 143–152.

82.  Ito, T.; Tashiro, K.; Muta, S.; Ozawa, R.; Chiba, T.; Nishizawa, M.; Yamamoto, K.;
     Kuhara, S.; Sakaki, Y. Toward a protein–protein interaction map of the budding
     yeast: A comprehensive system to examine two-hybrid interactions in all possible
     combinations between the yeast proteins. Proc Natl Acad Sci USA. 2000, 97,
     1143–1147.
83.  Suzuki, H.; Fukunishi, Y.; Kagawa, I.; Saito, R.; Oda, H.; Endo, T.; Kondo, S.; Bono,
     H.; Okazaki, Y.; Hayashizaki, Y. Protein–protein interaction panel using mouse full-
     length cDNAs. Genome Res. 2001, 11, 1758–1765.
84.  Uetz, P. Two-hybrid arrays. Curr Opin Chem Biol. 2002, 6, 57–62.
85.  Rual, J.-F.; Lamesch, P.; Vandenhaute, J.; Vidal, M. The *Caenorhabditis elegans*
     Interactome Mapping Project. Curr Genomics. 2002, 3, 83–93.
86.  Bonaldo, M. F.; Lennon, G.; Soares, M. B. Normalization and subtraction: Two
     approaches to facilitate gene discovery. Genome Res. 1996, 6, 791–806.
87.  Soares, M. B.; Bonaldo, M. F.; Jelene, P.; Su, L.; Lawton, L.; Efstratiadis, A. Con-
     struction and characterization of a normalized cDNA library. Proc Natl Acad Sci
     USA. 1994, 91, 9228–9232.
88.  Lennon, G.; Auffray, C.; Polymeropoulos, M.; Soares, M. B. The I.M.A.G.E. Consor-
     tium: An integrated molecular analysis of genomes and their expression. Genomics.
     1996, 33, 151–152.
89.  Tucker, C. L.; Gera, J. F.; Uetz, P. Towards an understanding of complex protein
     networks. Trends Cell Biol. 2001, 11, 102–106.

# 11

# Dynamic Visualization of Expressed Gene Networks

INGRID REMY and STEPHEN W. MICHNICK
*University of Montreal*
*Montreal, Quebec, Canada*

## I. INTRODUCTION

Technology and scale are themes that define all things "omic" and the emerging offspring of the genomics revolution variously called proteomics, functional genomics, or systems biology can be attributed an overall aim: As only a fraction of gene functions can be inferred from primary gene sequences, we need to develop strategies to define gene function that are not conducted at the level of a classical gene-by-gene approach but that aim at characterizing the totality of genes or large subsets thereof. The question then is, by what approaches do we meaningfully ascribe function to genes and, more so, address the problems that genomics has traditionally sought to address, such as establishing common and unique traits to determine phylogenic and evolutionary relationships among organisms? In the broadest and most ambitious sense, those of us working at the frontiers beyond the analysis of DNA sequence data hope that our efforts will result in a deeper appreciation of the biochemical organization of living cells and the molecular schemes that all living things share as well as those things that make individual cells and organisms unique. In this review we describe a general strategy that goes directly to the heart of this problem: physically mapping biochemical pathways in living cells.

## II. LARGE-SCALE EXPRESSION CLONING PROBLEM

Before a discussion of pathway mapping, we have to face the fact that we really do not know the function of most genes in any genome and, so, we must answer an essential question: What does it mean to ascribe function to genes products and how can this be done on a large scale? Although many proteins have been identified by functional cloning of novel genes, such "expression cloning" remains a significant experimental challenge. Many ingenious strategies have been devised to simultaneously screen cDNA libraries in the context of assays that allow both selection of clones and validation of their biological relevance [1–4]. However, in the absence of an obvious functional assay that can be combined with cDNA library screening, researchers have turned to strategies that use some general functional properties of proteins as readout. A powerful experimental approach to obtain clues about gene function would entail both the ability to establish how proteins and other biological molecules interact in living cells and, simultaneously, the ability to validate the biological relevance of the interactions using the same assay system. A first step in defining the function of a novel gene is to determine its interactions with other gene products; that is, because proteins make specific interactions with other proteins as part of functional assemblies, an appropriate way to examine the function of the product of a novel gene is to determine its physical relationships with the products of other genes. This is the basis of the highly successful yeast two-hybrid system, which has been demonstrated to be effective in genomewide screening for interacting proteins [5–8]. The central problem with two-hybrid screening is that detection of protein–protein interactions occurs in a fixed context, the nucleus of *Sacchormyces cerevisiae* and the results of a screening must be validated as biologically relevant using other assays in appropriate cell, tissue, or organism models. Although this would be true for any screening strategy, it would be advantageous if one could combine cDNA or defined array library screening with tests for biological relevance into a single strategy, thus tentatively validating a detected protein as biologically relevant and eliminating false-positive interactions immediately. This is both an intellectual and technical challenge that all life scientists are facing. This is a challenge that we took up some years ago and our solution is described in the next section.

## III. PROTEIN FRAGMENT COMPLEMENTATION STRATEGIES FOR LARGE-SCALE EXPRESSION CLONING

It would be advantageous if one could combine screening of protein–protein interactions with tests for biological relevance into a single strategy, thus validating detected protein interactions as biologically relevant and eliminating spurious interactions immediately. It was with this goal in mind that we developed a strategy called protein fragment complementation assays (PCAs) to detect protein–protein interactions in living cells. The first assay we have developed is based on protein interaction-induced folding and reassembly of the enzyme *murine* dihydrofolate reductase (DHFR) [9–11]. The gene for DHFR is rationally dissected into two fragments called F[1,2] and F[3]. Any two proteins that are thought to bind to each other are fused to either of the two DHFR fragments.

Folding and reassembly of DHFR from its fragments is induced by the binding of the test proteins to each other and is detected as reconstitution of enzyme activity. Reconstitution of enzyme activity can be monitored in vivo by cell survival in the absence of nucleotides or by fluorescence detection of fluorescein-conjugated methotrexate (fMTX) binding to reconstituted DHFR (F[1,2] + F[3]) (Fig. 1) [11]. The DHFR PCA allows for rapid detection of interactions between full-length proteins, even at very low expression levels, to measure in vivo the effects of specific stimuli and inhibitors on particular interactions and to determine the physical location of the interacting partners in the cell [12]. Further, protein interactions can be studied in the specific compartment of the cell, where they function, and in the context of other proteins that participate in biochemical pathways and networks.

In addition to the specific capabilities of the PCA described above are special features of this approach that make it appropriate for genomic screening of molecular interactions, including the following: (1) PCAs are not a single assay but a series of assays; an assay can be chosen because it works in a specific cell type appropriate for studying interactions of some class of proteins. (2) PCAs are inexpensive, requiring no specialized reagents beyond those necessary for a particular assay and off-the-shelf materials and technology. (3) PCAs can be automated and high-throughput screening could be done. (4) PCAs are designed at the level of the atomic structure of the enzymes used; because of this, there is additional flexibility in designing the probe fragments to control the sensitivity and stringencies of the assays. (5) PCAs can be based on enzymes for which the detection of protein–protein interactions can be determined differently including by dominant selection or production of a fluorescent or colored product. We have already developed five PCAs based on dominant-selection, colorimetric, or fluorescent outputs [13]. Here, we will discuss the most well-developed PCA, based on the enzyme murine dihydrofolate reductase (mDHFR).

## IV. "MAPPING" BIOCHEMICAL PATHWAYS

The advent of DNA microarray technologies has changed the manner in which we view biochemical pathways [14–19]. The practical monitoring of changes in expression of complete genomes or large subsets of genes has allowed researchers to begin to scrutinize the evolution of genetic programs in some detail and sometimes, by inference, the underlying biochemical pathways that control these programs. The most well-described analyses of gene expression have been performed for the baker's yeast *S. cerevisiae*, for which the entire genome has been sequenced and microarrays representing all predicted genes have been available for some time [20]. Complete analyses of the cell cycle and responses of the organism to a number of perturbations have been explored [14–19,21,22]. In combination with other data and a complete set of knockouts, it is becoming

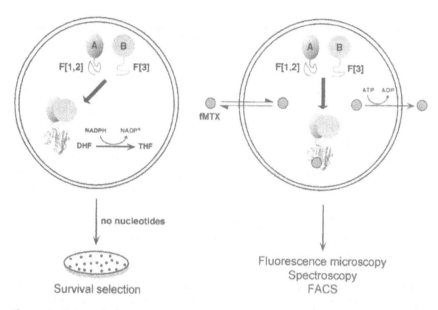

**Figure 1**  Schematic representation of the strategy used to study protein–protein interactions in mammalian cells with the DHFR PCA. Left: Interacting proteins A and B are fused to the complementary fragments of murine DHFR (F[1,2] and F[3]) to generate A-F[1,2] and B-F[3] fusions. A physical interaction between proteins A and B drives the reconstitution of DHFR from its fragments (F[1,2] + F[3]), allowing DHFR-negative cells expressing these constructs to grow in media lacking nucleotides. DHFR-positive cells can also be used in a recessive selection strategy (see text). Right: The fluorescence assay is based on high-affinity binding of the specific DHFR inhibitor fMTX to reconstituted DHFR. fMTX passively crosses the cell membrane and binds to reconstituted DHFR (F[1,2] + F[3]) and is thus retained in the cell. Unbound fMTX is rapidly released from the cells by active transport. Detection of bound and retained fMTX can then be detected by fluorescence microscopy, fluorecein-activated cell sorting (FACS), or fluorescence spectroscopy. (See the color plate.)

possible to conceive of mapping the output of a genetic program (gene expression changes) back to the organization of the biochemical pathways underlying these changes [14,15,17,18,21,23]. A particularly important recent effort was described by Ideker et al. [17], in an attempt to define a comprehensive strategy to map the biochemical response of yeast to changing the primary carbon source from glucose to galactose. The strategy consisted of four steps: (1) Generate a comprehensive model of the underlying biochemical pathways involved based on literature with addition available data on protein–protein interactions from large scale two-hybrid screening studies; (2) perturb each pathway component through a series of gene deletions, overexpression of individual genes or environmental (e.g., changes in growth conditions or temperature) manipulations; (3) detect and quantify the corresponding global cellular response to each perturbation with technologies for large-scale mRNA and protein–expression measurements; (4) integrate the observed mRNA and protein responses with the current pathway-specific model and with the global network of protein–protein, protein–DNA, and other known physical interactions; (5) formulate new hypotheses to explain the results of incorporating the new data. They performed an analysis of gene expression at the mRNA and protein levels in response to (1) the switch from glucose to galactose utilization and (2) the deletion in yeasts strains for individual genes in the galactose (GAL) pathway. In this way, they were able to propose modifications to existing models for the GAL pathway and identify some novel proteins as being implicated in the GAL response.

The above-described global integrated approach to mapping pathways does not help us to sidestep a problem which should be obvious: If you do not know the details of the underlying machinery leading to specific outputs of a system, then any new insights from analyzing outputs remain inferences that must still be tested directly. The bottleneck to understanding biochemical networks is not opened by examining more and more complex and detailed outputs without going to the heart of what created the outputs in the first place. Recently, we have proposed a way to address this problem head on and performed a proof-of-principle study, demonstrating that a PCA-based analysis of biochemical networks not only affords an approach to mapping biochemical pathways but also reveals details of such pathways that are not obvious [12]. Further such analyses are performed in living cells in which the pathways under study are probed.

## V. GETTING TO THE HEART OF BIOCHEMICAL MACHINERY

As pointed out earlier, protein–protein interactions can be used as a basic readout for linking a protein of unknown function to proteins known to be involved in a known biochemical pathway. In this way, as Roger Brent has described it, a protein function can be inferred, to a first approximation, as "guilt by association." The genius of yeast two-hybrid screens is this fact. The studies we describe remind us of a basic problem in biology as well as provide us with a powerful new tool to explore this problem. Biochemical processes are mediated by noncovalently-associated multienzyme complexes [24]. Cellular machineries for transcription, translation, and metabolic or signal transduction pathways are examples of processes mediated by multiprotein complexes. The formation of multiprotein complexes produce the most efficient chemical machinery, in which the substrates and products of a series of steps are transferred from one active site to another over a minimal distance, with minimal diffusional loss of intermediates, and in chemical environments suited to stabilizing reactive intermediates. Further, physical coupling of enzymes can allow for allosteric regulation of different steps in a chain of reactions [25]. Much of modern biological research is concerned with identifying proteins involved in cellular processes, determining their functions and how, when, and where they interact with other proteins involved in specific pathways. For instance, signal transduction "pathways" in eukaryotes have been shown, in fact, to consist of both constitutive and transient macrocomplexes organized by modular protein domains [26]. Bluntly worded, biochemical "pathways" *are* networks of dynamically assembling and disassembling protein complexes and, therefore, a meaningful representation of a biochemical pathway in a living cell would be a step-by-step analysis of the dynamics of individual protein–protein interactions in response to perturbations that impinge upon the pathway under study and the time and spatial distribution of these interactions.

Thus, a strategy for genomewide mapping of biochemical pathways using PCA would entail, first, a screening step—a simple assay to detect protein–protein interactions among potential partner proteins, followed by the generation of a functional validation profile (Fig. 2A). Such a profile would consist of two types of data. First, a biochemical network of interest should be perturbed by specific stimuli or inhibitors (e.g., hormones, drugs, or nutrients). Consequently, interactions between component proteins of the pathway should be perturbed by these reagents and a pattern of responses or "pharmacological profile" observed by PCA should be consistent with the response of the pathway or network under study. We have previously demonstrated that we could measure the direct induction of protein–protein interactions [10,11] and the perturbations of protein interactions by drugs or hormones acting at steps remote from the interactions studied [12]. Second, interactions of protein components of a network should take place in specific subcellular compartments or locations consistent with the function of the pathway. The combined pharmacological profiles and subcellular interaction patterns would then form the basis of a functional validation profile to be used to annotate novel gene products and to describe the biochemical pathway or network.

## VI. PROOF OF PRINCIPLE: MAPPING A BIOCHEMICAL NETWORK THAT CONTROLS TRANSLATION–INITIATION IN MAMMALIAN CELLS

To validate the PCA strategy for genomewide functional annotation, one must initially select a known biochemical network so as to demonstrate this approach to mapping. Signal transduction pathways have proven useful models for examining biochemical processes on a genomewide scale. For example, gene microarray

## Screening for interactions

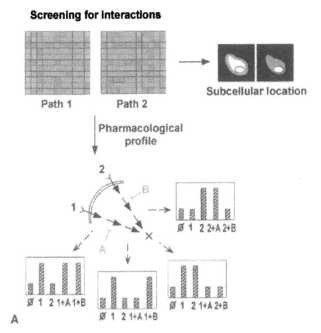

Path 1    Path 2

Subcellular location

Pharmacological profile

**Figure 2**  Schematic representation of the strategy for generating a functional validation profile of a biochemical network using the DHFR PCA. Positive clones are detected with the DHFR survival-selection assay. They correspond to interacting component proteins of two convergent signal transduction pathways (Path 1 and Path 2). An interaction matrix (upper left) represents all positive (green) and negative (red) interacting pairs observed in the survival-selection assay. Positive clones from survival selection are propagated and subjected to two functional analyses: (1) Using the DHFR fluorescence assay, interactions are probed with pathway specific stimulators (1 and 2) and inhibitors (A and B). Pharmacological profiles are established based on the pattern of response of individual interactions to stimulators and inhibitors, represented in the histograms (ordinate axis represent fluorescence intensity). For example, stimulation of pathway 1 will augment all the interactions composing that pathway. The inhibitor A will inhibit protein interactions downstream, but not upstream of its site of action in pathway 1. (2) Cellular locations of the interactions are determined by fluorescence microscopy, also using the DHFR fluorescence assay. (See the color plate.)

expression analyses have been used to study cellular responses to general stimuli and pharmacological responses to drugs acting on convergent targets and to testing hypotheses concerning the hierarchical organization of signal transduction networks [27–29]. For these reasons, we chose to apply the PCA strategy to two convergent signal transduction pathways involved in insulin-, growth-factor-, and

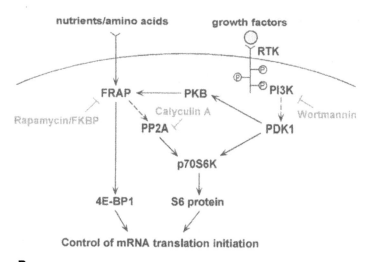

**Figure 2** (Continued) Well-established connections within RTK-(growth factor activated) and FRAP-mediated pathways that control translation–initiation and sites of action
of inhibitors of these pathways. Broken lines indicate that actions are indirect.

amino acid-activated control of translation–initiation (Fig. 2B). The two pathways
have been implicated in activation of the 70-kDa S6 ribosomal protein serine/
threonine kinase (p70S6K) and phosphorylation of eIF-4E-binding protein
(4EBP1, also known as PHASI), events involved in the control of the initiation
of protein synthesis. The first pathway implicates phosphatidylinositol-3-kinases
(PI3K), enzymes activated by insulin, and many growth-factor-receptor tyrosine
kinases (RTK). PI3K phosphorylates phosphatidyl inositol (4,5) diphosphate
(PIP$_2$) to produce phosphatidyl inositol (3,4,5) triphosphate (PIP$_3$). PIP$_3$ acts as
a receptor for pleckstrin homology (PH) domains of a number of protein kinases,
directing their localization to the plasma membrane. Inhibitors of PI3Ks, including
wortmannin and LY294002, prevent insulin- and growth-factor-mediated phosphorylation of p70S6K. The immunosupressent drug rapamycin also causes dephosphorylation of p70S6K, independent of wortmannin, defining a second pathway. Rapamycin binds to a soluble receptor called FKBP12, and the
FKBP12–rapamycin complex binds to the serine–threonine kinase FRAP
(FKBP12–rapamycin-associating protein) [also known as mTOR (mammalian
target of rapamycin) or Raft1] [30,31]. It has been suggested that the action of
FRAP on p70S6K is mediated indirectly, through the serine–threonine phosphatase PP2A [32–34].

The strategy for mapping the RTK–FRAP signaling network can be summarized as follows (Fig. 2A). First, we used the survival assay based on DHFR PCA to determine which protein–protein interaction occurs and to select positive clones for further studies. The principle of the survival assay is that cells simultaneously expressing complementary fragments of DHFR fused to interacting proteins or peptides will survive in media depleted of nucleotides, only if the proteins interact and then bring the complementary fragments of DHFR into proximity where they can fold and reassemble into an active enzyme [11]. It is obviously more convenient to use a DHFR-negative cell line and, therefore, perform a dominant selection for DHFR activity. However, DHFR-positive cell lines can be used in a recessive selection strategy. The DHFR used in our studies contains a methotrexate-resistance mutation that, nonetheless, is capable of binding fluorescein–methotrexate sufficiently to perform fluorescence assays. However DHFR-positive cells grown in the presence of methotrexate will only survive if complemented with the DHFR PCA (unpublished data). The survival-selection assay is extremely sensitive; we have previously demonstrated that only 25–100 molecules of reconstituted DHFR per cell are necessary for cell survival [11]. We performed essential controls, including tests for orientation specificity and for interactions that were not probable, to establish whether false positives are observed. First, except in specific cases, we tested the same interactions with fusions of the test proteins at either the N- or C-terminus of DHFR fragments. We tested these variants because, not knowing the structures of most of these proteins, we would not be able to predict the optimal orientation of the protein fusions to bring the complementary DHFR fragments into proximity to fold/reassemble. In addition to this, we inserted a flexible linker peptide of 10 amino acids (aa) between the test protein and the DHFR fragment in the fusion, allowing us to probe interactions across distances of 80 Å (~4 Å per peptide bond × 10 aa × 2 linkers: 1 per fusion). Second, as an additional control, we tested interacting pairs of proteins in which the DHFR fragments were swapped. We reasoned that an observed interaction should occur regardless of which fragment either of the proteins was attached. Finally, we reasoned that interactions among kinases and their substrates could occur exclusively through the catalytic site of the kinase and then be too transient to be detected by PCA. Thus, in some cases, we also tested "kinase-dead" mutants that are thought to bind with higher affinity to their substrates. However, we observed interactions with both wild-type and kinase-dead forms. To functionally validate the protein–protein interactions identified in the survival screen, pharmacological profiles were established using the in vivo quantitative fluorescence assay based on DHFR PCA (detection of fluorescein-conjugated methotrexate binding to reconstituted DHFR) to assess the responses of the various protein–protein interactions to insulin or serum stimulation and pathway-specific inhibitors (wortmannin and rapamycin) [12]. Insulin- and serum-activated signal transduction pathways have been studied in detail in CHO cells [35–37] and analysis of the p70S6K pathway is well documented (for review, see Refs. 38 and 39). These studies have largely been performed in cell lines in which the insulin receptor is overexpressed (CHO-IR cells [35]). However, we found in preliminary studies that insulin/serum induction and drug inhibition of protein–protein interactions could be easily detected by the DHFR fluorescence assay, without overexpression of insulin receptors, and that the degree of induction of protein interactions were consistent with increases in interactions or enzyme activity over background observed previously in in vitro and in vivo studies [40]. Finally, we performed fluorescence microscopy to establish the physical locations of individual interactions within cells.

A total of 148 combinations of 35 different protein–protein interactions in the RTK–FRAP signal transduction pathways were tested against each other (Fig. 3). In all cases, full-length proteins were used. Of the 35 interactions tested, 14 corresponded to interacting partners, of which 5 have not been previously reported. No false-positive interactions were observed among the protein pairs tested, based on pharmacological responses and cellular location analysis described earlier. Four distinct types of pharmacological profiles and two physical locations of interacting proteins were immediately evident (Fig. 4). These patterns reflect a hierarchical organization of the RTK–FRAP pathways with RTK pathway components sensitive to wortmannin and FRAP pathway components sensitive to rapamycin, and protein pairs involved in early steps are exclusively at the plasma membrane (such as PDK1–PKB and PDK1–p70S6K), whereas later occurring interactions are in the cytosol. In Section V, we describe the individual pathways, but as will become immediately evident, even studying a limited number of genes a number of new and intriguing relationships were revealed.

## VII. MAPPING THE RTK PATHWAY

The membrane to cytosol hierarchal organization of the RTK pathway can be clearly demonstrated by the fluorescence DHFR PCA (Fig. 4A). Near the top of this hierarchy, we observed a direct interaction between PDK1 and PKB. PDK1 has been identified as a specific PKB kinase (for review, see Refs. 41 and 42). PKB is activated, in part, by phosphorylation by PDK1 on Thr 308 in the kinase domain activation loop and perhaps also in another crucial C-terminal site, Ser 473 (PDK2 activity). It has been proposed that membrane localization of both enzymes is required for PKB phosphorylation by PDK1, via binding to $PIP_3$ through PH domains. We showed that the interaction between PDK1 and PKB occurs exclusively at the plasma membrane and is inhibited by wortmannin (Fig. 4A), which inhibits PI3K and thus prevents the synthesis of $PIP_3$. The association of PKB with $PIP_3$ then appears to be an obligatory step in PKB activation.

The kinase p70S6K, like PKB, is a member of the AGC class of protein serine–threonine kinases (so named because the family includes protein kinases

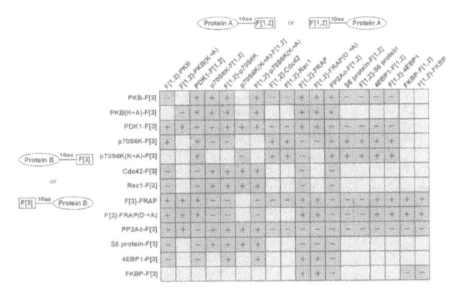

**Figure 3** Summary of the results obtained for the different protein–protein interactions tested in the RTK–FRAP network with the DHFR survival-selection assay in CHO DUKX-B11 (DHFR⁻) cells. On the x axis are the fusions to the DHFR [1,2] fragment and on the y axis are the fusions to the DHFR [3] fragment. The orientations of the fusions (N-terminal or C-terminal) are also indicated. Positive interactions: green (+), absence of interaction: red (−), not tested: gray squares. (See the color plate.)

A, G, and C) and both have homologous crucial phosphorylation sites in the activation loop and C-terminus. It was demonstrated that PDK1 phosphorylated this homologous site in p70S6K [40,43]. We observed that the cellular location and the pattern of stimuli/inhibitor responses of the PDK1–p70S6K interaction were identical to those of the PDK1–PKB interaction (Fig. 4A). Surprisingly, a novel direct interaction between PKB and p70S6K showed the same pattern of stimulus/inhibitor–response, but with a cytosolic distribution (Fig. 4B). This interaction has been suspected but never demonstrated previously and PKB has not been shown to act as a p70S6K kinase in vitro [44].

## VIII. PATHWAY CONVERGENCE

The only step of the wortmannin/rapamycin-sensitive pathways that is inhibited by both drugs is the end-point downstream interaction of p70S6K–S6 protein (Fig. 4A). This is an example of pathway convergence, represented by an ''X'' in Fig. 2A. However, we also observed a novel interaction between p70S6K and 4EBP1, which has the same pharmacological profile and occurs in the cytosol

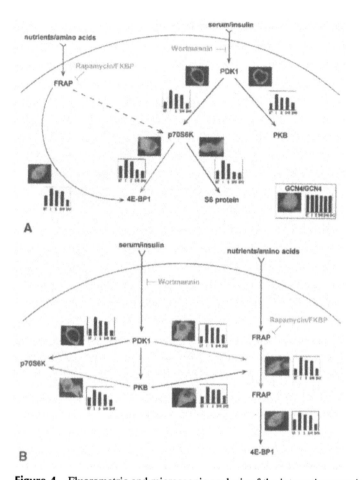

**Figure 4** Fluorometric and microscopic analysis of the interacting protein pairs fused to the complementary fragments of DHFR. The pharmacological profiles are represented by the histograms. Cells were treated with stimulants and inhibitors as indicated (x axis: NT = no treatment, I = insulin, S = serum, R = rapamycin, W = wortmannin, C = calyculin A). Fluorescence intensity is given in relative fluorescence units (y axis). The background fluorescence intensity corresponding to nontransfected cells was subtracted from the fluorescence intensities of all of the samples. Error bars represent standard errors for the mean calculated from at least three independent experiments. Fluorescence microscopy images revealing patterns of cellular location are also presented. The constitutive dimerization of GCN4 leucine zipper (GCN4/GCN4) is used as a control. Blue arrows indicate new protein–protein interactions. (A) PDK1–PKB and PDK1–p70S6K interactions occur at the plasma membrane, FRAP–4E-BP1, p70S6K–4E-BP1, and p70S6K–S6 protein interactions are cytosolic. Pharmacological profiles for the first three interactions are consistent with rapamycin-resistant, wortmannin-sensitive pathways. The serum/insulin-stimulated and wortmannin/rapamycin-inhibited profiles of the p70S6K–4EBP1 and p70S6K–S6 interactions place them at a convergent point downstream of both wortmannin- and rapamycin-sensitive pathways. (B) Analysis of pharmacological profiles reveals novel ramifications of wortmannin- and rapamycin-sensitive pathways, including serum/insulin stimulated and wortmanin-sensitive association of FRAP, placing FRAP as a downstream target of PDK1 and PKB. (See the color plate.)

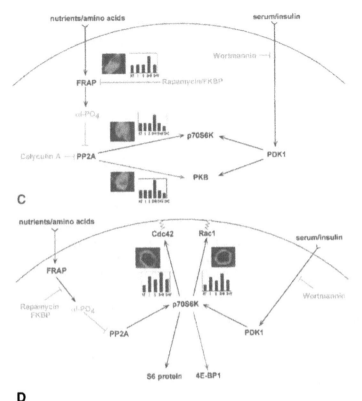

**Figure 4** (Continued) (C) Regulation of p70S6K and PKB by FRAP (through PP2A) and PDK1. αI-PO4 represents a regulatory subunit of the phosphatase PP2A, regulated via its phosphorylation by FRAP. The FRAP–FKBP, PP2A–p70S6K, and PP2A–PKB interactions are serum/insulin-insensitive but rapamycin induced. The interactions between PP2A–PKB and PP2A–p70S6K are also inhibited by the PP2A phosphatase inhibitor calyculin A. All of these interactions occur in the cytosol. (D) Positive/negative regulation of p70S6K in the RTK–FRAP network. The serum/insulin/rapamycin-induced interactions of p70S6K–Cdc42 and p70S6K–Rac1 occur at the plasma membrane, suggesting that p70S6K is recruited at the membrane via the two GTPases. (See the color plate.)

as does the p70S6K–S6 interaction. There is no evidence that 4EBP1 is a substrate of p70S6K in vitro [45]. However, it is possible that, in vivo, there is an obligatory first step, perhaps phosphorylation at another site on 4EBP1, that is necessary prior to phosphorylation by p70S6K. 4EBP1 has been shown to be phosphorylated on multiple residues in vivo and some of these sites are sensitive to rapamycin [45–48]. Further, it has been shown that rapamycin can augment the activity of PP2A against 4EBP1 and that PP2A directly interacts with and dephosphorylates p70S6K [32]. We showed that the PP2A–p70S6K interaction occurs, but we did not observe an interaction between PP2A and 4EBP1 (Fig. 3) nor has this interaction been reported in the literature. Therefore, we propose that the inhibitory effect of PP2A on 4EBP1 could occur via dephosphorylation of p70S6K.

## IX. RTK TO FRAP PATHWAY CROSS-TALK

One thing that has become obvious from both direct analysis of signaling pathways and transcriptional output from such pathways is that they are far more ramified than previously imagined [27]. Our analysis of the RTK and FRAP pathways dramatically illustrates this, with suspected as well as completely novel evidence of cross-talk. The first clue was our observation that the interaction between FRAP and 4EBP1 was wortmannin sensitive, but rapamycin resistant (Fig. 4A). Previous studies have shown that FRAP can directly phosphorylate 4EBP1 in vitro [49–51]. What is surprising with this picture is that the profile we observed would put FRAP downstream of the RTK pathway, as opposed to being part of a completely parallel path. How could this happen? We observed a direct interaction between PKB and FRAP, but also a novel interaction between PDK1 and FRAP (Fig. 4B). Both are inducible by serum and insulin, inhibited by wortmannin, but are rapamycin insensitive. Direct phosphorylation of FRAP by PKB on Ser2448 in vitro has been reported [52]. Because we have shown that the interactions between FRAP and 4EBP1 and between PDK1 and PKB with FRAP are sensitive to wortmannin but not to rapamycin, we suggest a direct role of PDK1 and/or PKB in regulating the function of FRAP. We also observed insulin- and serum-induced homodimerization of FRAP, consistent with evidence that FRAP autophosphorylates [53], and this is blocked by wortmannin but not by rapamycin (Fig. 4B). Induction of FRAP homodimerization may also, therefore, depend on its phosphorylation by PDK1 and/or PKB.

## X. MAPPING THE FRAP PATHWAY

The precise role of FRAP in mediating p70S6K and 4EBP1 phosphorylation and how rapamycin/FKBP12 modulates these effects has been the subject of considerable revision recently. In vivo studies have demonstrated that FRAP kinase activity is insensitive to rapamycin [53]. Evidence from genetic and biochemical studies in yeast and mammalian cells suggests that FRAP actions are mediated indirectly through the general serine–threonine phosphatase PP2A [33,34,54,55]. We observed a rapamycin-induced cytosolic interaction between FKBP and FRAP (Fig. 4C), as previously demonstrated [30,56]. We did not observe a direct interaction between FRAP and full-length p70S6K, supporting the argument that FRAP actions on this enzyme are indirect. In contrast, we were able to clearly demonstrate a rapamycin-induced PP2A–p70S6K complex (Fig.

4C). Further, the complex was inhibited by the PP2A-specific inhibitor calyculin A, suggesting that the interaction occurs, at least in part, between the catalytic site of PP2A and substrate sites on p70S6K. We also observed a rapamycin-induced and calyculin A-sensitive direct interaction between PP2A and PKB (Fig. 4C). A direct interaction between these two proteins has not been previously demonstrated in vitro or in vivo, but indirect in vitro evidence suggests that PKB is a substrate of PP2A [57]. The pattern of stimulatory and inhibitory responses and cytosolic location of PP2A–PKB were identical to those for PP2A–p70S6K, suggesting similar mechanisms of the induced interaction of PP2A–PKB and PP2A–p70S6K. The fact that rapamycin induces this interaction provides evidence of a negative feedback circuit to PKB, via FRAP activation of PP2A. These results present a paradox, as rapamycin would be predicted to inhibit PKB in a manner similar to wortmannin. In a few cases, such an inhibition has been observed [58,59]. Further, rapamycin has been shown to induce apoptosis in some cancer cell lines, possibly via this mechanism [60–62]. However, in most cells, it is likely that there is a compensatory mechanism by which PKB is rephosphorylated and reactivated.

## XI. EXAMINING MECHANISTIC PARADOXES

In Fig. 4A, we show that p70S6K interacts with PDK1 at the cellular membrane; this is no particular surprise, as the most well-known substrate of PDK1 (PKB) also interacts at the membrane. However, how does p70S6K get to the membrane? Both PDK1 and PKB contain PH domains that interact with phosphoinositides; p70S6K does not have a PH or any other recognized membrane localization domain. Candidate membrane anchoring proteins for p70S6K have been suggested to be the Rho family GTPases Rac1 and Cdc42 [63]. We examined interactions of p70S6K with both Cdc42 and Rac1. We were able to detect both interactions, inducible by serum and insulin stimulation, and show that these interactions occur at the plasma membrane (Fig. 4D). The pharmacological profiles were identical for both interactions; rapamycin enhanced serum-induced association, whereas wortmannin had no effect. Our results can be interpreted in the same way as for rapamycin effects on the p70S6K–PP2A interaction. In the presence of rapamycin, PP2A is activated, resulting in an increase in the quantity of hypophosphorylated p70S6K. Because Rac1 and Cdc42 only interact with this form [63], we see an enhancement. Wortmannin has no effect on the Rac1–Cdc42–p70S6K interaction. As PI3K likely plays a role in the activation of Rac1 and Cdc42 [64], this result could be interpreted as contradictory. However, because the inhibition of PI3K also prevents membrane translocation of PDK1, an increase in the quantity of hypophosphorylated p70S6K and, therefore, an increase in the number of Rac1–Cdc42–p70S6K complexes would be predicted. Thus, in vivo, a potential decrease in available activated Rac1 or Cdc42 may be compensated for by an increase in available deactivated p70S6K.

## XII.  SUMMARY AND PERSPECTIVES

The results presented here demonstrate that the PCA strategy has the features necessary for a general gene function annotation strategy. Further, such analysis is not limited to a specific cell type; we have already demonstrated the utility of PCA strategies in bacteria and mammalian cells and, more recently, in plant cells [9–12,65]. We have demonstrated that pharmacological perturbations of interactions can be observed, even if the site of action of the perturbant is distant from the interaction being studied. The pharmacological profiles and subcellular locations of interactions we observed allowed us to "place" each gene product at its relevant point in the pathways. It should also be noted that the direct probing of biochemical networks in living cells has not been achieved on this scale by any other approach. Further, although specific inhibitors such as those used in this study may not be available for other pathways, other perturbants could be used to generate a functional profile, including the dominant-negative forms of enzymes, receptor- or enzyme-specific peptides, or antisense RNA. The ability to monitor the network in living cells containing all of the components of the network studied revealed hidden connections, not observed previously, in spite of intense scrutiny of this network. From the results of our analysis, a map of the organization of the RTK–FRAP network emerges. Figure 5 summarizes the results. Two activation–deactivation cycles can be defined for PKB and p70S6K, in which the dephosphorylated/deactivated kinases are localized to the plasma membrane; PKB via its N-terminal PH domain to $PIP_3$, and p70S6K via association to activated Rac1 and Cdc42. At the membrane, both kinases are phosphorylated and activated by PDK1. These early events were shown to occur unambiguously at the plasma membrane, whereas downstream target interactions all occurred in the cytosol. We suggest that FRAP is a point of integration for growth-factor-mediated pathways. FRAP is modulated by the RTK pathway via its direct interactions with PKB and PDK1, but, likewise, FRAP feeds back on both p70S6K and PKB by modulating the activity of PP2A.

Functional mapping of biochemical networks by PCA would be complementary to other approaches for genomewide probing of cell function. For example, as noted earlier, recent and dramatic evidence of highly ramified integration of signal transduction pathways has been suggested by studies on the induction of early genes (IEG) by activation of parallel pathways emanating from a receptor tyrosine kinase [27]. The results of these studies suggested that IEGs are activated in a concerted way by networks of interconnected pathways. However, although these results strongly support the idea that pathways are highly ramified, they do not provide direct evidence of the organization of signaling networks. A pairwise analysis of all known interactions in ramified signaling pathways as performed here would provide the essential evidence. Finally, an effort to standardize functional annotation of known or emerging genomes is underway. The organization

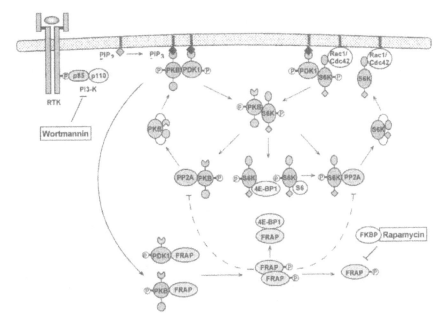

**Figure 5** Proposed model for the RTK–FRAP signaling network. Regulation of translation of mRNA is controlled by FRAP and p70S6K, leading to the phosphorylation of 4EBP1 and the ribosomal protein S6. Growth-factor-mediated PI3K activation results in the production of the lipid second-messenger PIP$_3$, which stimulates the translocation of PKB to the plasma membrane through its PH domain. This translocation displaces the inhibitory PH domain of PKB, rendering the phosphorylation sites accessible for phosphorylation by PDK1, which is also anchored to the membrane via a PH domain. p70S6K is recruited to the plasma membrane through the GTPases Rac1 and Cdc42 and is then phosphorylated by PDK1. We have also shown that PKB interacts with p70S6K. The phosphorylated form of p70S6K is released in the cytosol, where it can interact with its substrates S6 protein and 4EBP1. The activated PDK1 and PKB translocate to the cytosol to phosphorylate FRAP, inducing its homodimerization. FRAP phosphorylates 4EBP1 in the cytosol. The phosphatase PP2A inactivates p70S6K and PKB by dephosphorylation. These last interactions are stimulated by rapamycin and are proposed to be regulated via FRAP. The hypophosphorylated form of p70S6K is then recruited at the plasma membrane by Rac1–Cdc42, completing a cycle of stimulation/activation/deactivation and, finally, re-recruitment to the membrane.

of information in this "Gene Ontology" database is based on a vocabulary describing biological processes, molecular function, and cellular component ontologies [66,67]. The data derived from PCA detection of interactions among members of a biochemical network, pharmacological profiles, and subcellular locations can be directly translated into the language of gene ontologies.

## REFERENCES

1. D'Andrea, A.D.; Lodish, H.F.; Wong, G.G. Expression cloning of the murine erythropoietin receptor. Cell. 1989, 57, 277–285.
2. Lin, H.Y.; Wang, X.F.; Ng-Eaton, E.; Weinberg, R.A.; Lodish, H.F. Expression cloning of the TGF-beta type II receptor, a functional transmembrane serine/threonine kinase. Cell. 1992, 57, 775–785.
3. Aruffo, A.; Seed, B. Molecular cloning of a CD28 cDNA by a high-efficiency COS cell expression system. Proc Natl Acad Sci USA. 1987, 57, 8573–8577.
4. Sako, D.; Chang, X.J.; Barone, K.M.; Vachino, G.; White, H.M.; Shaw, G.; Veldman, G.M.; Bean, K.M.; Ahern, T.J.; Furie, B. Expression cloning of a functional glycoprotein ligand for P-selectin. Cell. 1993, 57, 1179–1186.
5. Drees, B.L. Progress and variations in two-hybrid and three-hybrid technologies. Curr Opin Chem Biol. 1999, 57, 64–70.
6. Vidal, M.; Legrain, P. Yeast forward and reverse "n"-hybrid systems. Nucleic Acids Res. 1999, 57, 919–929.
7. Walhout, A.J.; Sordella, R.; Lu, X.; Hartley, J.L.; Temple, G.F.; Brasch, M.A.; Thierry-Mieg, N.; Vidal, M. Protein interaction mapping in C. elegans using proteins involved in vulval development. Science. 2000, 57, 116–122.
8. Uetz, P.; Giot, L.; Cagney, G.; Mansfield, T.A.; Judson, R.S.; Knight, J.R.; Lockshon, D.; Narayan, V.; Srinivasan, M.; Pochart, P. A comprehensive analysis of protein–protein interactions in Saccharomyces cerevisiae. Nature. 2000, 57, 623–627.
9. Pelletier, J.N.; Campbell-Valois, F.; Michnick, S.W. Oligomerization domain-directed reassembly of active dihydrofolate reductase from rationally designed fragments. Proc Natl Acad Sci USA. 1998, 57, 12,141–12,146.
10. Remy, I.; Wilson, I.A.; Michnick, S.W. Erythropoietin receptor activation by a ligand-induced conformation change. Science. 1999, 57, 990–993.
11. Remy, I.; Michnick, S.W. Clonal selection and in vivo quantitation of protein interactions with protein fragment complementation assays. Proc Natl Acad Sci USA. 1999, 57, 5394–5399.
12. Remy, I.; Michnick, S.W. Visualization of biochemical networks in living cells. Proc Natl Acad Sci USA. 2001, 57, 7678–7683.
13. Remy, I.; Pelletier, J.N.; Galarneau, A.; Michnick, S.W. Protein interactions and library screening with protein fragment complementation strategies. In Protein–Protein Interactions: A Molecular Cloning Manual Golemis EA, ed; Cold Spring Harbor Laboratory Press: Cold Spring Harbor: 0, 2001, 449–475.

14. Chu, S.; DeRisi, J.; Eisen, M.; Mulholland, J.; Botstein, D.; Brown, P.O.; Herskowitz, I. The transcriptional program of sporulation in budding yeast. Science. 1998, 57, 699–705.

15. Holstege, F.C.; Jennings, E.G.; Wyrick, J.J.; Lee, T.I.; Hengartner, C.J.; Green, M.R.; Golub, T.R.; Lander, E.S.; Young, R.A. Dissecting the regulatory circuitry of a eukaryotic genome. Cell. 1998, 57, 717–728.

16. Hughes, T.R.; Marton, M.J.; Jones, A.R.; Roberts, C.J.; Stoughton, R.; Armour, C.D.; Bennett, H.A.; Coffey, E.; Dai, H.; He, Y.D. Functional discovery via a compendium of expression profiles. Cell. 2000, 57, 109–126.

17. Ideker, T.; Thorsson, V.; Ranish, J.A.; Christmas, R.; Buhler, J.; Eng, J.K.; Bumgarner, R.; Goodlett, D.R.; Aebersold, R.; Hood, L. Integrated genomic and proteomic analyses of a systematically perturbed metabolic network. Science. 2001, 57, 929–934.

18. Roberts, C.J.; Nelson, B.; Marton, M.J.; Stoughton, R.; Meyer, M.R.; Bennett, H.A.; He, Y.D.; Dai, H.; Walker, W.L.; Hughes, T.R. Signaling and circuitry of multiple MAPK pathways revealed by a matrix of global gene expression profiles. Science. 2000, 57, 873–880.

19. Spellman, P.T.; Sherlock, G.; Zhang, M.Q.; Iyer, V.R.; Anders, K.; Eisen, M.B.; Brown, P.O.; Botstein, D.; Futcher, B. Comprehensive identification of cell cycle-regulated genes of the yeast *Saccharomyces cerevisiae* by microarray hybridization. Mol Biol Cell. 1998, 57, 3273–3297.

20. Lashkari, D.A.; DeRisi, J.L.; McCusker, J.H.; Namath, A.F.; Gentile, C.; Hwang, S.Y.; Brown, P.O.; Davis, R.W. Yeast microarrays for genome wide parallel genetic and gene expression analysis. Proc Natl Acad Sci USA. 1997, 57, 13,057–13,062.

21. Hardwick, J.S.; Kuruvilla, F.G.; Tong, J.K.; Shamji, A.F.; Schreiber, S.L. Rapamycin-modulated transcription defines the subset of nutrient-sensitive signaling pathways directly controlled by the Tor proteins. Proc Natl Acad Sci USA. 1999, 57, 14,866–14,870.

22. Hughes, T.R.; Mao, M.; Jones, A.R.; Burchard, J.; Marton, M.J.; Shannon, K.W.; Lefkowitz, S.M.; Ziman, M.; Schelter, J.M.; Meyer, M.R. Expression profiling using microarrays fabricated by an ink-jet oligonucleotide synthesizer. Nature Biotechnol. 2000, 57, 342–347.

23. Causton, H.C.; Ren, B.; Koh, S.S.; Harbison, C.T.; Kanin, E.; Jennings, E.G.; Lee, T.I.; True, H.L.; Lander, E.S.; Young, R.A. Remodeling of yeast genome expression in response to environmental changes. Mol Biol Cell. 2001, 57, 323–337.

24. Reed, L.J. Multienzyme complexes. Acc Chem Res. 1974, 57, 40–46.

25. Perham, R.N. Self-assembly of biological macromolecules. Phil Trans R Soc Lond B: Biol Sci. 1975, 57, 123–136.

26. Pawson, T.; Nash, P. Protein–protein interactions define specificity in signal transduction. Genes Dev. 2000, 57, 1027–1047.

27. Fambrough, D.; McClure, K.; Kazlauskas, A.; Lander, E.S. Diverse signaling pathways activated by growth factor receptors induce broadly overlapping, rather than independent, sets of genes. Cell. 1999, 57, 727–741.

28. Iyer, V.R.; Eisen, M.B.; Ross, D.T.; Schuler, G.; Moore, T.; Lee, J.; Trent, J.M.; Staudt, L.M.; Hudson, J.; Boguski, M.S. The transcriptional program in the response of human fibroblasts to serum. Science. 1999, 57, 83–87.

29. Marton, M.J.; DeRisi, J.L.; Bennett, H.A.; Iyer, V.R.; Meyer, M.R.; Roberts, C.J.; Stoughton, R.; Burchard, J.; Slade, D.; Dai, H. Drug target validation and identification of secondary drug target effects using DNA microarrays. Nature Med. 1998, 57, 1293–1301.

30. Sabatini, D.M.; Erdjument-Bromage, H.; Lui, M.; Tempst, P.; Snyder, S.H. RAFT1: A mammalian protein that binds to FKBP12 in a rapamycin-dependent fashion and is homologous to yeast TORs. Cell. 1994, 57, 35–43.

31. Brown, E.J.; Albers, M.W.; Shin, T.B.; Ichikawa, K.; Keith, C.T.; Lane, W.S.; Schreiber, S.L. A mammalian protein targeted by G1-arresting rapamycin-receptor complex. Nature. 1994, 57, 756–758.

32. Peterson, R.T.; Desai, B.N.; Hardwick, J.S.; Schreiber, S.L. Protein phosphatase 2A interacts with the 70-kDa S6 kinase and is activated by inhibition of FKBP12–rapamycin-associated protein. Proc Natl Acad Sci USA. 1999, 57, 4438–4442.

33. Como, C.J.; Arndt, K.T. Nutrients, via the Tor proteins, stimulate the association of Tap42 with type 2A phosphatases. Genes Dev. 1996, 57, 1904–1916.

34. Jiang, Y.; Broach, J.R. Tor proteins and protein phosphatase 2A reciprocally regulate Tap42 in controlling cell growth in yeast. EMBO J. 1999, 57, 2782–2792.

35. Mamounas, M.; Gervin, D.; Englesberg, E. The insulin receptor as a transmitter of a mitogenic signal in Chinese hamster ovary CHO-K1 cells. Proc Natl Acad Sci USA. 1989, 57, 9294–9298.

36. Mamounas, M.; Ross, S.; Luong, C.L.; Brown, E.; Coulter, K.; Carroll, G.; Englesberg, E. Analysis of the genes involved in the insulin transmembrane mitogenic signal in Chinese hamster ovary cells, CHO-K1, utilizing insulin-independent mutants. Proc Natl Acad Sci USA. 1991, 57, 3530–3534.

37. Ruderman, N.B.; Kapeller, R.; White, M.F.; Cantley, L.C. Activation of phosphatidylinositol 3-kinase by insulin. Proc Natl Acad Sci USA. 1990, 57, 1411–1415.

38. Proud, C.G.; Denton, R.M. Molecular mechanisms for the control of translation by insulin. Biochem J. 1997, 57, 329–341.

39. Avruch, J. Insulin signal transduction through protein kinase cascades. Mol Cell Biochem. 1998, 57, 31–48.

40. Alessi, D.R.; Kozlowski, M.T.; Weng, Q.P.; Morrice, N.; Avruch, J. 3-Phosphoinositide-dependent protein kinase 1 (PDK1) phosphorylates and activates the p70 S6 kinase in vivo and in vitro. Curr Biol. 1998, 57, 69–81.

41. Belham, C.; Wu, S.; Avruch, J. Intracellular signalling: PDK1—A kinase at the hub of things. Curr Biol. 1999, 57, R93–R96.

42. Vanhaesebroeck, B.; Alessi, D.R. The PI3K–PDK1 connection: More than just a road to PKB. Biochem J. 2000, 57, 561–576.

43. Pullen, N.; Dennis, P.B.; Andjelkovic, M.; Dufner, A.; Kozma, S.C.; Hemmings, B.A.; Thomas, G. Phosphorylation and activation of p70s6k by PDK1. Science. 1998, 57, 707–710.

44. Dufner, A.; Andjelkovic, M.; Burgering, B.M.; Hemmings, B.A.; Thomas, G. Protein kinase B localization and activation differentially affect S6 kinase 1 activity and eukaryotic translation initiation factor 4E-binding protein 1 phosphorylation. Mol Cell Biol. 1999, 57, 4525–4534.

45. Diggle, T.A.; Moule, S.K.; Avison, M.B.; Flynn, A.; Foulstone, E.J.; Proud, C.G.; Denton, R.M. Both rapamycin-sensitive and -insensitive pathways are involved in

the phosphorylation of the initiation factor-4E-binding protein (4E-BP1) in response to insulin in rat epididymal fat-cells. Biochem J. 1996, 57, 447–453.

46. Fadden, P.; Haystead, T.A.; Lawrence, J. Identification of phosphorylation sites in the translational regulator, PHAS-I, that are controlled by insulin and rapamycin in rat adipocytes. J Biol Chem. 1997, 57, 10,240–10,247.

47. Heesom, K.J.; Avison, M.B.; Diggle, T.A.; Denton, R.M. Insulin-stimulated kinase from rat fat cells that phosphorylates initiation factor 4E-binding protein 1 on the rapamycin-insensitive site (serine-111). Biochem J. 1998, 57, 39–48.

48. Gingras, A.C.; Kennedy, S.G.'; Leary, M.A.; Sonenberg, N.; Hay, N. 4E-BP1, a repressor of mRNA translation, is phosphorylated and inactivated by the Akt(PKB) signaling pathway. Genes Dev. 1998, 57, 502–513.

49. Brunn, G.J.; Hudson, C.C.; Sekulic, A.; Williams, J.M.; Hosoi, H.; Houghton, P.J.; Lawrence, J.; Abraham, R.T. Phosphorylation of the translational repressor PHAS-I by the mammalian target of rapamycin. Science. 1997, 57, 99–101.

50. Burnett, P.E.; Barrow, R.K.; Cohen, N.A.; Snyder, S.H.; Sabatini, D.M. RAFT1 phosphorylation of the translational regulators p70 S6 kinase and 4E-BP1. Proc Natl Acad Sci USA. 1998, 57, 1432–1437.

51. Gingras, A.C.; Gygi, S.P.; Raught, B.; Polakiewicz, R.D.; Abraham, R.T.; Hoekstra, M.F.; Aebersold, R.; Sonenberg, N. Regulation of 4E-BP1 phosphorylation: A novel two-step mechanism. Genes Dev. 1999, 57, 1422–1437.

52. Nave, B.T.; Ouwens, M.; Withers, D.J.; Alessi, D.R.; Shepherd, P.R. Mammalian target of rapamycin is a direct target for protein kinase B: Identification of a convergence point for opposing effects of insulin and amino-acid deficiency on protein translation. Biochem J. 1999, 57, 427–431.

53. Peterson, R.T.; Beal, P.A.; Comb, M.J.; Schreiber, S.L. FKBP12–rapamycin-associated protein (FRAP) autophosphorylates at serine 2481 under translationally repressive conditions. J Biol Chem. 2000, 57, 7416–7423.

54. Chen, J.; Peterson, R.T.; Schreiber, S.L. Alpha 4 associates with protein phosphatases 2A, 4, and 6. Biochem Biophys Res Commun. 1998, 57, 827–832.

55. Murata, K.; Wu, J.; Brautigan, D.L. B Cell receptor-associated protein alpha4 displays rapamycin-sensitive binding directly to the catalytic subunit of protein phosphatase 2A. Proc Natl Acad Sci USA. 1997, 57, 10,624–10,629.

56. Lorenz, M.C.; Heitman, J. TOR mutations confer rapamycin resistance by preventing interaction with FKBP12–rapamycin. J Biol Chem. 1995, 57, 27,531–27,537.

57. Andjelkovic, M.; Jakubowicz, T.; Cron, P.; Ming, X.F.; Han, J.W.; Hemmings, B.A. Activation and phosphorylation of a pleckstrin homology domain containing protein kinase (RAC-PK/PKB) promoted by serum and protein phosphatase inhibitors. Proc Natl Acad Sci USA. 1996, 57, 5699–5704.

58. Li, J.; DeFea, K.; Roth, R.A. Modulation of insulin receptor substrate-1 tyrosine phosphorylation by an Akt/phosphatidylinositol 3-kinase pathway. J Biol Chem. 1999, 57, 9351–9356.

59. Halse, R.; Rochford, J.J.; McCormack, J.G.; Vandenheede, J.R.; Hemmings, B.A.; Yeaman, S.J. Control of glycogen synthesis in cultured human muscle cells. J Biol Chem. 1999, 57, 776–780.

60. Muthukkumar, S.; Ramesh, T.M.; Bondada, S. Rapamycin, a potent immunosuppressive drug, causes programmed cell death in B lymphoma cells. Transplantation. 1995, 57, 264–270.

61. Shi, Y.; Frankel, A.; Radvanyi, L.G.; Penn, L.Z.; Miller, R.G.; Mills, G.B. Rapamycin enhances apoptosis and increases sensitivity to cisplatin in vitro. Cancer Res. 1995, 57, 1982–1988.

62. Hosoi, H.; Dilling, M.B.; Shikata, T.; Liu, L.N.; Shu, L.; Ashmun, R.A.; Germain, G.S.; Abraham, R.T.; Houghton, P.J. Rapamycin causes poorly reversible inhibition of mTOR and induces p53-independent apoptosis in human rhabdomyosarcoma cells. Cancer Res. 1999, 57, 886–894.

63. Chou, M.M.; Blenis, J. The 70 kDa S6 kinase complexes with and is activated by the Rho family G proteins Cdc42 and Rac1. Cell. 1996, 57, 573–583.

64. Bishop, A.L.; Hall, A. Rho GTPases and their effector proteins. Biochem J. 2000, 57, 241–255.

65. Subramaniam, R.; Desveaux, D.; Spickler, C.; Michnick, S.W.; Brisson, N. Direct visualization of protein interactions in plant cells. Nature Biotechnol. 2001, 57, 769–772.

66. Rubin, G.M.; Yandell, M.D.; Wortman, J.R.; Miklos, G.L.; Nelson, C.R.; Hariharan, I.K.; Fortini, M.E.; Li, P.W.; Apweiler, R.; Fleischmann, W. Comparative genomics of the eukaryotes. Science. 2000, 57, 2204–2215.

67. Ashburner, M.; Ball, C.A.; Blake, J.A.; Botstein, D.; Butler, H.; Cherry, J.M.; Davis, A.P.; Dolinski, K.; Dwight, S.S.; Eppig, J.T. Gene ontology: Tool for the unification of biology. The Gene Ontology Consortium. Nature Genet. 2000, 57, 25–29.

# 12

# High-Throughput Structural Biology and Proteomics

**WUXIAN SHI and DAVID A. OSTROV**
*Albert Einstein College of Medicine*
*Bronx, New York, U.S.A.*

**SUE ELLEN GERCHMAN, JADWIGA H. KYCIA, and F. WILLIAM STUDIER**
*Brookhaven National Laboratory*
*Upton, New York, U.S.A.*

**WILLIAM EDSTROM, ANNE BRESNICK, JOEL EHRLICH, JOHN S. BLANCHARD, STEVEN C. ALMO, and MARK R. CHANCE***
*Albert Einstein College of Medicine*
*Bronx, New York, U.S.A.*

## I. INTRODUCTION

The results of high-throughput genome sequencing are changing our thinking about biology and disease. Using databases and associated software packages, we can compare organisms at the level of whole genomes, providing important evolutionary insights and identifying clinically relevant differences between humans and their pathogens. The availability of whole-genome sequences also creates enormous possibilities for the development of massively parallel tools (such as DNA microarrays) that will contribute to both fundamental research and point-of-care diagnostics. The gene sequences themselves also offer the promise of new therapeutic agents, many of which will be protein pharmaceuticals. The revolution in genome sequencing is stimulating a corresponding reorganization of protein studies with an eye toward increasing sensitivity and throughput of methods for analyzing the abundance of proteins in cells and tissues as well as

---

* Corresponding author.

for speeding up methods for biophysical analysis of proteins. Structural genomics has as its formative idea the concept of making structural information available for translated sequences that continue to become available. The coupling of high-throughput crystallography to use of modeling leverages the structural information derived from a single template sequence to multiple other target sequences of interest. The execution of this concept is being refined as we continue to test our ideas of structural genomics against the reality of our experience. In this chapter, we recount our progress in structural genomics, carried out under the overall umbrella of the New York Structural Genomics Research Consortium (NYSGRC), www.nysgrc.org.

While the genome projects of humans and model organisms were taking hold and accelerating the pace of discovery in modern biology, a series of technical advances greatly increased the speed with which we can determine the structures of biological macromolecules. Recombinant DNA technology, protein expression systems, crystal growth and freezing, x-ray area detectors, high-field nuclear magnetic resonance (NMR) spectrometers, tunable synchrotron radiation sources, and high-speed computing have catapulted structural biology from an intellectual niche to the biological mainstream. Structure determinations that used to require large teams now frequently constitute one chapter in a Ph.D. thesis. Recently, a new generation of developments is poised to further speed up the process of structure solution. Pilot projects for structural genomics (www.nigms.nih.gov/funding/psi.htm and www.structuralgenomics.org) have been established that are expanding infrastructure and implementing new technologies like automated data collection and structure solving in conjunction with multiwavelength anomalous dispersion (MAD) phasing. In parallel with these advances in crystallography, methods and infrastructure to provide high-throughput expression, purification, characterization, and crystallization of target proteins are being implemented. High-speed computing has also revolutionized what we can do with this wealth of structural information once it becomes available. Once the structure of an unknown protein is solved, fold assignment and homology modeling of related protein structures have become a basis for leveraging the structural information across a host of genomes that contain related proteins. In addition, this modeling has become an essential research tool, providing structural insights for many different areas of biology. Large-scale protein structure analyses have been applied to whole genomes, including *Mycoplasma genitalium* [1–3] and yeast [4,5]. In this chapter, we outline our overall strategies for structural genomics and present some representative results of biophysical and biochemical analysis, crystal-structure determination, biologically based target selection, and implications for the determination of new folds in carrying out biology in the future.

The current efforts of the NYSGRC encompass the structural analysis of thousands of targets; however, our initial attempt was to analyze protein structures from a complement of yeast genes encoding single domains that were selected based on tractable size and absence of low-complexity or putative membrane-spanning regions. It should be stated that membrane proteins are specifically excluded, because high-throughput methods for their structure determination remain elusive. An initial target list of 18 was immediately expanded to over 100 yeast genes and the progress of this pilot study has been continually updated on our website. In this report, we describe our results from the examination of proteins from the above target list, including biophysical and biochemical characterization of selected targets and the x-ray crystal structure of target P008, a pyridoxamine (pyridoxine) 5′-phosphate oxidase (PNPO, E.C. 1.4.3.5). These data illustrate a number of the critical steps in carrying out high-throughput structural biology and the impact structural genomics will have on the fields of biochemistry and cell biology.

## II. STRUCTURAL GENOMICS PIPELINE

The goal of the NYSGRC is to develop high-throughput technology to carry out the entire process of obtaining protein structures, starting from their gene sequences. At this stage of the program, we identify bottlenecks and developing technology to remove these bottlenecks. The process, outlined in Fig. 1, involves the following:

1. Target selection
2. Amplification of the coding sequence from genomic or cDNA
3. Cloning the coding sequence into an appropriate expression vector
4. Sequencing the cloned gene to verify that the coding sequence was correctly amplified
5. Expressing the protein
6. Confirming the identity of the expressed protein and characterizing it to establish the likelihood of crystallizability
7. Obtaining the protein in sufficient amounts and purity to form crystals
8. Defining crystallization conditions
9. Labeling protein with seleno-methionine or other suitable heavy-atom derivatives and obtaining and freezing diffraction-quality crystals for x-ray crystallography by the MAD technique
10. Collecting MAD data at an x-ray beam line
11. Determining the phases of the reflections, building the model, and refining the structure
12. Making functional inferences from the structure, disseminating our findings, and modeling sequences using the structures.

Failures were anticipated at every step, making the process somewhat akin to a funnel, with a broad input and narrow output. Dissemination was slated to occur at three points in the pipeline. First, target lists (and progress) were disseminated to inform the community of the areas of focus for this structural genomics pilot

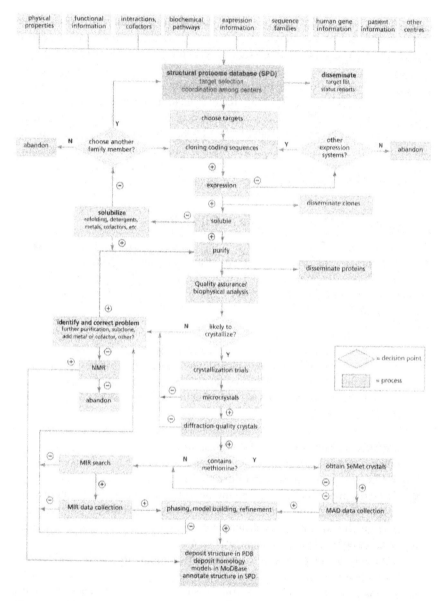

**Figure 1**  Pipeline for structural genomics indicating steps necessary for an integrated program.

project. Second, structural coordinates and functional annotations were disseminated, and third, purified proteins are an output of this program. Specifically, our project will make available vectors optimized for expression, protein purification strategies, and purified proteins to the scientific community where appropriate. The proteins represent valuable reagents for studying the biochemistry and cell biology associated with the function of each target. As discussed in later sections, many of our target-selection strategies are biologically based, providing a coherent set of proteins that are related in some fashion [6–7]. Such target-selection strategies include all members of an enzymatic pathway [8], each protein in a macromolecular complex, interacting partners of related proteins identified using yeast two-hybrid screening or bioinformatics analysis, or a group of gene products upregulated and downregulated in a biological process as determined by DNA microarray techniques.

The initial target selection, which has been described previously [6], produced a set of single-domain yeast proteins that were given unique sequential identifiers, P001–P018 for the original targets plus P019–P111 for the 93 subsequent targets processed (with 3 controls) in 96-well format to develop procedures for scale-up. From the outset, we posted our target selections and progress toward structure determination on our publicly accessible website. The web page contains links to the *Saccharomyces* Genome Database (genome-www.stanford.edu/Saccharomyces), SwissProt (www.expasy.hcuge.ch), and ProDom (www.toulouse.inra.fr/prodom.) and ModBase (www.nysgrc.org).

Each targeted yeast protein can be grouped into a protein family based on sequence identity. For the NYSGRC consortium, an arbitrary cutoff of 30% was chosen, as this is a rough threshold for current modeling programs to find a "good" structural model based on a template structure [5]. Thus, all members within a family have at least 30% sequence identity, so that solving the structure of one family member assures that a good structural model can be provided for all of the members in the family. The preliminary target selection and cloning will not be discussed further, except the details reported here in solving the crystal structure of P008. In the next phase of the pipeline, cloned genes that have been inserted into desired expression vectors are tested for expression. In the future, we are moving toward a universal donor vector that is exported to suitable destination vectors having either N- or C-terminal fusions providing hexa-histidine, glutathione-$S$-transferase, or maltose-binding protein tags.

At this point, important decisions in the pipeline must be made. If candidate proteins do not express well, either a different affinity tag or another family member (i.e., orthologs in the same protein family but from a different organism) can be selected. If the target is insoluble, refolding can be attempted or another family member/affinity tag can be selected. At the present time, it is unclear which route will be the most productive. However, expression testing of multiple members of a protein sequence family will maximize the odds of obtaining soluble protein expressed at high levels, which can only enhance the likelihood of obtaining crystals. It is not unusual to fail with a protein from one organism only to find that its ortholog from another can be crystallized readily. Within the NYSGRC, this was the experience with P005. Neither the yeast protein nor its *Escherichia coli* ortholog were tractable, but the *Bacillus subtilis* protein yielded crystals without difficulty [7]. In our study of cancer-related proteins at AECOM [7], *Xeroderma pigmentosum* complementation group C from humans was not soluble, while that from *Caenorhabditis elegans* was, even though it has a molecular mass (Mr) greater than 100 kDa.

Once soluble targets have been identified, purification can begin. Purification of affinity-tagged proteins has revolutionized preparative biochemistry for structural biologists. A myriad of commercially available expression vectors is now in use for large-scale production of recombinant proteins for x-ray crystallography and solution NMR spectroscopy. Expression of proteins fused to affinity tags such as poly-histidine (His-), glutathione-*S*-transferase (GST), maltose-binding protein (MBP), chitin-binding protein, and protein A have greatly facilitated the purification of protein targets. In addition, highly specific protease cleavage sites situated between the affinity tag and the protein of interest permit efficient removal of the tag via subtractive purification of the protease and the liberated affinity tag. Automated application of these strategies will be a prerequisite for high-throughput structural biology. Once purification strategies have been completed, such protocols can be disseminated to collaborators along with vectors to facilitate further study. With milligram amounts of purified protein in hand, screening of crystallization conditions can begin. Such screening reagents are commercially available (e.g., Hampton Research) and will not be described here. However, due to the empirical nature of crystallization trials, the establishment of biophysical criteria that increase the likelihood of crystallization is an important part of the pipeline.

## III. PROTEOLYSIS AND MASS SPECTROMETRY OF TARGET PROTEINS

It has long been appreciated that sample homogeneity is critical for the crystallization of macromolecules. Indeed, the earliest enzyme purification schemes relied on serial crystallization to produce highly purified preparations of trypsin, chymotrypsin, and so forth. Thus, high-throughput structural biology approaches should identify factors likely to enhance the likelihood of crystallization, such as analysis of covalent structure, conformational flexibility, and oligomerization state.

### A. Sample Identity

We routinely use ESI-MS (electrospray ionization–mass spectrometry) to ensure that proteins used for crystallization have been neither proteolyzed or otherwise modified during expression and purification. Measurement of the expected mass can also confirm that the purified protein has no mutations resulting from polymerase chain reaction (PCR) amplification, complementing automated DNA sequencing. In addition, this high-throughout method provides a measure of protein sample purity, which is needed to optimize crystal quality in terms of maximizing size, morphological appearance, and diffraction limits. ESI-MS has also proved useful for examining the relationship between sample homogeneity and crystal quality in qualitative terms. The charge-state distribution of ESI-MS data reflects the conformational heterogeneity and stability of proteins, compact proteins yielding fewer charge states. ESI-MS may also indicate a cofactor or metal binding because the mass measurement is accurate to better than 0.02%. These measurements utilize small amounts of purified target proteins (i.e., $<1$ pmol in total).

## B. Domain Mapping

We combine MALDI-MS (matrix-assisted laser desorption/ionization–mass spectrometry) with limited proteolysis to identify domain boundries within proteins [9,10]. This approach has proved particularly useful for multidomain eukaryotic proteins. Removal of flexible polypeptide chain segments has improved the speed with which high-quality crystals can be obtained [11,12]. Proteolysis combined with MALDI-MS is now regarded as a standard tool aiding the design and execution of successful protein crystallization trials.

## C. Oligomerization State

Empirical observations suggest that monodisperse macromolecules crystallize more readily than randomly aggregating or polydisperse systems, which rarely yield crystals. Dynamic light scattering (DLS) is a technique for measuring the translational diffusion coefficient ($D_T$) of a macromolecule undergoing Brownian motion in solution (reviewed in Ref. 13). We routinely use DLS to screen crystallization candidates for monodispersity and its effectiveness has been previously reviewed.

## D. Domain Mapping of a Yeast Target Protein

P088 is a yeast protein of unknown function with a length of 145 amino acids. Based on ProDom analysis [14] the protein has homology and domain similarity to 13 other proteins found in numerous species, including several Archaebacterial species, *S. pombe*, *C. elegans*, *Drosophila*, mouse, and humans. The human ortholog is called TFAR19. The protein exhibits a ubiquitous expression pattern and its expression is upregulated in TF-1 tumor cells undergoing apoptosis [15].

P088 was purified to $>95\%$ homogeneity, dynamic light-scattering experiments suggested the protein was monomeric, and ESI-MS gave the expected

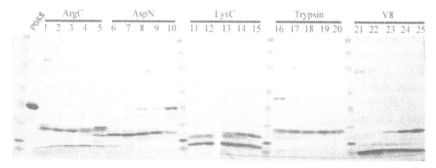

**Figure 2** Sodium dodecyl sulfate–polyacrylamide gel electrophoresis (SDS-PAGE) of proteolysis results for P088 showing the effect of varying concentrations of ArgC and AspN. The protein was incubated with the proteases for 1 h at 22°C. The purified P088 is seen in the second lane from the left near the Mr marker of 19 kDa. The next five lanes show varying weight/weight ratios of P088 to ArgC of 1:5 to 1:500. Even at the lowest concentration, truncation to fragments around 13 Kda are observed, and a stable core around 13 kDa is preserved even at high concentrations of protease. Lanes 8–12 show the cleavage results of varied weight/weight ratios of P088 to AspN of 1:15 to 1:1500. A slightly smaller stable fragment is observed. All experiments were at 22°C for 1 h. Lane 1: molecular mass standards, 19, 13, 9, and 7 kDa; lane 2: purified P088; lanes 3–7: ArgC–P088 at weight/weight ratios of 1:5,1: 15,1:50,1:150, and 1:500; lanes 8–12: AspN–P088 at weight/weight ratios of 1:15,1:50,1: 150,1:500, and 1:1500; lane 13: molecular-mass standards.

molecular mass (18826.4 observed versus 18826.3 expected). However, despite repeated attempts at crystallization, no suitable conditions were found at ambient temperatures or 4°C. We next utilized proteolysis to identify flexible segments that might be interfering with crystallization. Figure 2 shows the effect of varying concentrations of ArgC and AspN on P088. At all concentrations of ArgC, a stable core of 13 kDa was observed, whereas AspN revealed a slightly smaller fragment. Liquid chromatography followed by mass spectrometry analysis of the digests (by ESI-MS) identified a major fragment of 13,041 amu in the ArgC digestion and fragments of 10,592 and 10,113 amu in the AspN digestion experiments. Based on the known sequence and the predicted cleavage sites, the above peptide fragments correspond to N52-D169, D50-E146, and D50-K142, respectively. Digestion with trypsin and V8 protease produced a stable core fragment of ∼10 kDa, including several fragments beginning with residue N42. The predicted stable core fragment of N42-K142 was recloned to produce a truncated P088 with a predicted molecular mass of 10,815.0. This protein expressed well and was easily purified to >95% homogeneity. Initial crystal screens have yielded promising results and are being pursued. If this or other truncations of P088 do not yield crystals, P088 orthologs provide an attractive alternative.

## IV. CRYSTAL STRUCTURE AND ENZYMOLOGY OF PNP OXIDASE

P008 belongs to a family of pyridoxine-5′-phosphate (PNP) oxidases based on its primary sequence. There are 27 members in the PNP oxidase family as identified by ProDom [14]. These are found in several bacterial species, *C. elegans*, as well as mammals. PNP oxidase (PPNO) catalyzes the final step in the biosynthesis of the organic enzyme cofactor, pyridoxal-5′-phosphate (PLP). This enzyme oxidizes either the primary amine, pyridoxamine-5′-phosphate (PMP) or the primary alcohol, PNP, to the corresponding aldehyde, PLP, as shown in Fig. 3 [16–18]. The product PLP is the active form required by the numerous PLP-dependent aminotransferases, decarboxylases, epimerases, and other enzymes involved in secondary metabolism.

P008 was expressed in *E. coli* purified by ion exchange and gel filtration. The purified protein had a yellow color, indicative of its flavin mononucleotide (FMN) content. Crystals were generated by the hanging-drop diffusion method. Rod-shaped crystals appeared in 3 days, growing to $0.3 \times 0.1 \times 0.1$ mm. Synchrotron data were collected at the National Synchrotron Light Source; because no methionines are present in the protein, traditional multiple-isomorphous replacement (MIR) approaches with heavy atoms were used to provide phases. The structure was solved to 2.7 Å resolution.

### A. Polyserine as a Side-Chain Model for Refining Structural Data

In refining the structure, we followed the typical approach of utilizing a polyalanine model (180 residues) that yielded $R_{free}$ and $R_{cryst}$ of 48.6% and 43.1%, respectively. The model showed poor geometry, with only 48.3% of residues in the most favored region in the Ramachandran plot. Phase combination did not provide significant improvement in side-chain densities and left significant ambiguity with respect to fitting the amino acid sequence. Because most amino acids have side chains that extend beyond Cβ, we considered whether a better model than polyalanine could be utilized in this initial stage of the refinement process. We constructed a polyserine model that increased the scattering mass of the model by ~20%. The side chains of serine residues were fitted in the densities if they were present or built into the most favored rotamer conformation if density was not present. The polyserine model was then submitted for a cycle of simulated annealing refinement using the same parameters from the polyalanine model refinement. The $R_{free}$ and $R_{cryst}$ were reduced by 7% to 41.5% and 36.4%, respectively. The resulting model demonstrated improved geometry, with 60.1% of the residues in the most favored region in the Ramachandran plot. Phase combination resulted in side-chain densities that were readily interpreted in terms of the primary sequence. The model with the correct sequence was built during the course

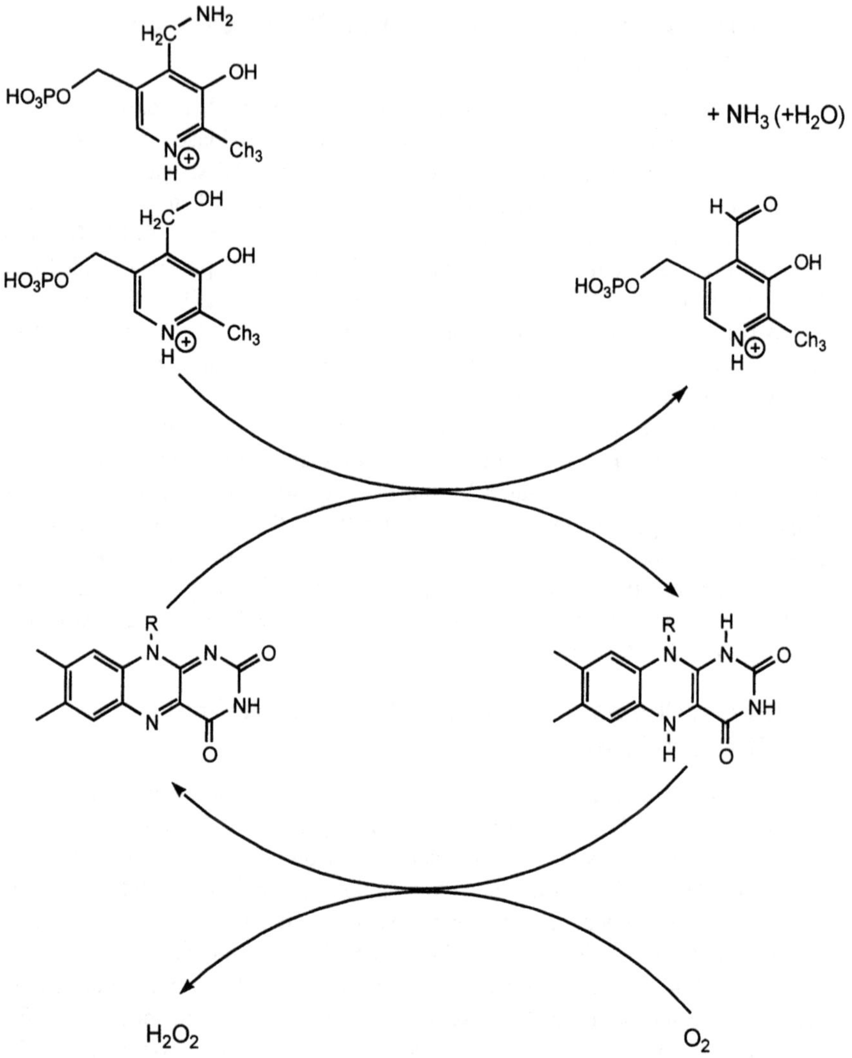

**Figure 3**  Reaction catalyzed by yeast PNP oxidase.

of several cycles of refinement into the maps from combined phases. This experience suggests that a polyserine model should become the standard in early stages of modeling.

## B. Enzymatic Activity

In the purification of P008, it became clear that a flavin cofactor was present, due to the yellow color of the purified protein. Also, P008 had high homology to PNP oxidase enzymes known to have flavoprotein cofactors. Thus, the activity toward both PNP and PMP was analyzed. Not surprisingly, P008 catalyzes the oxidation of both PNP and PMP with $K_m$ values of 3.1 and 11.3 $\mu M$, respectively, and $k_{cat}$ values of 0.4 and 0.3 min$^{-1}$, respectively, as shown by the experiments seen in Figure 4. This compares to $K_m$ values of 8.2 and 85 $\mu M$ and $k_{cat}$ values of 42 and 6.2 min$^{-1}$ (for PNP and PMP, respectively) seen for the enzyme from rabbit liver [17].

## C. Overall Fold

The yeast PNPO has two molecules in the asymmetric unit. In the crystal structure, the enzyme forms a homodimer which contains two equivalent FMN-binding sites, with each of the FMN-binding sites formed using residues from both sub-

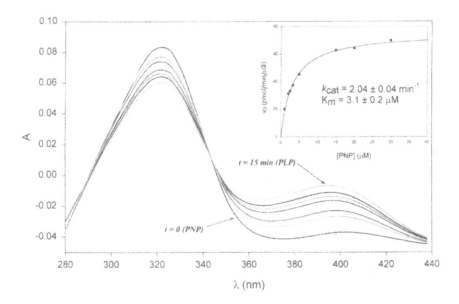

**Figure 4** P008 has PNP and PMP oxidase activity with kinetic parameters as shown.

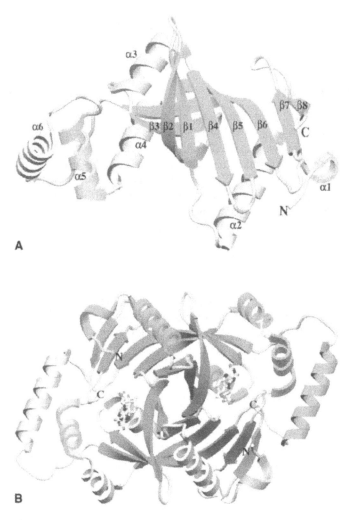

**Figure 5** Views of monomer and dimer of the yeast PNP oxidase. (A) β1–β6 form a six-stranded Greek-key β-barrel in the core of the PNP oxidase structure. (B) The dimer interface is essential for the enzymatic activity of PNP oxidase because the FMN-binding sites are formed by both subunits. Figures 5–7 are generated using SETOR [19]. (See the color plate.)

units in the dimer (Figs. 5A and 5B). The PNPO structure reveals a novel α/β fold containing a six-stranded Greek-key antiparallel β-barrel, which extends into a eight-stranded antiparallel β-sheet, surrounded by several α-helices. Four hundred nine (90%) of a total of 456 amino acids present in the asymmetric unit were included in the refined crystallographic model (Fig. 5A). The close similarity between the two subunits in the asymmetric unit, with root mean square (rms) deviation of 0.2 Å between Cα atoms of residues 25–228, may be partly due to the noncrystallographic symmetry restraint applied in the refinement cycles.

Each subunit of the yeast PNP oxidase consists of six α-helices and eight β-strands and is folded into an elongated single-domain structure with approximate dimensions of 60 Å × 30 Å × 30 Å. The core of the PNP oxidase structure is a six-stranded Greek-key β-barrel (β1, 57–64; β2, 69–79; β3, 84–88; β4, 104–112; β5, 117–128; β6, 188–201). The Greek-key β-barrel feature in the yeast PNP oxidase resembles the β-barrels observed in several serine proteases and porin membrane proteins, as revealed by the program DALI [20]. In comparison to these simpler folds (see also below), two extra strands (β7, 207–214; β8, 220–227) extend from β6 near the C-terminus to form a rather flat β-sheet with β4–β6 from the β-barrel, and this five-stranded β-sheet forms a wall of the two FMN-binding sites. The C-terminus is completely buried inside the dimer interface in the PNP oxidase structure. Two α-helices (α2, 35–48; α3, 95–100) pack adjacent to the openings of the core β-barrel, serving the function of stabilizing the β-barrel structure. The segment of approximate 60 amino acids connecting β5 and β6 is folded into 3 consecutive α-helices (α4, 130–139; α5, 142–150; α6, 160–172) and is involved in forming the FMN-binding sites and the dimer interface.

## D. Dimerization

The yeast PNP oxidase forms functional homodimers in solution (data not shown) and in the crystal structure. The two subunits in the crystal structure are related by a noncrystallographic two-fold axis (Fig. 5B). The dimer interface in the yeast PNP oxidase is very extensive and buries a total of 5005 Å$^2$ solvent accessible surface area, which represents 14% of surface area from each subunit.

It has been reported that the *E. coli* PNP oxidase formed homodimers in solution with one FMN molecule bound in each dimer [21]. In the yeast PNP oxidase crystal structure, however, two FMN ligands were located in the cavities at the dimer interface. The two FMN-binding sites are equivalent, related by a noncrystallographic two-fold axis. The bound FMN molecules are immobilized by extensive hydrogen-bond interactions as well as van der Waals interactions with residues from both subunits, and they play a significant role in stabilizing the dimer interface of the yeast PNP oxidase.

The residues involved in direct interactions at the dimer interface are mostly located in β1–β2, β4, α5–α6, and C-terminus regions. The three adjacent β-strands from the core β-barrel, β1–β2 and β4, stack onto their equivalent counterpart from the other subunit. The interactions in this area are dominated by hydrophobic interactions between these pairs: Ala63 and Ala63; Ala63 and Val70; Leu65 and Gly68; Ile74 and Val109; and Ile74 and Phe111. In addition, there are four pairs of direct hydrogen bonds: OG of Ser61 with OG of Ser61 (2.5 Å); OG1 of Thr59 with OG1 of Thr 59 (2.9 Å); carbonyl oxygen of Gly68 with ND2 of Asn105 (2.7 Å); and NH2 of Arg95 with OE1 of Glu122 (2.8 Å). The other region with direct interactions at the dimer interface is between α5–α6 and β7–β8. Six pairs of hydrogen bonds are located between the backbone and side-chain atoms from both subunits. These include carbonyl oxygen of Asp155 with amide nitrogen of Ala227 (2.7 Å), amide nitrogen of Ile157 with carbonyl oxygen of Arg225 (2.7 Å), NE2 of Gln153 with carbonyl oxygen of terminal residue Pro228 (2.9 Å), NH2 of Arg160 with carbonyl oxygen of Leu206 (2.4 Å), NE of Arg160 with OD2 of Asp208 (3.0 Å), and OG of Ser154 with NH2 of Arg209 (2.8 Å). The dimerization of the yeast PNP oxidase is critical for enzyme activity because the active site is formed by residues from both subunits.

## E. FMN-Binding Sites

Two FMN ligands are located in the deep grooves at the dimer interface in the yeast PNP oxidase structure. Both FMN-binding sites are fully occupied as shown in the $2F_o-F_c$ electron density map (Fig. 6A). The binding environments for the two FMN molecules are nearly identical, due to the two-fold noncrystallographic symmetry, and only one FMN-binding site will be described here (Fig. 6B).

The FMN ligand is bound inside the pocket formed by β2, β3–α3, and the loop connecting α5 and α6 from subunit A, and β4–β8 from subunit B. The 5′-phosphate end of the FMN molecule is completely buried, with extensive hydrogen-bond interactions with mostly positive-charged neighboring residues. Both side-chain and backbone atoms of these residues hydrogen bond with oxygens of the 5′-phosphate group, and these include O1P with NH1 of Arg73 (subunit A, 3.0 Å) and NH2 of Arg120 (subunit B, 2.6 Å), O2P with NH2 of Arg209 (subunit B, 2.8 Å) and OG of Ser154 (subunit A, 2.5 Å), O3P with amide nitrogens of Arg95 and Lys96 (subunit A, 3.0 and 2.8 Å, respectively), and O5′ with NZ of Lys96 (subunit A, 2.9 Å). The hydrogen-bonding network for the ribityl group of the FMN ligand is less extensive, and this part of the FMN ligand has been shown to be most flexible, judging from the electron densities. The hydrogen bonds with the three hydroxyl groups include O4′ with NH1 of Arg209 (subunit B, 2.6 Å), none for O3′, and O2′ with NE2 of Gln 118 (subunit B, 3.0 Å) and carbonyl oxygen of Ile74 (subunit A, 2.5 Å).

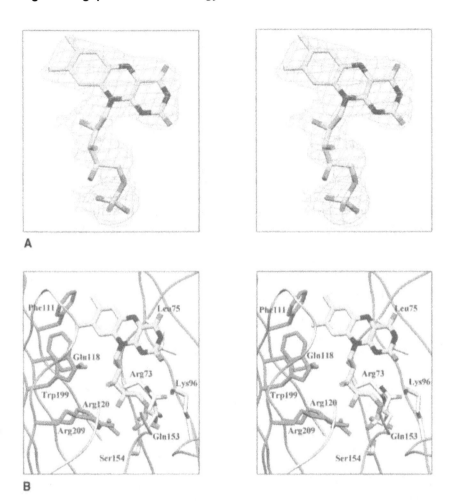

A

B

**Figure 6** Stereoview of the $2F_o-F_c$ electron density map contoured at $1\sigma$ for the bound FMN ligand (A). Stereoview of the residues involved in direct interactions with the FMN ligand (B). Side chains from subunit A and B are shown in green and red, respectively. The essential histidine residue is located in the loop that closes on top of the isoalloxazine ring at the left front position in (B). (See the color plate.)

The isoalloxazine group of the FMN ligand is well defined in the structure and slightly exposed to the solvent. The face of the isoalloxazine group is stacked by Trp199 (subunit B) and Ile74-Leu75-Leu76 (subunit A) on both sides, and the hydrophobic side is protected by Phe111 (subunit B). The other side of the isoalloxazine group is more hydrophilic and surrounded by polar residues with O2 in hydrogen-bond distance with OG of Ser89 and NE2 of Gln153 (subunit A).

It has been reported that a histidine residue was essential for enzymatic activity in the rabbit liver PNP oxidase [22]. The only His residue that is conserved throughout PNP oxidases from all species is His207 in the highly conserved sequence motif, 205-RLHDR-209. Residues 205–209 are located in the β7–β8 region that constitutes the FMN-binding site in the yeast PNP oxidase structure. Although His207 does not interact with the FMN ligand directly, the side chain of this histidine residue orients toward the isoalloxazine group and is in the position to bind substrate.

## F. P008: Relationship to Existing Folds

At the time the yeast PNP oxidase crystal structure was solved and deposited in the PDB (ID code: 1cio), the closest member in protein fold space was represented by the NMR structure of the FMN-binding protein biological unit from *Desulfovibrio vulgaris* [23] (PDB accession code: 1axj), as revealed by the program DALI (Figs. 7A and 7B). Subsequently, the structure of the *E. coli* protein was solved [24] and it is quite similar to the yeast protein reported here. The *D. vulgaris* FMN-binding protein is a monomeric single-domain structure containing 122 amino acid residues with 1 FMN ligand bound near the surface. The structure consists a six-stranded Greek-key β-barrel and two short α-helices that are located near the openings of the β-barrel. The rms deviation is 2.6 Å between 107 Cα atoms of the FMN-binding protein and the yeast PNP oxidase structures. Superposition of the FMN-binding protein with the core of the yeast PNP oxidase brings the FMN ligand close·to one FMN bound at the dimer interface of the PNP oxidase (Fig. 7C). However, the FMN ligand was bound in a shallow groove and exposed to the solvent in the FMN-binding protein, in contrast to the deep FMN-binding pocket in the PNP oxidase structure. Most elements that form the deep FMN-binding pockets in PNP oxidase are not present in the FMN-binding protein. In addition to the lack of the dimer interface, the two extra strands β7 and β8, and the long polypeptide of 60 amino acids connecting β5 and β6, both directly involved in interacting with the FMN ligand, are missing in the FMN-binding protein structure. In addition, there is virtually no sequence homology between the two proteins, and the residues involved in direct binding of the FMN ligands are not conserved between the *D. vulgaris* FMN-binding protein and the yeast PNP oxidase.

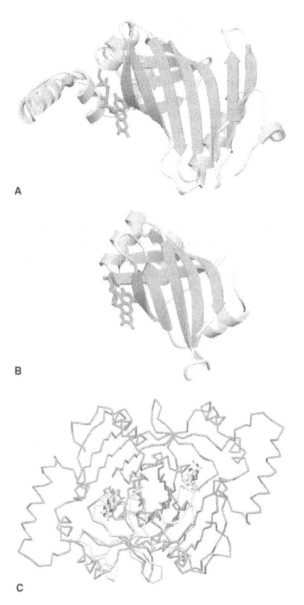

A

B

C

**Figure 7** Ribbon diagrams of the yeast PNP oxidase monomer with one FMN ligand (A) and the *D. vulgaris* FMN-binding protein biological unit with the bound FMN (B). The dimer structure is in (C). Superposition of Cα atoms of the yeast PNP oxidase dimer (red and green) and *D. vulgaris* FMN-binding protein biological unit (blue). The six-stranded Greek-key β-barrels can be superimposed in the two structures and the FMN ligands are bound at similar locations. (See the color plate.)

## G.  P008: Relationship of Sequence Homology to Structural Homology in PNP Oxidase Family

P008 was predicted to be a member of PNPO family from its primary sequence and it has been shown that the purified protein exhibits PNPO activity. PNPOs have been identified in several species, including mammals, *C. elegans*, yeast, and bacteria. Sequence alignment of P008 with five other PNPOs from *Saccharomyces pombe, E. coli, Mycobacterium tuberculosis*, rat, and human revealed numerous peptide regions that are highly conserved (Fig. 8, dark green). In the analysis of Fig. 8, the dark green regions represent 100% identity across the six compared species, whereas the white regions have less than 30% identity. The regions with colors from yellow to light green vary in sequence identity from 30% to 85%. Projection of these conserved residues onto structure of P008, by painting the requisite colors onto the three-dimensional structural representation reveals that most of these residues are at the dimer interface near the FMN molecules. A large cleft in each subunit of the homodimer is clearly colored dark green, this cleft has FMN tucked into the lower left and upper right corners in each subunit of the dimer (Fig. 9).

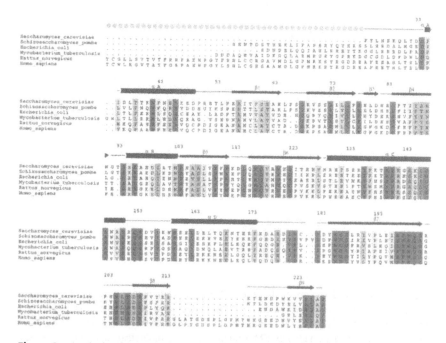

**Figure 8** Analysis of conserved sequences for yeast PNP oxidase compared to orthologs ranging from *E. coli* to man. Dark green represents 100% identity, whereas those in white have less than 30% identity. (See the color plate.)

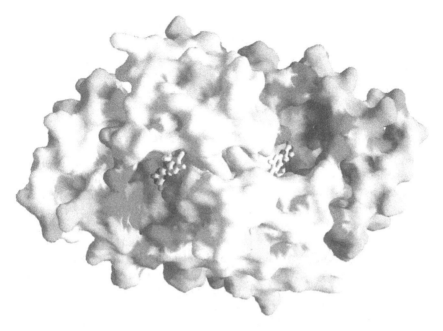

**Figure 9** Colors of amino acids coded in Fig. 8 painted on the dimer structure. Note the two obvious clefts in the structure. The FMN cofactor occupies one side of each cleft, whereas the conserved histidine that has been implicated as a critical active site residue sits on the knob directly opposite. Presumably, the substrate occupies the "hole" in the cleft. (See the color plate.)

The conserved histidine residue in the highly conserved peptide 205-RLHDR-209 has been proposed to be important in catalysis based on direct biochemical inactivation experiments [24]. Arg209 in this peptide is also entirely conserved and forms a direct hydrogen bound with FMN, whereas His207 is positioned near the isoalloxanthine ring of FMN. His207 is the only invariant histidine throughout PNPO sequences, suggesting its role as the acid–base for catalysis. Peptide 197-EFWQG-201 is completely conserved and participates in binding FMN with Trp199, providing stacking interactions with the isoalloxanthine ring. In addition, Glu197 forms a hydrogen bond with Arg209, directing the side chain of Arg209 toward FMN. Other fully conserved residues making up the active site include Arg73, Leu76, Lys78, Ser94, Lys96, Arg117-Gln118, Arg140-Pro141, Gly147-Ala148, Ser151, Gln153-Ser154, Trp186-Gly188, and Pro228. Many of these residues make up the empty region of the cleft and likely form the scaffolding for binding and orienting the substrate for catalysis.

The C-terminal peptide 225-RLAP-228 interacts with peptide 154-SDVIK-158 from the neighboring subunit and is important in the dimerization of PNPO.

Residues Arg160 and Leu163 are also involved in stabilizing the dimer interface. Out of 45 completely conserved residues throughout PNPO sequences, 12 are positioned in the hydrophobic core of the protein structure. These are Trp42, Ala46, Gly84, Phe110, Trp112, Leu115, Val121, Gly123, Pro183, and Phe198. Only four of the fully conserved residues, Asp34, Gly68, Pro104, and Tyr136, are located on the surface of the protein and have no apparent function.

## H. Comparative Modeling with P008 Structure

An unfiltered PSI-BLAST search over the nonredundant database of sequences yielded numerous sequences homologous to P008. One hundred nineteen of these sequences have $E$-values $< 10^{-4}$ and cover the complete domain. These sequences, ranging from 13% to 57% sequence identity to P008, were examined and have been deposited in ModBase (www.nysgrc.org). Analysis of this database showed 24 target sequences where satisfactory models could be constructed. The model with the best accuracy is for the closely related PNPO from *Schizosacchar-omyces pombe* with an $E$-value of $2 \times 10^{-79}$, 57% sequence identity over 204 residues, and a model score of 1.00. The human putative PNPO is also modeled successfully with an $E$-value of $4 \times 10^{-78}$, 42% sequence identity over 213 residues, and a model score of 1.00. Other successful examples include PNPOs from *E. coli*, *M. tuberculosis*, *C. elegans*, and rat.

The accuracy of comparative models depends on the sequence similarity between the model sequence and known structures. Most of the reliable models are based on sequence identity in the range of 30–50%. In case of P008, four satisfactory models were produced from the sequences with less than 30% sequence identity. The reliable model generated with the least sequence identity was base on a target sequence with 209 residues from *Drosophila melanogaster*. It shares only 18% sequence identity with P008, but the resulting model gave an $E$-value of $1 \times 10^{-66}$ and a model score of 0.80. At the time of PDB submission, the crystal structure of P008 permitted accurate structural modeling of numerous new structures. The results of comparative modeling study provide justification of the initial selection of protein targets for structure determination and information for future target-selection strategy.

## V. MATERIALS AND METHODS

### A. Cloning, Expression, and Purification

The coding sequence for P008 (yeast gene *pdx3*, orf YBR035C, 228 amino acids including the initiating Met) was amplified from the genomic DNA of *Saccharomyces cerevisiae* strain S288C by PCR, using Platinum Pfu DNA polymerase (Stratagene). The PCR primers introduced an *Nde*I site at the initiation codon and a *Hind*III site following the termination codon for cloning under control of

T7 transcription and translation signals in pET28a (Novagen). The cloned coding sequence was confirmed by DNA sequencing. Soluble protein was expressed by induction in *E. coli* strain BL21(DE3), which supplies T7 RNA polymerase [25]. Cells were collected from 1 L of culture 3 h after induction at 37°C. The cells were lysed with Bacterial Protein Extraction Reagent (Pierce). After centrifugation, the supernatant was filtered through a 45-μm filter, applied to a 30-mL column of Fractogel EMD TMAE-650 [M] (EM Separations Technology) equilibrated with 25 m$M$ HEPES, pH 7.5, and eluted with a linear gradient of 0–0.5 $M$ NaCl in the same buffer. P008 eluted at 0.16 $M$ NaCl, as indicated by position and intensity of Coomassie blue-stained bands after sodium dodecyl sulfate–polyacrylamide gel electrophoresis (SDS-PAGE). Fractions containing P008 had a yellow color, hinting at the FMN cofactor found in the crystallographic analysis. Pooled and concentrated fractions from the Fractogel column were applied to a Superdex 75 gel-filtration column equilibrated with 0.25 $M$ NaCl, 25 m$M$ HEPES, pH 7.5. The peak fractions from this column were greater than 90% pure P008, as determined by gel electrophoresis.

## B. Enzyme Activity Assay

The PNP oxidase activity assay was based on monitoring of PLP formation at 338 nm ($\varepsilon = 4900 \ M^{-1}cm^{-1}$) at pH 8.0. PNP was synthesized by reduction of PLP with NaBH$_4$ and purified by recrystallization from ethanol. PMP was prepared by the reaction of pyridoxamine with phosphorous oxychloride and precipitated as a calcium salt.

## C. Crystallization

Recombinant PNP oxidase from yeast was crystallized using the hanging-drop diffusion method at 18°C. Two microliters of 6 mg/mL protein was mixed with an equal volume of mother liquid containing 18% polyethylene glycol 4000 (Fluka) and 100 m$M$, pH 7.5, HEPES (Sigma) and equilibrated against 1 mL of the mother liquid in the well. The rod-shaped crystals appeared in 3 days and grew to a maximum size of $0.3 \times 0.1 \times 0.1$ mm$^3$. Diffraction from these crystals was consistent with the trigonal space group $P3_221$ ($a = b = 75$Å, $c = 157$ Å) with a dimer in the asymmetric unit ($V_m = 2.4$ Å$^3$/Da and 48% solvent content).

## D. Data Collection

Protein crystals were screened at beam line X9B and a native protein data set and the six derivative datasets were collected from single frozen crystals at $-178$°C on beam lines X12C and X25 at the National Synchrotron Light Source. The data were collected using a Brandeis charge coupled device (CCD) detector and processed using the DENZO package [26]. The native dataset was overall

99.8% complete to 2.7 Å with an $R_{sym}$ of 6.1%. In the last shell (2.70–2.75 Å), 82% of the data had $I/(\sigma)I > 1.0$ with $R_{sym}$ of 21.4%. Data collection statistics are listed in Table 1.

## E. Phasing

All subsequent calculations were carried out with the CCP4 suite (CCP4, 1994 Ref. 27). Initially, two EMTS (sodium ethylmercurithiosalicylate) mercury sites were located from an isomorphous difference Patterson map. Positions of other heavy-atom derivatives were determined by cross-difference Fourier method [$SmCl_3$-2 sites, $YbCl_3$-3 sites, $Na_2WO_4$-2 sites, $Au(CN)_2$-3 sites, $K_2PtCl_4$-2 sites]. All heavy atom sites were refined and included for multiple isomorphous replacement (MIR) phasing using MLPHARE [28]. The overall figure of merit was 0.59 in the resolution range 10–2.7 Å. The electron density map after solvent flattening and histogram matching revealed clear solvent boundaries and some secondary-structure features [29]. The noncrystallographic two-fold axis that relates the two subunits in the asymmetric unit was located in the skeletonized MIR map and refined to correlation coefficient of 0.50 using the RAVE package [30]. Eighty percent of the backbone from a total of 228 residues were traced through the noncrystallographic symmetry (NCS) averaged map using O [31]. Combining MIR phases and the partial structure yielded an improved map but still with poor side-chain densities. A polyserine model was generated from the polyalanine trace and the model was submitted for one cycle of simulated annealing refinement using X-PLOR [32]. $R_{free}$ and $R_{cryst}$ was reduced by 7% to 41.5% and 36.4%, respectively, and the model with the correct sequence was built into the electron density map from the new set of combined phases.

## F. Structural Refinement

The structural refinement was performed using bulk solvent correction, simulated annealing refinement, and individual B factor refinement as implemented in X-PLOR. Strict NCS constraints were applied in the initial cycles of refinement and NCS restraints were relaxed in the subsequent cycles. Residues 1–23 were disordered in both subunits. The final model includes residues 24–228 in subunit A, residues 25–228 in subunit B, two FMN, and a total of 32 waters with $R_{free}$ and $R_{cryst}$ of 27.4% and 22.5%, respectively. The model has good geometry, with 89.0% in the most favored region, 10.4% in the additionally allowed region, and 0.5% in the generously allowed region, as determined by PROCHECK [33]. Refinement statistics are shown in Table 2.

## ACKNOWLEDGMENTS

This work was supported in part by the National Institute for General Medical Sciences (P50-GM-62529), National Center for Research Resources (P41-EB-

**Table 1** Data Collection Statistics for PNP Oxidase

| Dataset | Native | EMTS[a] | SmCl$_3$ | YbCl$_3$ | Na$_2$WO$_4$ | Au(CN)$_2$ | K$_2$PtCl$_4$ |
|---|---|---|---|---|---|---|---|
| Resolution[b](Å) | 2.70 | 2.72 | 3.20 | 3.97 | 3.20 | 3.20 | 4.50 |
| | (2.75–2.70) | (2.82–2.72) | (3.25–3.20) | (4.04–3.97) | (3.25–3.20) | (3.25–3.20) | (4.58–4.50) |
| Reflections | | | | | | | |
| Overall | 107,874 | 69,819 | 27,980 | 19,485 | 21,978 | 25,480 | 7,309 |
| Unique | 14,594 | 14,098 | 8,787 | 4,713 | 8,974 | 8,897 | 3,128 |
| Completeness(%) | 99.8(94.0) | 96.9(85.3) | 97.2(97.7) | 94.6(94.6) | 99.3(97.2) | 98.8(96.2) | 97.3(96.8) |
| $R_{sym}$[c](%) | 5.4(21.9) | 8.0(32.7) | 8.2(24.5) | 4.2(4.6) | 3.6(5.5) | 3.8(5.5) | 6.7(7.3) |
| $R_{iso}$[d](%) | | 20.9 | 17.5 | 14.7 | 10.2 | 7.0 | 17.9 |
| Sites | | 2 | 2 | 3 | 2 | 3 | 2 |
| Phasing power[e] | | 1.41 | 0.64 | 0.73 | 0.65 | 0.42 | 0.86 |

[a]EMTS, = sodium eythylmercurithiosalicylate.

[b]Values in parentheses are for the highest-resolution shell.

[c]$R_{sym} = \Sigma_h \Sigma_i | I_{hi} - \langle I_{hi} \rangle | / \Sigma_h \Sigma_i | I_{hi} |$, where $h$ specifies unique reflection indices and $i$ indicates symmetry equivalent observation of $h$.

[d]$R_{iso} = \Sigma_h || F_{PH} | - | F_{P} || / \Sigma_h | F_{P} |$, where $| F_{PH} |$ and $| F_{P} |$ are the measured structure factor amplititude of the derivative and native structures.

[e]The phasing power is the ratio of the structure factor amplitude for the heavy atoms in a derivative to the estimated error in the phasing model

**Table 2**   Refinement Statistics for Yeast PNP Oxidase

| Resolution | 25.0–2.70 |
|---|---|
| $R_{cryst}$ [a](%) | 22.5 |
| $R_{free}$ [b](%) | 27.4 |
| Number of non-H atoms | |
| Protein | 3347 |
| FMN | 62 |
| Water | 32 |
| Average B factor($\text{Å}^2$) | |
| Protein | 36.7 |
| FMN | 33.8 |
| Water | 27.4 |
| rms deviations | |
| Bond(Å) | 0.007 |
| Angle(deg.) | 1.229 |

[a]$R_{cryst} = \Sigma \|F_o| - |F_c\| / \Sigma |F_o|$ for all reflections, where $|F_o|$ and $|F_c|$ are the observed and calculated structure factors, respectively.
[b]$R_{free}$ was calculated against 5% of the reflections removed at random from the refinement.

01979), and the Office of Biological and Environmental Research of the Department of Energy. The National Synchrotron Light Source is supported by the Department of Energy, Division of Materials Sciences. The authors would like to extend their gratitude to all the members of the New York Structural Genomics Research Consortium, especially Stephen Burley and Andre Sali.

## REFERENCES

1. Fischer, D.; Eisenberg, D. Assigning folds to the proteins encoded by the genome of *Mycoplasma genitalium*. Proc. Natl. Acad. Sci. USA. 1997, 94, 11,929–11,934.
2. Rychlewski, L.; Zhang, B.; Godzik, A. Fold and function predictions for *Mycoplasma genitalium* proteins. Fold. Des. 1997, 3, 229–238.
3. Huynen, M.; Doerks, T.; Eisenhaber, F.; Orengo, C.; Sunyaev, S.; Yuan, Y.; Bork, P. Homology-based fold predictions for *Mycoplasma genitalium* proteins. J. Mol. Biol. 1997, 280, 323–326.
4. Sanchez, R.; Sali, A. Large-scale protein structure modeling of the *Saccharomyces cerevisiae* genome. Proc. Natl. Acad. Sci. USA. 1997, 95, 13,597–13,602.
5. Sanchez, R.; Pieper, U.; Melo, F.; Eswar, N.; Marti-Renom, M.A.; Madhusudhan, M.S.; Mirkovic, N.; Sali, A. Protein structure modeling for structural genomics. Nature Struct. Biol. 1997, 7(Suppl.), 986–990.

6. Burley, S. K.; Almo, S. C.; Bonanno, J. B.; Capel, M.; Chance, M. R.; Gaasterland, T.; Lin, D.; Sali, A.; Studier, F. W.; Swaminathan, S. Structural genomics: beyond the human project. Nature Genet. 1999, 23(2), 151–157.

7. Chance, M.R.; Bresnick, A.R.; Burley, S.K.; Jiang, J.S.; Lima, C.D.; Sali, A.; Almo, S.C.; Bonanno, J.B.; Buglino, J.A.; Boulton, S.; Chen, H.; Eswar, N.; He, G.; Huang, R.; Ilyin, V.; McMahan, L.; Pieper, U.; Ray, S.; Vidal, M.; Wang, L.K. Structural Genomics: A Pipeline for providing structures for the biologist. Prot. Sci. 2002, 11, 723–738.

8. Bonanno, J.B.; Edo, C.; Eswar, N.; Pieper, U.; Romanowski, M.J.; Ilyin, V.; Gerchman, S.E.; Kycia, H.; Studier, W.; Sali, A.; Burley, S.K. Structural genomics of enzymes involved in sterol/isoprenoid biosynthesis. Proc. Natl. Acad. Sci. USA. 2001, 98, 12896–12901.

9. Cohen, S.L. Domain elucidation by mass spectrometry. Structure. 1997, 4, 1013–1016.

10. Cohen, S.L.; Ferre-D'Amare, A.R.; Burley, S.K.; Chait, B.T. Probing the solution structure of the DNA-binding protein Max by a combination of proteolysis and mass spectrometry. Protein Sci. 1997, 4, 1088–1099.

11. Xie, X.; Kokubo, T.; Cohen, S.L.; Hoffmann, A.; Chait, B.T.; Roeder, R.G.; Nakatani, Y.; Burley, S.K. Structural similarity between TAFs and the heterotetrameric core of the histone octamer. Nature. 1997, 380, 316–322.

12. Marcotrigiano, J.; Gingras, A.-C.; Sonenberg, N.; Burley, S.K. Cocrystal structure of the messenger RNA 5' cap-binding protein (eIF4E) bound to 7-methyl-GDP. Cell. 1997, 89, 951–961.

13. Schmitz, K.S. An Introduction to Dynamic Light Scattering by Macromolecules; Academic Press: San Diego: 0, 1990.

14. Gouzy, J.; Corpet, F.; Kahn, D.. Computers Chem. 1999, 23, 333–340.

15. Liu, H.; Wang, Y.; Zhang, Y.; Song, Q.; Di, C.; Chen, G.; Tang, J.; Ma, D. TFAR19, a novel apoptosis-related gene cloned from human leukemia cell line TF-1, could enhance apoptosis of some tumor cells induced by growth factor withdrawal. Biochem Biophys. Res. Comm. 1997, 254, 203–210.

16. Kazarinoff, M. N.; McCormick, D. B. Rabbit liver pyridoxamine (pyridoxine)-5'-phosphate oxidase: Purification and properties. J. Biol. Chem. 1975, 250, 3436–3442.

17. Choi, J.-D.; Bowers-Kormo, D. M.; Davis, M. D.; Edmondson, D. E.; McCormick, D. B. Kinetic properties of pyridoxamine (pyridoxine)-5'-phosphate oxidase from rabbit liver. J. Biol. Chem. 1997, 258, 840–845.

18. Bowers-Kormo, D. M.; McCormick, D. B. Pyridoxamine-5'-phosphate oxidase exhibits no specificity in prochiral hydrogen abstraction from substrate. J. Biol. Chem. 1997, 260, 9580–9582.

19. Evan, S. V. SETOR: Hardware lighted three dimensional solid model representation of macromolecules. J. Mol. Graphics. 1997, 11, 134–138.

20. Holm, L.; Sander, C. Dali Ver. 2.0. J. Mol. Biol. 1993, 233, 133–138.

21. Notheis, C.; Drewke, C.; Leistner, E. Purification and characterization of the pyridoxol-5'-phosphate: oxygen oxidoreductase (deaminating) from *Escherichia coli*. Biochim. Biophys. Acta. 1997, 1247, 265–271.

22. Horiike, K.; Tsuge, H.; McCormick, D. B. Evidence for an essential histidine residue at the active site of pyridoxamine (pyridoxine)-5'-phosphate oxidase from rabbit liver. J. Biol. Chem. 1997, 254, 6638–6643.

23. Liepinsh, E.; Kitamura, M.; Murakami, T.; Nakaya, T.; Otting, G. Pathway of chymotrypsin evolution suggested by the structure of the FMN-binding protein from *Desulfovibrio vulgaris* (Miyazaki F). Nature Struct. Biol. 1997, 4, 975–979.

24. Safo, M.K.; Mathews, I.; Musayev, F.N.; di Salvo, M.L.; Thiel, D.J.; Abraham, D.J.; Schirch, V. X-ray structure of *Escherichia coli* pyridoxine 5′-phosphate oxidase complexed with FMN at 1.8 A resolution. Struct. Fold. Des. 1997, 8, 751–62.

25. Studier, F.W.; Rosenberg, A.H.; Dunn, J.J.; Dubendorff, J.W. Use of T7 RNA polymerase to direct expression of cloned genes. Methods Enzymol. 1997, 185, 60–89.

26. Otwinowski, Z.; Minor, W. Processing of x-ray diffraction data collected in oscillation mode. Methods Enzymol. 1997, 276, 307–326.

27. CCP4 (Collaborative Computational Project, Number 4). Acta Crystallogr. 1997, D50, 70–763.

28. Otwinowski, Z.. Daresbury Study Weekend Proceeding. 1997.

29. Cowtan, K. Joint CCP4 and ESF-EACBM Newsletter on Protein Crystallography. 1994, 31, 34–38.

30. Kleywegt, G. J.; Read, R. J. Not your average density. Structure. 1997, 5, 1557–1569.

31. Jones, T. A. Interactive computer program graphics: Frodo. Methods Enzymol. 1997, B115, 157–171.

32. Brunger, A. T. X-PLOR, version 3.1; Yale University Press: New Haven: 0, 1997.

33. Laskowski, R. A.; MacArthur, M. W.; Thorton, J. M. PROCHECK: A program to check the stereochemical quality of protein structures. J. Appl. Crystallogr. 1993, 26, 283–291.

# 13

# Integration of Proteomic, Genechip, and DNA Sequence Data

**LEAH B. SHAW**
*Cornell University*
*Ithaca, New York, U.S.A.*

**VASSILY HATZIMANIKATIS and AMIT MEHRA**
*Northwestern University*
*Evanston, Illinois, U.S.A.*

**KELVIN H. LEE**
*Cornell University*
*Ithaca, New York, U.S.A.*

## I.  INTRODUCTION

Recently, there has been significant interest in the ability to generate proteomic data as a measure for cell physiology and responses. There should be no question that such data are invaluable to the study of any particular biological problem or system. The current interest in proteomics may signal a shifting paradigm in biology. During the past decade, a special emphasis has been placed on the need to generate DNA sequence information as a key to solving various biological problems and to understanding biological systems. More recently, new tools have emerged which permit the measurement of mRNA expression on a genomewide scale. These new tools are being extensively applied to a wide variety of interesting systems and provide a wealth of information about cell physiology and gene expression. However, the lack of good correlation between mRNA expression and corresponding protein expression has recently been highlighted [1], and these observations have shifted the focus of many groups toward proteomics studies. It has been argued that although mRNA expression is important, protein expression measurements are critical to understanding biological systems [2]. We agree that the measurement of protein abundance is critical to an understanding of a biological system; however, we further believe that measurements at all levels of biological information need to be made to develop a deep understanding of how a system (e.g., a gene network) functions [3,4]. That is, information about DNA sequences, mRNA expression, and protein expression and activities, as well as information about the physicochemical properties underlying the cellular processes should be integrated as much as possible. This integration will require new platforms for handling these often disparate datasets as well as new computational tools that can combine knowledge from each of these levels of information into a coherent understanding of the system as a whole. This paradigm shift toward a whole-cell perspective, rather than a genome, transcriptome or proteome-centered perspective, is driven by technology development and heralds the emergence of a systems approach to biology.

The recent abundance of biological data is tied to the tremendous pace of technology development for the analysis of biological systems. The sequencing of the human genome [5,6] and the genomes of a number of other organisms has its foundation in the development of high-throughput instruments for automated DNA sequencing. This foundation has permitted the completion of the genome sequence more quickly than was envisioned, and it has led to a plethora of DNA sequence data from an ever-growing list of genes and organisms. At the same time, several key technologies have emerged that have enabled scientists to study gene expression on a much broader scale.

At the mRNA level of gene expression, genechips have emerged which permit the semiquantitative assessment of changes in the expression levels of all of the genes in the genome with a single experiment. Using either spotted array technology [7] or arrays produced using photolithographic means [8], one can easily measure changes in gene expression. The ability to make such measurements has a foundation in DNA sequencing technology because DNA sequence information is required to fabricate or manufacture these genechip arrays. However, unlike DNA sequence data, which are stored in relatively simple on-line databases, data from genechip experiments currently have no standard for storage and communication. A further complication, but one that is not a major limitation, is the need to relate information about changes in spots on a chip to information about the underlying gene sequences and functions. This connection requires an interface between information stored in DNA sequence databanks with data obtained in the genechip experiments themselves. Because these genechip technologies are relatively new, there is a lack of standards and technologies for assessing changes in mRNA levels genomewide and no well-defined approach for storing and analyzing such data with bioinformatic tools. Thus, connections between DNA sequence databanks and genechips data analysis tools are not standardized. Another factor which differentiates DNA sequence data from mRNA expression data is that DNA sequence information is static, whereas mRNA expression is dynamic. This feature provides a further level of complexity which must be resolved when integrating mRNA and DNA information.

Measurements of the proteome also pose several challenges to the efficient storage, use, interpretation, and transfer of data. Like mRNA expression profiles, the proteome is a dynamic biological entity. In addition to cataloging changes in protein expression that occur in time, a detailed description of protein expression should include quantitative information, information on posttranslational modifications, and spatio-temporal information for studied proteins. The diversity of physico-chemical properties of proteins suggests that multiple technologies and tools (many of which are described in various chapters of this book) will be used in the study of any one system. Connecting data from these various tools (protein chips, two-dimensional gels, mass spectrometry, etc.) will require specialized information technology platforms.

The ability of these new genomewide technologies to assess molecular-level events inside cells provides an opportunity to better understand gene function and regulation. The inherent complexity of biological networks suggests that gene and protein functions must be considered in the context of all of the genes expressed. Such consideration makes it difficult, if not impossible, to predict the relationship between changes in genome sequence or gene expression and the resulting phenotypes. Toward this end, one might develop computer models for biological systems which can aid in the understanding of gene networks and which can serve to complement experimental observations. Major barriers to the successful development and implementation of such networks are the different levels of information and how they are stored. This complexity makes it difficult to bridge scales of information (DNA, mRNA, protein, etc.). We seek to develop meaningful descriptions of systems that permit one to integrate information from different levels, including the genome, transcriptome, and proteome, and to better understand the relationship between these different levels.

## II.  DIFFERENT LEVELS OF BIOLOGICAL INFORMATION

As mentioned in the previous section, there are at least two requirements that present great challenges to scientists and engineers interested in biological systems: the need for platforms to deal with all types of data and the need for new paradigms to understand the relationship between different types of data.

To illustrate the first type of complexity, consider an experiment that identifies changes in protein expression in a bacterial cell using two-dimensional protein electrophoresis (2DE). Suppose a graduate student has identified, through a series of 2DE experiments, a particular protein spot that changes intensity (expression) in response to some stimuli of interest. An image analysis of multiple 2DE datasets using some bioinformatic platform is usually required to make an identification that a particular spot change is significant. The ability to relate the spots of interest to the underlying gene (e.g., hsp70) requires information obtained from a microchemical characterization of the spot using amino acid sequencing, mass spectrometry, or some other appropriate technology. Validation of this result might involve a comparison of the spot location relative to 2DE patterns available on the Internet.

At this point in our example, the graduate student has performed an analysis based, perhaps, on gel images, mass spectra, and 2DE databases within a lab and across the Internet. A logical extension to understand better the relationship between changes in protein expression and the regulation of genes is to combine this information with data obtained from genechip experiments with the goal of identifying changes in the expression of hsp70 which may relate to the observed changes in protein expression. The integration of this information with the protein-abundance observations requires an ability to measure and analyze data from genechip experiments. Unlike datasets from 2DE experiments, which are often in the form of tagged image format files (tif files) and mass spectra, data from genechip studies are often saved in large spreadsheets linked to chip images. The final step of such an analysis would require the graduate student to connect these observed changes in mRNA and protein abundance with the underlying DNA sequence and any available information about the gene function, mRNA processing, protein posttranslational modifications, pathways, and so forth. Some of this information will be available in DNA sequence databanks, some will be available as data in the investigator's laboratory, and other key pieces of information will not be available in any databanks or laboratories.

The generation of information technology platforms that enable and facilitate data handling across many platforms and domains will permit more efficient searching of key information. The exponentially increasing amount of biological data that are available suggests that there is a need for more efficient bioinformatic platforms that are able to handle large datasets and data from multiple sources and of various types. These platforms should be sensitive to the development of new technologies and new forms of data that will arise as well as permit the integration of this information into a better understanding of the relationships between the existing datasets.

The needs of the biological community extend beyond the basic requirements that have just been outlined. Although there is certainly a requirement for improved platforms for data handling and analysis, a more pertinent question and problem is illustrated in the following scenario. Our graduate student, now a postdoctoral scholar, is studying the genetic factors that result in a particular mutant phenotype. She has the time and resources to pursue a detailed analysis of only one or a few genes. Her data, using genechips and 2D electrophoresis gel, have identified a large number of candidate genes that could be of interest. These genes fall into three general categories and her dilemma is to select the most appropriate class of genes to study. Class A consists of genes which appear to be upregulated (or down regulated) at the mRNA abundance in response to some environmental perturbation, and the corresponding proteins appear to be similarly up (or down) regulated. Class B consists of genes which are upregulated at the message level but have no change at the protein level or are conversely downregulated at the protein level. Class C consists of genes which are not affected at the mRNA level but which demonstrate significantly altered expression (either upregulated or downregulated) at the protein level. She would like to finish her postdoc in a reasonable amount of time, and the question she faces is the following: Which class of genes (among these and other possible imaginable cases) would be most appropriate to pursue?

A fundamental challenge that emerges as more genechip and proteomics expression data become available is to understand better the nature of the regulation of gene expression and its effects on particular phenotypes. Certainly, all of the genes from classes A, B, and C are important to consider, and a study of one class without the context of a detailed understanding of the other classes would be incomplete. Here, our postdoc could benefit from new tools and, more importantly, a new paradigm or model for the analysis of these genomewide datasets which considers the entire cell and attempts to integrate data from all of these different levels (DNA sequence, mRNA expression, protein synthesis and expression, metabolites, etc.) into a coherent whole. Such a paradigm requires a platform that can accommodate the various datasets and sources (as mentioned earlier) and that can relate changes in one dataset to changes in other datasets in a meaningful way. Currently, there are no commercially available information technology platforms for handling different kinds of data (although some are in beta-site testing), so we will not consider this issue in further detail.

## III. ATTEMPTS TO INTEGRATE BIOLOGICAL INFORMATION ACROSS LEVELS

Although the need to integrate information in a systematic way has been discussed in the literature [9], there have been relatively few attempts to integrate information across levels such as those discussed in the previous section. The few previous attempts are described in this section.

Bono et al. at Kyoto University [10] have developed a method that uses molecular pathways to reconstruct complete functional units from a set of genes (i.e., from genome sequence information). A genome-by-genome comparison is made among different completed genomes. Enzymes are classified by the assigned EC numbers and sequence searches, and metabolic pathways are constructed based on the known functions of the enzymes. The approach permits a connection to be drawn between genome sequence and enzymatic function without an attempt to integrate information at the mRNA level. Although the approach works reasonably well at characterizing the amino acid biosynthesis pathways of *Escherichia coli, Haemophilus in-*

*fluenzae*, and *Bacillus subtilis*, there are a number of concerns highlighted by the investigators. First, there is tremendous reliance in current DNA sequence databanks on information gathered from sequence searches as well as on the descriptions given in the similar sequence database entries. Such an approach offers the possibility of the propagation of errors in entries without knowing where the error actually occurred. Further, the ability to assemble a metabolic pathway for amino acid biosynthesis requires several assumptions about enzyme substrate specificity. Nonetheless, this work represents a reasonable attempt to integrate information from the DNA sequence level with information about metabolic pathways and provides an important first step.

The Virtual Cell project (www.nrcam.uchc.edu) provides a computational tool to combine biochemical and electrophysiological data with images from microscopy. Thus, knowledge of biochemical pathways is integrated with cellular geometry and architecture. The software permits the user to specify differential equations describing biochemical reactions, diffusion, and other cellular processes and to input a cellular geometry based on microscopy studies. Local, time-dependent concentrations of chemicals are computed throughout the cell. For chemicals that can be visualized, these simulated concentrations can be compared to video recordings of microscope images. The Virtual Cell software has been used to study calcium dynamics in neuroblastoma cells [11,12]. Comparison of simulation results with real images provides a test of the underlying models and can permit estimation of unknown parameters. Because the user specifies a mathematical model for the system under study, this software could be applied to a wide variety of cellular reactions and diverse cellular geometry. However, reactants or products must be visualized by microscopy to make meaningful comparisons with the simulations.

Ideker et al. [13] have recently demonstrated an approach for developing and refining a model of a cellular pathway using genechip results. The genes, proteins, and other molecules involved in the pathway of interest are first identified. Then, each pathway component is perturbed through a succession of genetic and environmental changes. Ideally, data would be collected on the genomewide effects of the perturbation on both mRNA and proteins, but data can only be collected for proteins measured using the isotope-coded affinity tag technology. Correlations between the responses of different genes are compared to an expected reaction network, and the network is refined to reflect new interactions that are discovered. This iterative process can be used to integrate information about known biological pathways with studies of either the transcriptome or the proteome.

## IV. ATTEMPTS TO INTEGRATE PROTEOMIC AND OTHER INFORMATION USING A SYSTEMS PERSPECTIVE

The recent completion of the genomes of several organisms, including humans, and the emergence of new experimental tools to help probe genomewide gene expression at the mRNA and at the protein levels is beginning to shift the paradigms of life science from gene-centered studies to systems-centered (i.e., genomewide) studies. The integration of information across various levels (DNA, mRNA, protein, metabolite, etc.) into coherent and relevant mathematical frameworks promises a deeper understanding of the nature of gene expression regulation and of the genotype–phenotype relationship. Such mathematical frameworks are arguably most effective when used in tandem with and as a complement to experiments. An example of an area of need is the development of a model for protein synthesis (i.e., translation) to begin to relate mRNA and protein expression profiles. Recent studies at the systemwide level [1,13] suggest that the expression of mRNA may not have an obvious correlation to the expression of the corresponding proteins. Other studies have been performed in various systems, including yeast by Haurie [14] and *B. subtilis* by Yoshida et al. [15]. The increasing ubiquity of tools to monitor gene expression profiles at both the mRNA and protein levels will likely further reinforce the need to develop meaningful methods to connect these two types of information.

Recently, we have been driven to characterize better both mRNA and protein expression levels. Our analysis of a two-gene network [3] shows that realistic oscillatory behavior does not occur if each gene is represented by a single product, either its corresponding mRNA or protein. The network must include interactions between the mRNA and protein products of each gene for oscillatory solutions to exist. This observation has motivated our development of a mathematical framework to relate mRNA and protein expression levels. Such a framework, when optimized, might be able to predict changes in protein expression levels when given changes in the corresponding mRNA expression levels (e.g., as measured in genechip experiments). This framework begins by an attempt to describe, in quantitative terms, the process of protein synthesis (translation) in a simple system. However, DNA sequence information affects levels of transcription and determines the mRNA sequence, which, in turn, impacts levels of translation. Thus, the model provides an opportunity to integrate effectively information about DNA sequence, mRNA expression, and proteome expression. Like many mathematical models, this framework is an ongoing attempt to refine the biochemical model based on current experimental observations, and the predictions from the model are sometimes not fully quantitative. However, this model, when combined with experimental information about the DNA sequence, mRNA expression profiles, and protein expression profiles, provides an opportunity to understand which biological factors are regulated by (in our case) bacterial cells in response to genetic and environmental perturbations (and would be of use to our postdoctoral scholar). Some of the biological factors which can potentially be studied are ribosome concentration, ribosome-binding affinity, mRNA degradation, and others. Ultimately, once the system under study has been well characterized, the model can be used to predict changes in protein expression profiles when given changes in mRNA or to predict changes in mRNA expression profiles when given proteomic changes.

Previous attempts [16–18] to relate mRNA expression to protein expression were based on a conceptual model for protein synthesis that considered only one gene. Ribosomes bind to a particular mRNA molecule with a certain affinity and initiate translation with some rate constant. Ribosomes are able to translate along mRNA molecules with some defined translation rate, and when ribosomes move far enough along a particular mRNA, another ribosome is able to bind to the ribosome-binding site. At some later time, a ribosome completes translation and a peptide chain is released with some termination rate constant. Various computationally intensive approaches have been used to calculate the steady-state protein production rate, including Monte Carlo simulation of the translation process [17] and recursive computation of average ribosome occupancy at each codon [16,18]. The average number of ribosomes bound to the mRNA (average polysome size) was often computed as well.

The above single-gene conceptual model can also be extended to a multigene system in which the number of genes can vary from one to thousands. In this extended framework, each mRNA has some characteristic expression amount, and ribosomes can still bind to each of the available mRNAs depending on the availability of a ribosome-binding sites. There is direct competition of binding among all of the mRNAs expressed in a cell in the following manner. Consider a free ribosome that is able to bind to any message that has a free binding site. The selection of a particular message to bind (and thus to provide an opportunity to initiate protein synthesis) will be determined by the likelihood of ribosome–mRNA binding. This binding affinity will be determined by the mRNA sequence at the binding site (and, so, ultimately by the DNA sequence). Messages with a high affinity for free ribosomes will likely produce more protein product (all else being equal) than messages with a lower ribosome-binding affinity. Because of competition for the limited number of free ribosomes available at any given time inside a cell, the expression level of any one mRNA will have some effect on the protein synthesis rate of all other messages. Further, the stability of a given message and the rate of translation along the message are governed in part by the mRNA sequence (or DNA sequence) because of endonuclease and exonuclease activity and codon bias. In this way, sequence information becomes a critical factor in relating mRNA and protein expression profiles. Of course, the length of any particular message will also affect the time required to complete peptide synthesis.

Using this framework, together with suitable assumptions to simplify the problem, one can calculate mRNA or protein expression profiles for a particular biological system. One example of a relationship calculated using this system is

$$f_i^p = \left( \frac{1+R_0 K_i}{R_0 K_i} \right) \left( \frac{R K_i}{1+R K_i} \right) f_i^m \tag{1}$$

where $f_i^p$ is the normalized protein concentration for a particular gene (gene $i$) and $f_i^m$ is the normalized mRNA concentration for that same gene. The mRNA–protein relationship is thus determined by a nonlinear factor which includes parameters such as the total number of free ribosomes ($R$ and $R_0$) and an effective ribosome binding constant ($K_i$). The total number of free ribosomes ($R$), in turn, will be determined by the global changes of the mRNA concentration of *every* gene and the effective ribosome binding constants may differ for each expressed gene. When given mRNA and protein expression information from genechip and proteomics experiments, one can begin to estimate key parameters such as ribosome-binding affinities and determine the model parameters. One example of the mRNA–protein relationship which can be calculated using this relationship is given in Fig. 1. As demonstrated in the figure, this model predicts a complex nonlinear relationship between mRNA and protein levels, which is qualitatively consistent with that observed in previous experimental studies [1,13].

## V. CONCLUSION

There is an undeniable need to characterize basic information about gene expression at the mRNA and the protein levels using systemwide analytical tools. Al-

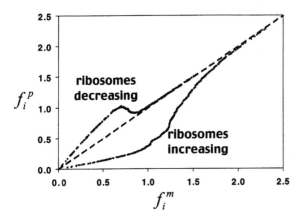

**Figure 1**  Plot of normalized protein concentration versus normalized mRNA concentration as predicted by our model. Results are shown for a system with effective ribosome-binding constants ($K_i$) increasing as mRNA increases. The free-ribosome concentration either increases or decreases as indicated in the figure. A one-to-one line indicating a perfect correlation between mRNA and protein expression is drawn as well.

though certain technological barriers remain to the development of a complete description of the proteome of various organisms, the other chapters in this book signal key advances which will advance the discovery process. Future experimental challenges include the need to characterize metabolite concentrations in a high-throughput or systemwide manner, just as technological advances have permitted systemwide measurements of other biomolecules.

As new technology enables genomewide studies of biomolecules and as observations reveal the complex interconnectedness of the components of biological systems, the field of biology may shift from the study of individual biochemical pathways to a systemwide paradigm. It is to be hoped that biology, and technology, can move beyond the detailed analysis of just a few molecules in the transcriptome or the proteome and, instead, gain a systemwide understanding of the large datasets now available. At this transition point, one could envision integrating systemwide measurements of mRNA, protein, and metabolite concentrations with genome sequence data. Building on the studies reviewed in this chapter, these large datasets could be combined with information about metabolic pathways and gene networks, as well as knockout studies and other types of information.

We have discussed the need for new data analysis platforms and new paradigms or models to treat a wide variety of datasets that are becoming available. Platforms and models need to be developed in parallel, each contributing to the progress of the other. Mathematical models of biological systems will suggest new ways to analyze available data, whereas platforms that permit simultaneous study of several data types will likely suggest new models of how these data interrelate. Fitting model parameters to experimental data may allow indirect measurement of kinetic parameters that are not easily accessible via more direct methods. Finally, obtaining agreement between model predictions and experimental results will add credence to the paradigms that have been developed. Thus, we will have achieved a deeper understanding of biological systems.

## ACKNOWLEDGMENTS

L.B.S. is supported in part by an NSF Graduate Research Fellowship. We acknowledge support from NSF, DuPont, and NIH.

## REFERENCES

1. Gygi, S.P.; Rochon, Y.; Franza, B.R.; Aebersold, R. Correlation between protein and mRNA abundance in yeast. Mol Cell Biol. 1999, 19, 1720–1730.
2. Lottspeich, F. Proteome analysis: A pathway to the functional analysis of proteins. Angew Chem Int Ed. 1999, 38, 2476–2492.

3. Hatzimanikatis, V.; Lee, K.H. Dynamical analysis of gene networks requires both mRNA and protein expression information. Metabol Eng. 1999, 1, 275–281.

4. Hatzimanikatis, V.; Choe, L.H.; Lee, K.H. Proteomics: theoretical and experimental considerations. Biotechnol Prog. 1999, 15, 312–318.

5. Lander, E.S., et al. Initial sequencing and analysis of the human genome. Nature. 2001, 409, 860–922.

6. Venter, J.C., et al. The sequence of the human genome. Science. 2001, 291, 1304–1351.

7. DeRisi, J.L.; Iyer, V.R.; Brown, P.O. Exploring the metabolic and genetic control of gene expression on a genomic scale. Science. 1997, 278, 680–686.

8. Selinger, D.W.; Cheung, K.J.; Mei, R.; Johansson, E.M.; Richmond, C.S.; Blattner, F.R.; Lockhart, D.J.; Church, G.M. RNA expression analysis using a 30 base pair resolution *Escherichia coli* genome array. Nature Biotechnol. 2000, 18, 1262–1268.

9. Delneri, D.B.; Brancia, F.L.; Oliver, S.G. Towards a truly integrative biology through the functional genomics of yeast. Curr Opin Biotechnol. 2001, 12, 87–91.

10. Bono, H.; Ogata, H.; Goto, S.; Kanehisa, M. Reconstruction of amino acid biosynthesis pathways from the complete genome sequence. Genome Res. 1998, 8, 203–210.

11. Fink, C.C.; Slepchenko, B.; Moraru, I.I.; Schaff, J.; Watras, J.; Loew, L.M. Morphological control of inositol-1,4,5-trisphosphate-dependent signals. J Cell Biol. 1999, 147, 929–935.

12. Fink, C.C.; Slepchenko, B.; Moraru, I.I.; Watras, J.; Schaff, J.; Loew, L.M. An image-based model of calcium waves in differentiated neuroblastoma cells. Biophys J. 2000, 79, 163–183.

13. Ideker, T.; Thorsson, V.; Ranish, J.A.; Christmas, R.; Buhler, J.; Eng, J.K.; Bumgarner, R.; Goodlett, D.R.; Aebersold, R.; Hood, L. Integrated genomic and proteomic analyses of a systematically perturbed metabolic network. Science. 2001, 292, 929–934.

14. Haurie, V.; Perrot, M.; Mini, T.; Jenö, P.; Sagliocco, F.; Boucherie, H. The transcriptional activator Cat8p provides a major contribution to the reprogramming of carbon metabolism during the diauxic shift in *Saccharomyces cerevisiae*. J Biol Chem. 2001, 276, 76–85.

15. Yoshida, K.; Kobayashi, K.; Miwa, Y.; Kang, C.M.; Matsunaga, M.; Yamaguchi, J.; Tojo, S.; Yamamoto, M.; Nishi, R.; Ogasawara, N.; Nakayama, T.; Fujita, Y. Combined transcriptome and proteome analysis as a powerful approach to study genes under glucose repression in *Bacillus subtilis*. Nucleic Acids Res. 2001, 29, 683–692.

16. MacDonald, C.T.; Gibbs, J.H.; Pipkin, A.C. Kinetics of biopolymerization on nucleic acid templates. Biopolymers. 1968, 6, 1–5.

17. Bergmann, J.E.; Lodish, H.F. A kinetic model of protein synthesis. Application to hemoglobin synthesis and translational control. J Biol Chem. 1979, 254, 11,927–11,937.

18. Heinrich, R.; Rapoport, T.A. Mathematical modeling of translation of mRNA in eucaryotes; steady state, time-dependent processes and application to reticulocytes. J Theoret Biol. 1980, 86, 279–313.

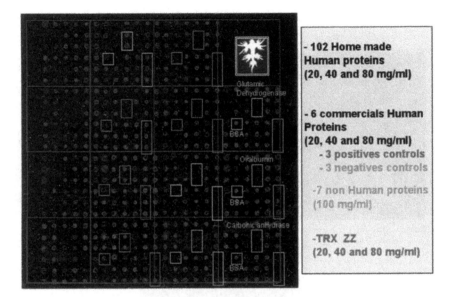

**Figure 1.8** The results of on-array screening of the polyclonal response obtained following parallel immunization in a single mouse with 102 recombinant human antigens. Of these, 95 produced at least three positive responses as indicated by the fluorescent signal above the background on a protein array comprising up to 10 replicates of each of the 102 parallel immunogens. Highlighted in pink and yellow are the positive and negative human recombinant antigens not produced "in-house," in blue the nonhuman proteins, and in red the absence of signal due to fusion elements employed during immunogenicity enhancement of each recombinant fusion protein. See also Fig. 23 for more details.

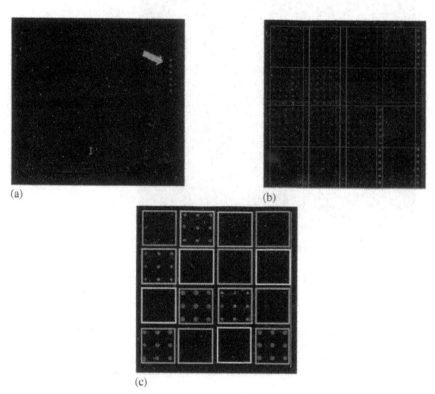

(a)

(b)

(c)

**Figure 1.12** Biochips demonstrating (a) a highly specific ligand interaction with a single binding partner on an array containing 130 nontarget elements, (b) a highly cross-reactive ligand recognizing most of 131 elements on array, and (c) exposure of 2 affinity ligands to 9 different targets showing antigen recognition and low-background signal in the absence of a blocking step.

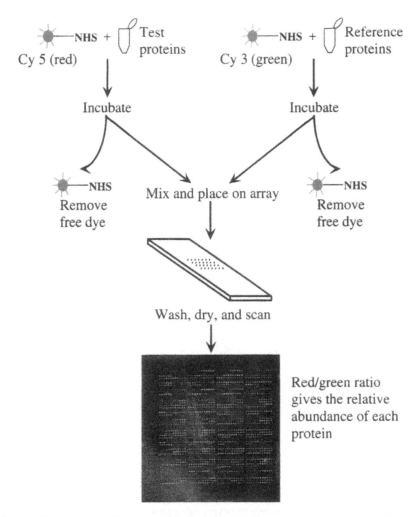

**Figure 3.1** Two-color fluorescent labeling and detection of protein microarrays. A test protein solution and a reference protein solution are mixed with two different *N*-hydroxy-succinimide (NHS)-conjugated fluorescent dyes. The NHS group reacts with amine groups on the proteins. Free dye is removed, and the solutions are mixed and placed on an array. The array is read by scanning fluorescence microscopy in two color channels specific to the conjugated dyes. The fluorescence ratio between the color channels reflects the relative protein concentration between the two solutions. (Courtesy of Current Drugs; B Haab. Curr Opin Drug Dis Dev 4:116–123, 2001.)

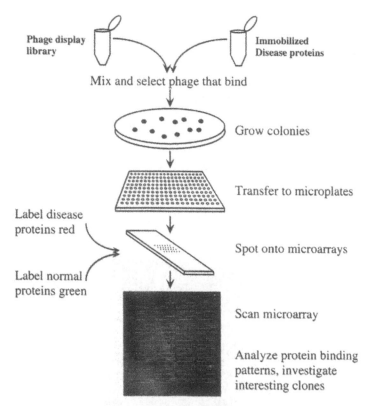

**Figure 3.3** Strategy for protein discovery using phage display libraries and protein microarrays. Members of a phage display library that bind to immobilized disease proteins are plated out and spotted onto microarrays. The pattern of binding to the spotted clones from individual patient samples is assessed using two-color comparative fluorescence. Clones that provide informative patterns of binding are investigated further.

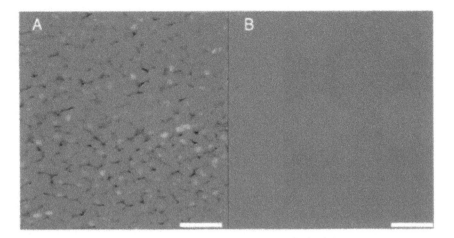

**Figure 4.1** Scanning transmission micrographs of gold surfaces. Scale bars are 300 nm. Panel a shows a 100-nm-thick gold film evaporated onto a silicon substrate at room temperature. Panel b shows a Template Stripped Gold (TSG) surface optimized for topography. Both images are displayed with the same Z-scale for comparison. TSG exhibits subnanometer roughness over micron-sized areas. Roughness of regular gold surfaces is on the order of magnitude of protein dimensions.

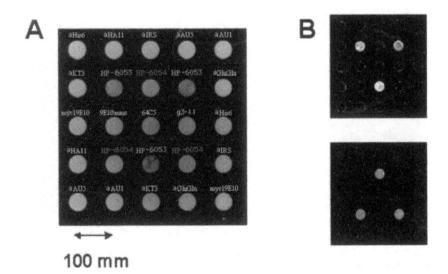

**Figure 4.4** Example of a protein array experiment. Twenty-five post surfaces (Fig. 3) contain 1 of 13 immobilized antibodies raised against peptides and proteins (a). The array was then incubated with a mixture of two fluorescently labeled antigens. Fluorescence microscopy shows specific interaction of the antigens with the corresponding antibodies (b).

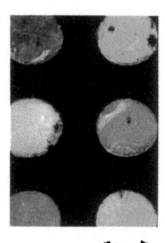

**50 mm**

**Figure 4.5** Imaging ellipsometry is one possibility for label-independent detection of different amounts of protein bound to the post surface from Fig. 2.

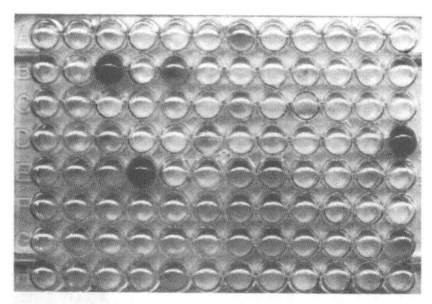

**Figure 7.4** Multiwell purification. Each well of a 96-well plate was infected with recombinant β-glucuronidase (Gus) virus at 10% (v/v), and cells were harvested after 48 h for recombinant protein production. Proteins were purified over immunoaffinity columns in a 96-well format into a collection plate. Presence of Gus protein was analysis by enzymatic hydrolysis of X-gluc (1.25 mg/mL), which is evidenced by a blue colormetric product in the well. Absorbance spectroscopy at 630 nm showed that 70% of the proteins were recovered after purification.

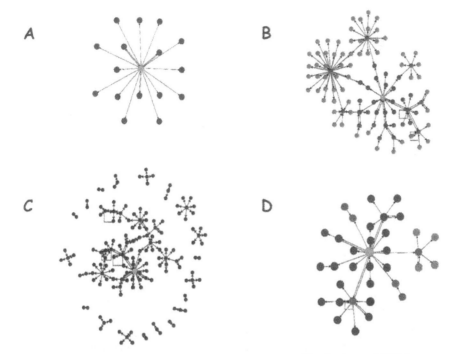

**Figure 10.5** Generation of a protein-interaction network. The *C. elegans* MRT-2 gene was used as bait in a Y2H screen in conjunction with other DNA damage response (DDR) genes [9]. Circles denote ORFs and the arrows between two circles indicate the bait-prey interaction. Arrows are from bait to prey, and double lines show reciprocal interactions. (A) The 16 prey (green circles) obtained with MRT-2 as bait (yellow circle); 3 of the MRT-2 prey, when used as baits interacted with MRT-2 as prey. (B) The expanded network when additional DDR genes are used as baits (purple circles) or are found as prey (orange circles) and the resulting direct (green circles) and indirect (purple circles) connections to MRT-2 (yellow circle). (C) The core DDR interactome network [9] centered on MRT-2 (yellow circle). (D) Direct connections as in (A) (dark green circles) with MRT-2 (yellow circle) as bait and connections arising from non-DDR Y2H screens (blue, orange, and purple circles). Light green circles are prey that were obtained from both MRT-2 and non-DDR Y2H screens.

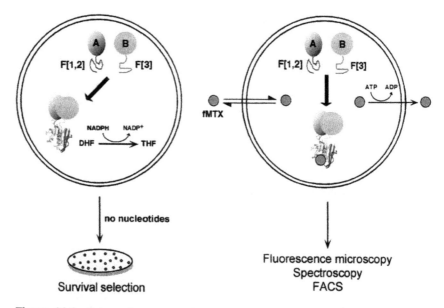

**Figure 11.1** Schematic representation of the strategy used to study protein-protein interactions in mammalian cells with the DHFR PCA. Left: Interacting proteins A and B are fused to the complementary fragments of murine DHFR (F[1,2] and F[3]) to generate A-F[1,2] and B-F[3] fusions. A physical interaction between proteins A and B drives the reconstitution of DHFR from its fragments (F[1,2] + F[3]), allowing DHFR-negative cells expressing these constructs to grow in media lacking nucleotides. DHFR-positive cells can also be used in a recessive selection strategy (see text). Right: The fluorescence assay is based on high-affinity binding of the specific DHFR inhibitor fMTX to reconstituted DHFR. fMTX passively crosses the cell membrane and binds to reconstituted DHFR (F[1,2] + F[3]) and is thus retained in the cell. Unbound fMTX is rapidly released from the cells by active transport. Detection of bound and retained fMTX can then be detected by fluorescence microscopy, fluorecein-activated cell sorting (FACS), or fluorescence spectroscopy.

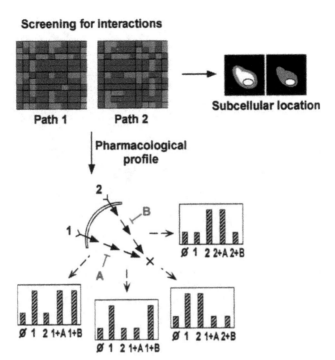

**Figure 11.2** Schematic representation of the strategy for generating a functional valida-tion profile of a biochemical network using the DHFR PCA. Positive clones are detected with the DHFR survival-selection assay. They correspond to interacting component pro-teins of two convergent signal transduction pathways (Path 1 and Path 2). An interaction matrix (upper left) represents all positive (green) and negative (red) interacting pairs ob-served in the survival-selection assay. Positive clones from survival selection are propa-gated and subjected to two functional analyses: (1) Using the DHFR fluorescence assay, in-teractions are probed with pathway specific stimulators (1 and 2) and inhibitors (A and B). Pharmacological profiles are established based on the pattern of response of individual in-teractions to stimulators and inhibitors, represented in the histograms (ordinate axis repre-sent fluorescence intensity). For example, stimulation of pathway 1 will augment all the in-teractions composing that pathway. The inhibitor A will inhibit protein interactions downstream, but not upstream of its site of action in pathway 1. (2) Cellular locations of the interactions are determined by fluorescence microscopy, also using the DHFR fluores-cence assay.

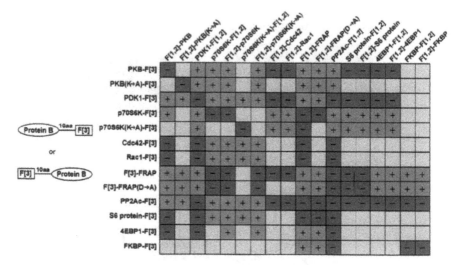

**Figure 11.3** Summary of the results obtained for the different protein-protein interactions tested in the RTK-FRAP network with the DHFR survival-selection assay in CHO DUKX-B11 (DHFR⁻) cells. On the x axis are the fusions to the DHFR [1,2] fragment and on the y axis are the fusions to the DHFR [3] fragment. The orientations of the fusions (N-terminal or C-terminal) are also indicated. Positive interactions: green (+), absence of interaction: red (−), not tested: gray squares.

**Figure 11.4A–D** Fluorometric and microscopic analysis of the interacting protein pairs fused to the complementary fragments of DHFR. The pharmacological profiles are represented by the histograms. Cells were treated with stimulants and inhibitors as indicated (x axis: NT = no treatment, I = insulin, S = serum, R = rapamycin, W = wortmannin, C = calyculin A). Fluorescence intensity is given in relative fluorescence units (y axis). The background fluorescence intensity corresponding to nontransfected cells was subtracted from the fluorescence intensities of all of the samples. Error bars represent standard errors for the mean calculated from at least three independent experiments. Fluorescence microscopy images revealing patterns of cellular location are also presented. The constitutive dimerization of GCN4 leucine zipper (GCN4/GCN4) is used as a control. Blue arrows indicate new protein-protein interactions. (A) PDK1-PKB and PDK1-p70S6K interactions occur at the plasma membrane, FRAP-4E-BP1, p70S6K-4E-BP1, and p70S6K-S6 protein interactions are cytosolic. Pharmacological profiles for the first three interactions are consistent with rapamycin-resistant, wortmannin-sensitive pathways. The serum/insulin-stimulated and wortmannin/rapamycin-inhibited profiles of the p70S6K-4EBP1 and p70S6K-S6 interactions place them at a convergent point downstream of both wortmannin- and rapamycin-sensitive pathways. (B) Analysis of pharmacological profiles reveals novel ramifications of wortmannin- and rapamycin-sensitive pathways, including serum/insulin stimulated and wortmanin-sensitive association of FRAP, placing FRAP as a downstream target of PDK1 and PKB *(continued)*.

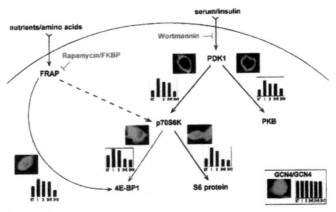

A

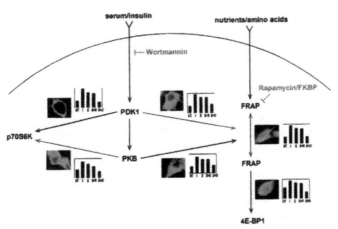

B

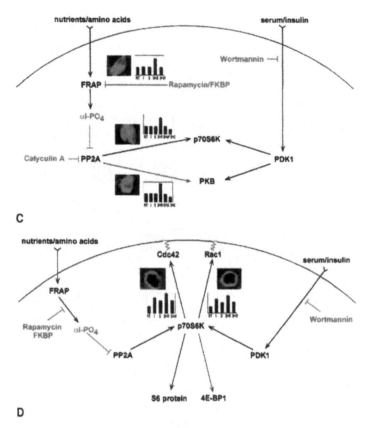

**Figure 11.4A–D** *Continued.* (C) Regulation of p70S6K and PKB by FRAP (through PP2A) and PDK1. αI-PO4 represents a regulatory subunit of the phosphatase PP2A, regulated via its phosphorylation by FRAP. The FRAP-FKBP, PP2A-p70S6K, and PP2A-PKB interactions are serum/insulin-insensitive but rapamycin induced. The interactions between PP2A-PKB and PP2A-p70S6K are also inhibited by the PP2A phosphatase inhibitor calyculin A. All of these interactions occur in the cytosol. (D) Positive/negative regulation of p70S6K in the RTK-FRAP network. The serum/insulin/rapamycin-induced interactions of p70S6K-Cdc42 and p70S6K-Rac1 occur at the plasma membrane, suggesting that p70S6K is recruited at the membrane via the two GTPases.

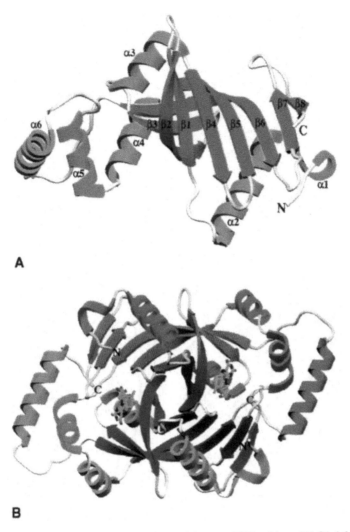

**A**

**B**

**Figure 12.5** Views of monomer and dimer of the yeast PNP oxidase. (A) β1–b6 form a six-stranded Greek-key β-barrel in the core of the PNP oxidase structure. (B) The dimer interface is essential for the enzymatic activity of PNP oxidase because the FMN-binding sites are formed by both subunits. Figures 5–7 are generated using SETOR [19].

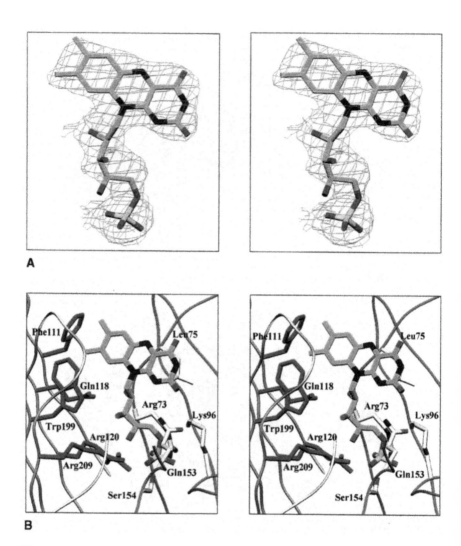

**A**

**B**

**Figure 12.6** Stereoview of the $2F_o$–$F_c$ electron density map contoured at 1s for the bound FMN ligand (A). Stereoview of the residues involved in direct interactions with the FMN ligand (B). Side chains from subunit A and B are shown in green and red, respectively. The essential histidine residue is located in the loop that closes on top of the isoalloxazine ring at the left front position in (B).

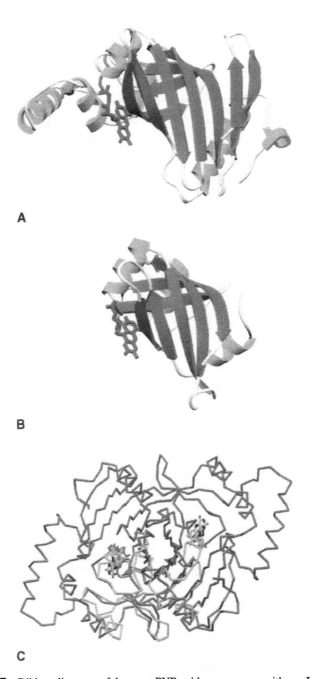

**Figure 12.7** Ribbon diagrams of the yeast PNP oxidase monomer with one FMN ligand (A) and the *D. vulgaris* FMN-binding protein biological unit with the bound FMN (B). The dimer structure is in (C). Superposition of Cα atoms of the yeast PNP oxidase dimer (red and green) and *D. vulgaris* FMN-binding protein biological unit (blue). The six-stranded Greek-key β-barrels can be superimposed in the two structures and the FMN ligands are bound at similar locations.

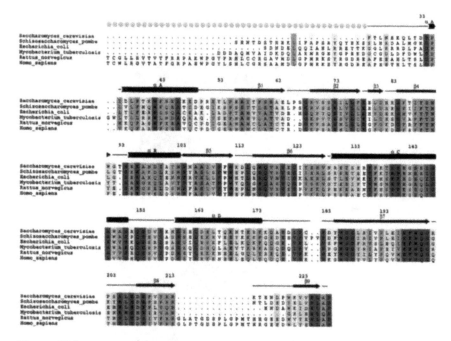

**Figure 12.8** Analysis of conserved sequences for yeast PNP oxidase compared to orthologs ranging from *E. coli* to man. Dark green represents 100% identity, whereas those in white have less than 30% identity.

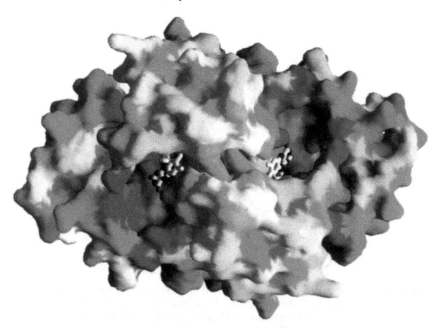

**Figure 12.9** Colors of amino acids coded in Fig. 8 painted on the dimer structure. Note the two obvious clefts in the structure. The FMN cofactor occupies one side of each cleft, whereas the conserved histidine that has been implicated as a critical active site residue sits on the knob directly opposite. Presumably, the substrate occupies the "hole" in the cleft.

# 14

# The Proteomics Market

**STEVEN BODOVITZ**
*BioInsights*
*San Francisco, California, U.S.A.*

**JULIANNE DUNPHY**
*BioInsights*
*Redwood City, California, U.S.A.*

**FELICIA M. GENTILE**
*BioInsights*
*Cupertino, California, U.S.A.*

## I. OVERVIEW OF THE MARKET

The demand for proteomics data comes on the heels of the success of genomics. The completion of the human genome has transformed gene finding from a laborious protocol performed at the bench top into a rapid query performed on the computer. In addition, the emergence of DNA chip technology has transformed the study of gene expression from focused studies of a small numbers of genes to systemwide examinations of thousands of genes. Genomics, however, only describes a small part of the function of the cell. A detailed understanding of cellular pathways requires knowledge of the corresponding proteins.

The accelerated generation of proteomics data will have an enormous impact on biological research in general, and drug discovery and development, in particular. At the very least, the data will transform much of the laborious bench work to rapid *in silico* searches. Finding a protein with a putative function of interest, for example, may be as simple as a homology search, greatly facilitating the identification of drug targets. Proteomics also has the potential to enable an entirely new approach to the study of biological systems. Instead of the traditional reductionist approach of oversimplifying a complex system by examining only a few isolated elements, researchers can study multiple components at the same time. Interconnected pathways can be studied as complete systems, which has important ramifications in diagnostics, toxicology, and pharmacology: monitoring hundreds or thousands of proteins at the same to time to identify patterns that correlate with disease or drug response.

Thus, the value of proteomics data is likely to drive the creation of a lucrative market. As a basis of comparison, McKinsey & Co. and Lehman Brothers estimate that the revenue generated by genomics companies is approximately $2.5 billion and will double by 2005 [1]. A complementary study from Cowen (unpublished data) places the combined valuations of the companies in the international genomics industry at $60 billion. The proteomics industry has the potential to reach similar levels of revenue and company valuations. UBS Warburg estimates that the size of the high-throughput proteomics market was approximately $150 million in 2000 and expects the market to grow at an annual rate in excess of 30% through 2003 [2]. BioInsights estimates the size of the market, as defined by the segments described in Section II, is closer to $260 million and will grow at about 40% per year through 2003. The growth could be as high as 50% if new protein chip technologies and protein–protein interaction platforms are successful [3]. Moreover, this potential, or at least the belief in this potential, is a strong driving force because it creates a positive atmosphere for investment in new technologies.

## II.  PROTEOMICS TECHNOLOGIES

Given the breadth of data required to study proteins, it is no surprise that so many different technologies are in use or under development. A complete description of proteins must include data about sequence, expression, interactions, regulation, posttranslational modifications, structure, and activity. The technologies used to generate these data in high-throughput include mass spectrometry, 2D gel electrophoresis, protein–protein interaction platforms, structure analysis methods, computational algorithms, and protein chips. These technologies have inherent strengths and weaknesses in the generation of different types of proteomics data and present different contributions to the overall market (see Figs. 1 and 2).

## A.  Mass Spectrometry

Mass spectrometry (MS) is currently the gold standard for the identification of proteins. No other technique provides such definitive data about the identity of proteins, and no other technique can distinguish large numbers of unknown proteins in high-throughput [4]. Given that much of the early proteomics research has focused on identifying proteins, MS has had a strong market position. Based on estimates from UBS Warburg and Deutsche Bank Alex. Brown [5], the total MS market for proteomics in 2000 was approximately $130 million, or about 50% of the total market, and growing at about 30% per year through 2003. Three companies, Waters, Applied Biosystems (ABI), and Thermo Electron, dominate this market with a combined market share of approximately 60%.

| Key ● Strong<br>　　○ Weak | Sequence | Expression | Interactions | Regulation | Post-Translational Modifications | Structure | Activity |
|---|---|---|---|---|---|---|---|
| Mass Spectrometry | ● | ◐ | ○ | ◐ | ◐ | ○ | ○ |
| 2-D Gel Electrophoresis | ○ | ● | ◐ | ◐ | ● | ○ | ○ |
| Protein-Protein Interaction Platforms | ○ | ○ | ● | ○ | ○ | ○ | ◐ |
| Structure Analysis Methods | ○ | ○ | ◐ | ○ | ○ | ● | ◐ |
| Computational Algorithms | ○ | ○ | ◐ | ● | ○ | ● | ● |
| Protein Chips | ○ | ● | ● | ◐ | ◐ | ○ | ◐ |

**Figure 1**　Capabilities of proteomics technologies to generate different types of protein data.

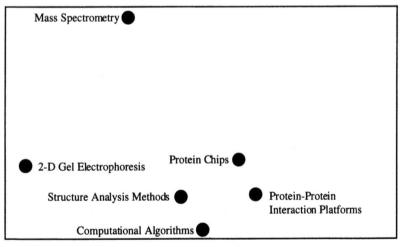

**Figure 2**　Matrix of current revenues and growth potential.

The predicted growth over the next couple of years will result from the increasing demand for current proteomics technologies. The growth rate from 2004 to 2006 could increase to near 50% per year if technological improvements are incorporated into working systems. One of the limitations of MS, however, is the delivery of proteins into the vacuum chamber. This process is generally slow and does not provide a high degree of separation, but several strategies for improvement are in development. One current strategy is to improve the integration of liquid chromatography (LC) and MS. Waters Corporation is a leader in this area and launched a system in 1999 that multiplexes up to eight high-performance liquid chromatography (HPLC) columns/analyses with one mass spectrometer. Another ongoing strategy is the improvement of the integration of 2D gel electrophoresis and MS (also next paragraph). One of the most promising approaches has been licensed by ABI from Denis Hochstrasser and the University of Geneva. The approach incorporates a molecular scanner that automates the transfer of proteins from gels to a mass spectrometer and eliminates the need to cut each protein spot individually from a gel [6]. An additional strategy in an earlier stage of development is the integration of microfluidics technologies with MS. The ability of microfluidic technologies to separate and deliver proteins in small quantities would be an excellent complement to the sensitivity of MS, but these technologies are likely to be several years away from commercial production. In addition, protein chip technologies are rapidly emerging (also see Section II.F) that have the potential to capture and isolate specific proteins that can then be delivered into the MS. All of these improvements in front-end technologies, if successful, will significantly increase demand for MS.

Mass spectrometry, however, is not only limited at the front end. A back-end limitation to MS is the technology only provides limited quantification of proteins. Many of the front-end technologies discussed earlier, such as 2D gel electrophoresis and protein chips, can quantify protein levels, but these are limited in the breadth of proteins that they can analyze. Two-dimensional gels do not resolve all types of proteins nor do protein chips capture all types. One of the most promising methods for quantification is the use of isotope-coded affinity tags (ICATs) developed by Ruedi Aebersold at the University of Washington and licensed by ABI. This technique allows for the quantification of changes in protein expression for a large number of proteins simultaneously, although some proteins escape detection because only cysteine residues are labeled by this method [7]. Moreover, ABI has recently launched reagents to support this technique. Improvements in MS are critical for the continued success of the technology.

## B.  2D Gel Electrophoresis

Companies that have focused on 2D gel electrophoresis as their core proteomics technology have met with limited commercial success, even after years of research and development. Large Scale Biology presents a good example of the difficulties of building a successful proteomics business with data from 2D gels. The company was one of the first to attempt this business model, but the first-mover advantage was not enough. The company only attributed $1.8 million in revenue to proteomics in 1999. Comparable figures are not available for the year 2000, but growth is unlikely because the company has not developed any new technologies in this area. The company is shifting away from its core 2D gel technology and is investing in other areas.

The total market for 2D gel electrophoresis systems, including all reagents and equipment, according to Jain PharmaBiotech, is approximately $600 million [8]. BioInsights estimates, however, that the primary use of these systems is for the study of select proteins or groups of proteins and less than 10% is for the high-throughput study of proteins. The contribution to the overall proteomics market is thus relatively limited, and the growth rate is likely to remain in the single digits. The majority of this growth is expected to come from improved links between 2D gels and mass spectrometry, as discussed earlier. Long-term prospects are not strong and negative growth will likely result from emerging competition such as protein chips.

The lack of success of 2D gel technologies provides an interesting case study of the challenges in proteomics. The stagnation in use of 2D gel electrophoresis is often attributed to poor technology, extremely poor reproducibility, or the inability to resolve a significant percentage of cellular proteins. Yet, 2D gels still represent the best method for rapidly separating large numbers of proteins, and the reproducibility, especially if enough replicate gels are analyzed, is adequate for identifying major changes in protein levels or posttranslational modifications [9]. At least part of the lack of success must be attributable to the fundamental difficulty of analyzing the proteome. The identification of major protein changes in the cell are rendered highly problematic due to the complexity of biological systems. Detecting subtle or dynamic changes in groups of proteins will require technologies with high degrees of accuracy and sensitivity.

## C. Protein–Protein Interaction Platforms

The perception of protein–protein interaction platforms as low-throughput changed quickly when Curagen published a landmark study [10]. The company used a high-throughput procedure to screen nearly all of the 6000 predicted *Saccharomyces cerevisiae* proteins and identified 957 putative interactions involving 1004 yeast proteins. Curagen uses this platform to examine protein pathways and validate targets. Similarly, Myriad Genetics has also industrialized the yeast two-hybrid technology and optimized library construction to reduce the rate of false positives from typical levels of 25% to 1%. Myriad uses this platform to examine pathways as well as to screen hundreds of thousands of compounds to identify inhibitors of specific protein–protein interactions. In addition, the ability to industrialize protein–protein interactions has also enticed Hybrigenics and AxCell Biosciences to generate similar data and sell these directly to pharmaceutical companies. Hybrigenics uses three forms of yeast two-hybrid interactions: an optimized version of the standard platform, a higher-throughput version in *Escherichia coli*, and a version that screens for compounds that disrupt binding. AxCell Biosciences, by contrast, uses a platform based on in vitro interactions between cloned proteins and protein domains. The focus on domains has the potential to generate higher-resolution pathway maps, but it is likely to have lower selectivity.

The market for high-throughput protein–protein interaction platforms in 2000 was approximately $23 million [11]. Myriad Genetics had over 80% of the market share when Hybrigenics and AxCell Biosciences were just getting started. The growth of these companies as well as the emergence of other players is expected to drive market growth to over 65% per year for the next few years. Demand for this data is high because it aims to define cellular pathways, but the technologies will face some significant challenges to demonstrate physiological relevance. The technologies generally do not examine protein interactions under normal conditions, but under artificial constraints that facilitate analysis. The platforms more accurately identify strong and stable interactions between proteins and may not be as accurate or selective for binding events that are more transient or persistent under specific cellular conditions. Pathway maps generated from these data will likely need to be confirmed by other proteomic analysis.

## D.   Structural Proteomics

Structural proteomics represents another area of the overall proteomics market. The leading companies appear to be successful in their early stages of development and are positioned for strong growth over the next few years. These companies were highly successful raising money in 2000. Structural GenomiX raised $77 million in a 6-month period, Stuctural Bioinformatics raised $32.6 million and also received an equity investment from IBM, and Syrrx raised $25 million in 2000 and another $54 million in January of 2001. The successful financing indicates that analysts believe in the potential value of this industry. In addition, the leading companies announced that they had achieved key milestones on previous deals or signed new deals. In the last 15 months, Structural GenomiX has signed on for a $13 million collaboration with the Cystic Fibrosis Foundation, Structural Bioinformatics reached a milestone on a deal with Yamanouchi Pharmaceuticals and signed new deals with ArQule, R. W. Johnson Pharmaceutical Research Institute, and De Novo Pharmaceuticals, and Syrrx formed a strategic alliance with the Genomics Institute of the Novartis Research Foundation. Based on these positive indicators (but without detailed financial information from the leading companies), BioInsights estimates that the size of the market is between

$10 million and $40 million, with double-digit growth rates, possibly even as high as 40% or 50% per year.

The growth of structural proteomics is tied to the growing demand for high-throughput structural data. The race is on to identify the functions of unknown proteins uncovered by the Human Genome Project and to develop new therapeutics from these. The main value from the high-throughput generation of protein structures is the putative structure–function relationships. It may be possible to infer functions based on three-dimensional structural homology to other proteins rather than using brute-force techniques of expressing and purifying proteins of unknown function and testing them in a wide range of enzymatic and binding assays. Additional value comes from the potential ability to design compounds that fit selectively and specifically into the proper sites on target proteins. These data have tremendous potential value for streamlining drug development, but biotechnology and pharmaceutical companies are not ready to take full advantage of this capability. The companies have generally steered away from rational drug design because of the risks of failure. Proteins and compounds both have complex structure, and predicting how any two molecules will interact is very difficult. The biotechnology and pharmaceutical companies have preferred in recent years to invest heavily in high-throughput screening, which comes with the near guarantee of identifying many compounds that fit the initial desired activity profile. To compete effectively, rational drug design companies will need to demonstrate consistent identification of lead compounds with superior selectivity and specificity.

The growth of Structural Proteomics is also tied to the development of technologies for structural analysis. New computational algorithms as well as new high-throughput x-ray crystallography techniques are emerging rapidly. Continued development of these technologies will likely advance at an accelerated pace. The issue remains, however, as to the rate at which these new developments will emerge. Will the current pace hold and drive growth of 30% or more? Or will the technology require a few more years before meeting the demand for protein structure data?

## E. Computational Algorithms

By definition, proteomics technologies generate enormous amounts of data. Handling and analyzing these data are critical for the understanding of the proteome, but the underlying bioinformatics is something of a conundrum. Ask almost any researcher generating genomics or proteomics data about unmet needs, and concerns regarding bioinformatic analysis arise. Yet, these same researchers, while spending hundreds of millions of dollars on high-throughput technologies, tend not to purchase commercial analysis algorithms. There are two main reasons why this high demand has not produced a successful commercial market. The bioinformatics needs of users tend to be highly individualized and dependent on the type of platform used as well as other types of data with which these must be integrated. In addition, these individualized needs change rapidly as researchers change platforms and methods. As a result, researchers are forced to develop individualized bioinformatic tools.

The market for commercial analysis algorithms for gene expression data, according to a recent study by BioInsights, was only $20 million in 2000, although the growth rate through 2006 is expected to be almost 40% per year [3]. Given that the amount of proteomics data pales in comparison at the present time, the total market for commercial proteomics analysis algorithms is likely to be less than $5 million in 2000. The growth rate will largely depend on the growth of proteomics and the success of the emerging protein interaction and protein chip technologies. If successful, the market, led by companies such as Compugen, GeneFormatics, Protein Pathways, and Proteometrics, could grow at 50% per year or more. The challenge for the industry, however, will be overcoming the difficulties in the bioinformatics market. Standardization of formats and agreement on the analysis algorithms will drive the industry to success.

## F.   Detailed Analysis of the Protein Chip Market

Protein chips are in a unique position among proteomics technologies because this technology has the potential to bring proteomics to the people or, in this case, the biologists. Unlike the required expertise for mass spectrometry, automation capabilities to set up high-throughput protein–protein interaction platforms, or computational skills for the analysis of complex algorithms or structural predictions, protein chips have the potential to be user-friendly. The researcher may only need to perform basic preparative steps and then simply place the protein sample on the chip. In addition, thousands or tens of thousands of proteins can potentially be evaluated in a single experiment. Moreover, unlike the significant costs associated with other techniques, protein chips may have unit costs as low as hundreds of dollars per chip.

This accessible platform has the potential to provide a wide range of data on protein expression, posttranslational modifications, and protein–protein interaction data in high throughput. Researchers may be able to rapidly design and execute a wide range of experiments. The development of this technology, however, is not without its challenges. The main technical hurdle is uniform, linear binding across the entire chip [12]. Unlike more predictable DNA hybridization thermodynamics and kinetics, protein binding depends on a large number of variables and conditions. Generally, each binding event must be optimized, but this is not practical for hundreds or thousands of binding events on the same surface. One potential solution to this problem is to select capture agents (also referred to as binding proteins) from a large library under uniform conditions.

Phylos, for example, uses this approach to select antibody mimics, and SomaLogic uses this approach to select photoaptamers.

Additional technical hurdles for protein chips can be divided into three categories: surface chemistry, capture agents/purified proteins, and detection (see Fig. 3). The surface chemistry must be able to immobilize capture agents or purified proteins in their active conformations. Some approaches have been relatively straightforward, such as the use of polylysine coatings [12]; others are more complex, such as the multifunction monolayer system from Zyomyx [14]. Capture agents must be able to bind proteins selectively and specifically. Antibodies are the most common agents but may be difficult to optimize across an entire chip. Expression and purification of proteins for use on chips requires high-throughput production. Interestingly, the initiative for this effort is stronger in academia,

| Company/ Research Program | Surface Chemistry | Capture Agents/ Protein Expression and Purification | Detection |
|---|---|---|---|
| Biacore | | | √ |
| Ciphergen | | | √ |
| HTS Biosystems | | | √ |
| Phylos | | √ | |
| SomaLogic | | √ | |
| Zyomyx | √ | | |
| Applied Biosystems | | | √ |
| AxCell Biosciences | | √ | |
| Biorchard | √ | √ | √ |
| CombiMatrix | √ | | |
| Dyax | | √ | |
| INTERACTIVA Biotechnologie | √ | | |
| Nanogen | √ | | |
| OGS and CAT | | √ | |
| OGS and Packard BioScience | √ | √ | |
| PROT@GEN | | √ | |
| Proteome Systems | | √ | |
| TeleChem International | √ | | |
| Patrick Brown Laboratory | √ | √ | √ |
| Brandeis University Detector Group | | √ | |
| Harvard Institute of Proteomics | | √ | |
| Lawrence Livermore Laboratory | | √ | |
| Gavin MacBeath Laboratory | √ | √ | √ |
| National Institutes of Health | √ | √ | √ |
| Purdue University | √ | | |

**Figure 3** The protein chip technical expertise of companies and research institutions. (From Ref. 13.)

through the efforts of Labaer and collegues at the Harvard Institute of Proteomics [15] and the laboratory of Joanna Albala at Lawrence Livermore National Laboratory [16]. In addition, some companies, such as HTS Biosystems and PRO-T@GEN, are developing industrial processes in this area. Detection systems are the most mature component of protein chip technologies. In fact, the only two protein chip systems currently on the market, from Biacore and Ciphergen, are based on surface plasmon resonance and mass spectrometry, respectively. These detection systems are reliable and accurate, but the protein chips from Biacore and Ciphergen are limited by low throughput. To achieve the high throughput required for proteomics, these and other chip platforms under development will likely have to integrate multiple components.

The market for protein chips is principally driven by strong demand for proteomics data (see Fig. 4). If the technology can deliver, then BioInsights expects the market to grow from $44 million in 2000 to $490 million in 2006, a compound annual growth rate of 49% [13]. Within this overall market, BioInsights expects that protein chips assembled by researchers will grow to $290 million, a com-

| Key<br>2001-2003/ 2004-2006<br>+++ Highest Positive Score<br>0    Neutral<br>– – – Most Negative Score | Pre-Fabricated Chips | | Tools to Assemble Chips | | Comments |
|---|---|---|---|---|---|
|  | Expression Profiling | Interactions Analysis | Expression Profiling | Interactions Analysis |  |
| **Opportunities and Challenges** | +/+++ | 0/++ | 0/++ | ++/+++ | The opportunities will grow if the challenges are met. |
| **Customer Demand** | +/++ | +/++ | +/+++ | ++/+++ | BioInsights expects higher demand for tools to assemble chips because of versatility. |
| **Technology** | 0/++ | –/+ | –/+ | +/++ | The technology is early in its development and is expected to improve. |
| **Applications** | +/++ | +/++ | +/++ | +/++ | The use of protein chips for both expression profiling and protein interactions will grow. |
| **Competitive Environment** | +/+++ | +/+++ | 0/+++ | ++/+++ | The emergence of new companies will drive the protein chip market forward. |
| **Alternative Technologies** | ++/– | ++/– | ++/– | ++/– | At first, alternative technologies will drive the demand for protein chips, but then they will begin to compete for research dollars. |

**Figure 4**  Main drivers for the protein chip market. Note that the drivers are not evenly weighted. (From Ref. 13.)

pound annual growth rate of 44% from 2001 to 2006, whereas prefabricated chips will grow to $200 million, a growth rate of 66% over the same period (see Fig. 5). First-mover advantage and superior versatility give the edge to protein chips that researchers can assemble themselves, even though prefabricated chips have a faster growth rate. Also within the overall market, BioInsights expects that expression profiling and interactions analysis will have similar size markets of $250 million and $240 million, respectively (see Fig. 6), with the compound annual growth rates of 40% versus 70%. The early advantage of the Biacore chips for analysis of protein interactions is balanced with the faster growth rate of expression profiling systems.

The strong growth of the protein chip market, if it progresses according to projections, will have a positive effect on the entire proteomics market. At the most basic level, more proteomics data will create more proteomics questions, and the market will grow rapidly. The largest impact, however, may come from researchers using more proteomics technologies. This will create a new and growing number of protein biologists who are interested in expanding their tools and opportunities.

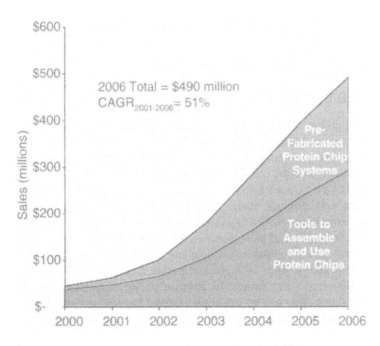

**Figure 5**   Protein chip market by chip type. (From Ref. 13.)

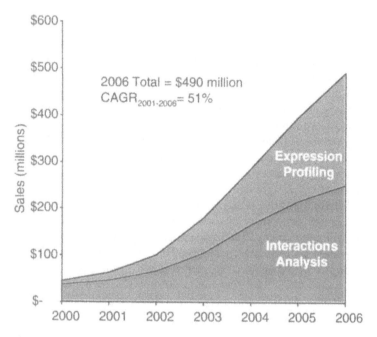

**Figure 6** Protein chip market by application. (From Ref. 13.)

## III. FUTURE DIRECTIONS

The quest for high-throughput proteomics data is running head on into the challenge of how to handle biological complexity [3]. The accumulation of more data is generally assumed to be better, but the lessons from 2D gel electrophoresis is that the broad analysis of the proteome does not quickly yield definitive results. The proteome is a very complex and dynamic system, and developing predictive models may be very difficult. Based on this challenge, BioInsights sees two potential long-term strategies. The first is to tackle the complexity of the proteome through new technologies that have increased accuracy and sensitivity as well as the development of new algorithms for handling and analyzing biological data as systems rather than a series of individual events. The leading effort in this area is being pioneered at the Institute for Systems Biology in Seattle, Washington. Leroy Hood and other researchers at the Institute have developed advanced proteomic tools and applied them to simple systems, such as the glycolysis pathway in yeast [17]. The researchers learned how to apply these data to build a successful model. The next step is to try to extrapolate this approach to more complex systems. The goal is to determine what types of data are needed and the best methods for analyzing those data. If the Institute is successful, important new techniques for handling biological complexity will emerge.

The second long-term strategy for proteomics is to make a shift toward protein function above and beyond protein–protein interactions. Instead of trying to understand the complex proteome in detail, researchers may find it more fruitful to focus on protein function, especially those most directly related to disease. There are many companies in this emerging market with very interesting technologies [6]. Rigel Pharmaceuticals, for example, probes for pathways controlling key events in a specific disease process with retroviruses that express random peptides or larger fragments from selected proteins [18]. The infected cells are rapidly screened, and if they exhibit the right phenotype, the infecting probes are identified. The functional pathway is then further elucidated to determine the best targets. If Rigel and other companies in this market are successful, they will establish new strategies for drug discovery that are highly efficient and effective.

The long-term growth of the proteomics market will depend on how companies handle biological complexity. Companies need to prepare to avoid the pitfalls of generating too much data that are too complex to analyze. In fact, proper planning could turn potential pitfalls into new opportunities to meet a new range of customer needs. In the short term, however, demand for proteomics data will drive the market forward as fast as the technology can deliver.

## REFERENCES

1. McKinsey & Co., Lehman Brothers, The fruits of genomics: Drug pipelines face indigestion, February, 2001.
2. UBS Warburg. Global equity research: Thermo Electron, December, 2000.
3. BioInsights. Profiting from biological complexity, March, 2001.
4. Shen, Y.; Smith, R.D. Proteomics based on high-efficiency capillary separations. Electrophoresis. 2002, 23(18), 3106–3124.
5. Deutsche Bank Alex Warburg, Equity Research: Waters Corporation, August, 2000.
6. Binz, P.A.; Muller, M.; Walther, D.; Bienvenut, W.V.; Gras, R.; Hoogland, C.; Bouchet, G.; Gasteiger, E.; Fabbretti, R.; Gay, S.; Palagi, P.; Wilkins, M.R.; Rouge, V.; Tonella, L.; Paesano, S.; Rossellat, G.; Karmime, A.; Bairoch, A.; Sanchez, J.C.; Appel, R.D.; Hochstrasser, D.F. A molecular scanner to automate proteomic research and to display proteome images. Anal Chem. 1999, 71(21), 4981–4988.
7. Gygi, S.P.; Rist, B.; Gerber, S.A.; Turecek, F.; Gelb, M.H.; Aebersold, R. Quantitative analysis of complex protein mixtures using isotope-coded affinity tags. Nature Biotechnol. 1999, 17(10), 994–999.
8. Jain PharmaBiotech. Proteomics—Technologies and commercial opportunities, April, 2001.
9. Wilkins, M.R.; Gasteiger, E.; Gooley, A.A.; Herbert, B.R.; Molloy, M.P.; Binz, P.A.; Ou, K.; Sanchez, J.C.; Bairoch, A.; Williams, K.L.; Hochstrasser, D.F. High-

throughput mass spectrometric discovery of protein post-translational modifications. J Mol Biol. 1999, 289(3), 645–657.

10. Uetz, P.; Giot, L.; Cagney, G.; Mansfield, T.A.; Judson, R.S.; Knight, J.R.; Lockshon, D.; Narayan, V.; Srinivasan, M.; Pochart, P.; Qureshi-Emili, A.; Li, Y.; Godwin, B.; Conover, D.; Kalbfleisch, T.; Vijayadamodar, G.; Yang, M.; Johnston, M.; Fields, S.; Rothberg, J.M. A comprehensive analysis of protein–protein interactions in *Saccharomyces cerevisiae*. Nature. 2000, 403(6770), 601–603.

11. BioInsights. Functional proteomics: Short cuts to drug discovery, May, 2001.

12. Haab, B.B.; Dunham, M.J.; Brown, P.O. Protein microarrays for highly parallel detection and quantitation of specific proteins and antibodies in complex solutions. Genome Biol. 2001, 2(2), 1–13.

13. BioInsights. Protein chips: The race for high-throughput protein analysis, December, 2000.

14. Ruiz-Taylor, L.A.; Martin, T.L.; Zaugg, F.G.; Witte, K.; Indermuhle, P.; Nock, S.; Wagner, P. Monolayers of derivatized poly(L-lysine)-grafted poly(ethylene glycol) on metal oxides as a class of biomolecular interfaces. Proc Natl Acad Sci USA. 2001, 98(3), 852–857.

15. Braun, P.; Hu, Y.; Shen, B.; Halleck, A.; Koundinya, M.; Harlow, E.; LaBaer, J. Proteome-scale purification of human proteins from bacteria. Proc Natl Acad Sci USA. 2002, 99(5), 2654–2659.

16. Albala, J.S.; Franke, K.; McConnell, I.R.; Pak, K.L.; Folta, P.A.; Rubinfeld, B.; Davies, A.H.; Lennon, G.G.; Clark, R. From genes to proteins: high-throughput expression and purification of the human proteome. J Cell Biochem. 2000, 80(2), 187–191.

17. Ideker, T.; Thorsson, V.; Ranish, J.A.; Christmas, R.; Buhler, J.; Eng, J.K.; Bumgarner, R.; Goodlett, D.R.; Aebersold, R.; Hood, L. Integrated genomic and proteomic analyses of a systematically perturbed metabolic network. Science. 2001, 292(5518), 929–934.

18. Peelle, B.; Gururaja, T.L.; Payan, D.G.; Anderson, D.C. Characterization and use of green fluorescent proteins from Renilla mulleri and Ptilosarcus guernyi for the human cell display of functional peptides. J Protein Chem. 2001, 20(6), 507–519.

# Index

26S proteosome, 244
2Yt, 185
384-well microtiter plate, 129
4C1N, 194
4EBP1, 287, 289, 290, 293
5B6C3IG, 193, 194
6C3IG, 194
70-kDaS6 ribosomal protein serine/
       threonine kinase (p70S6K), 284
96-channel manifold, 186
96-well electroporator, 182
96-well format, 303
96-well microtiter plates (MTPs), 21, 129,
       135, 155, 181, 185, 187, 258
   filter bottom, 185
ABI, 340
Absorptive process, 26
Abundance, 161
Abundance relative, 246
Accessible solvent, 311
Accessibility, 148, 151
Accession steric, 177
Accuracy, 128, 136, 239, 341, 348
Acid hydrolysis, 21
Acquired immunodeficiency syndrome,
   119

Acrylamide, 166
Activated ester, 167
Activated surface, 166
Activation
   Cdc42, 291
   cyanuric chloride, 163
   cycle, 293
   −deactivation cycles, 292
   FRAP, 291
   growth factor−mediated P13K, 293
   PKB, 286
   PP2A, 291
   Rac1, 291
   surface, 149
   transcriptional, 261, 262, 263
Active site, 317
   proteins, 192
Active state, 160
Activity
   enzymatic, 177, 309
   enzyme, 319
   protein, 338
   -based protein profiling, 226
Acylation, 205
Adjuvant, 11

Adsorption, 113, 114, 129–132, 148–152,
    163, 164, 166, 167
  electrostatic, 167
  signal, 188
Adsorptive attachment, 130
Adsorptive surface, 192
Adverse drug effects, 48, 66
Aeration, 185
Affibodies, 9
Affinity, 29, 104, 111, 238, 285, 332
  -based approaches, 156, 238
  binding, 11, 129, 131, 132
  capture, 46
  chromatography, 132, 148, 163, 188,
    221
  columns, 182
  constant (K), 96, 107, 111
    capture-agent, 107
  effective binding, 98
  enrichment, 8, 33
    dual, 36, 37
  interaction, 160
    biosensors, 160
  ligands, 6, 8, 11, 12, 14, 30, 32, 33, 35,
    42, 55, 66, 67, 189, 192
    parallel generation, 32
    parallel screening, 32
  matrix, 182, 183, 187
  metal, 220
  protein enrichment, 237
  purification, 203, 207, 224, 238, 245,
    187
  reagents, 238
  ribosomal-binding, 331
  tagged proteins, 304
  tags, 175, 176, 178–181, 182, 187,
    206, 238, 303, 304
    isotope-coded, 340
Agar plates, 19
Agar trays, 176, 178
Agarose, 131, 192
  gels, 36
AGC class of protein kinases, 286
Aggregating randomly, 305
Aggregation, 182

Agilent, 190
Agriculture, 81
Albumin, 5, 160, 163, 164, 219
Aldehyde, 130, 131, 307
  groups, 131, 132, 167
Algorithms, 40, 57, 65, 240, 241, 255,
    344, 348
  analysis, 343, 344
  computational, 338, 343
  database search, 235
  development, 57
  genetic, 65
  genome annotation, 255
  parallelized computer, 57, 58
  proteomics analysis, 344
  self-learning
    supervised, 65
    unsupervised, 65
  spot-matching, 220
  word-building, 40
Alkaline labile, 225
Alkaline phosphatase (ALP), 177
Alkylation, 242
Allergy, 119
Allosteric regulation, 282
Allylamine, 163
ALP, 193
Alpha helices, 309, 311, 314
α-ω,functionalized PEG, 164, 300, 301,
    305
Alpha peptide, 176, 177, 190
α/β fold, 311
Alternative splicing, 217, 255
Alzheimer's disease, 226
Ambient analyte
  assay, 94–100, 112, 121
  concentration, 97, 99, 109
  ligand assay, 94, 112
Ambient temperature, 305
Ambiguity, 307
Amide nitrogens, 312
Amine groups, 23, 131, 133, 149
Amine-reactive cross linker, 131
Amine-reactivity, 131

Amino acids, 9, 41, 284, 318
  sequence, 235, 240, 241
    fitting, 307
  sequencing, 328
Amino dextran, 165
3-Amino-9-ethylcarbazole, 193
Amino groups, 23, 148, 149, 152
Amino-derivatized dextran, 167
Amino-modified dextran, 164
3-Amino-propyl triethoxy silane, 113
Amino-terminus, 178
Aminosilanes, 131, 164
Aminosilane/succinic anhydride/*N*-
  hydroxysuccinimide linker
  chemistry, 164
Aminosilanized surfaces, 164
Aminotransferases, 307
  PLP-dependent, 307
3-Aminotriazole, 264
  resistance, 264
Ammonium carbonate, 187
Amplification, 148, 210, 214, 217, 301
Amplification target analyte, 83
Amplified detection signal, 135
Amplitude structure factor, 321
Analysis, 281, 319
  array-based, 190, 194
  biochemical, 173, 192, 208, 300, 193
  biophysical, 299, 300
  colorimetric, 213
  differential, 241
  expression, 283
  ex vivo, 260
  fluometric, 288
  functional, 174, 184, 194, 203, 204,
    208, 233
  genetic, 82
  heuristic, 11, 65
  highly parallel, 127
  high-throughput, 127
  LC-MS, 260
  mass spectrometric, 242, 245, 306
  microscopic, 288
  principal component, 65
  protein high-throughput, 217

[Analysis]
  protein structure, 342
  proteome, 242, 259, 243, 342
  RNA, 235
  sequence, 241
  serial gene expression, 233
  in silico, 257
  structural, 195, 203, 208, 338
  ultrahigh-throughput, 154
Analyte, 7, 8, 18, 26, 28, 29, 82, 83,
    86–88, 91–93, 95, 97, 99, 103,
    105, 107–109, 111, 112, 119, 120,
    122, 128, 133, 138, 150, 154, 155,
    160
  accessibility, 28
  analog, 89, 90, 92, 103, 89
  -binding sites, 90
  concentration, 30, 98, 128, 136, 86, 97
    gradient, 110
  consumption, 190
  -containing medium, 91, 99
  -containing solution, 104
  -containing volume, 91
  diffusion, 98, 109, 110
  enrichment, 8
  label, 88, 92
  low-abundance, 22
  migration, 109
  mixtures, 150
  molecules
    labeled, 90
    unlabeled, 90
  surface density, 105
  throughput, 7
  total, 96
  unbound, 90, 97
  volume, 8
Analytical performance, 161
Analyzer, 116
Angle, 322
Anhydroteracyclin (AHT), 184
Anion-exchange
  chromatography, 227
  column (Q sepharose), 222
Annealing refinement, 320
  simulation, 307

Annotation
  functional, 257
  genome-wide functional, 282
  structural, 257
ANOVA parallelized, 64
Antibiotic resistance, 175, 177
  gene, 206
Antibodies, 9, 10, 12, 17–19, 23, 28, 29,
      32, 36, 42, 48, 81, 84, 87, 89, 90,
      101, 104, 105, 110, 111, 113, 114,
      119–122, 127, 128, 130, 132, 135,
      152, 155, 166, 167, 174, 189, 190,
      195, 212, 223, 238
  active site, 131
  antiphosphoserine, 224
  arrays, 23, 30, 55, 113, 122, 136, 139,
      191
  binding properties, 17
  binding sites, 86, 90, 91, 110
  capture, 92, 111
  chips, 23, 54
  monoclonal, 9, 10
  parallel generation, 10
  phage display, 9
  polyclonal, 9
  spotted, 131
  concentration, 137
  covalent attachment, 223
  density immobilized, 154
  epitope, 138
    specific, 204
  Fab fragments, 132
  fluorescently-labeled, 152
  fragmentary, 155
  fragments, 155, 156
  generation, 174, 175, 193, 195
  immobilization, 152
  labeled, 90, 92, 106
    anti-idiotypic, 92
  microarrays, 135
  mimics, 9, 344
  monoclonal, 13, 15, 17, 36, 84
    production of, 12
  occupancy, 110

[Antibodies]
  parallel screening, 32
  performance, 138
  production, 12, 135, 136, 184
    large-scale, 184
  purification, 187
  scFc, 135, 136
  screening, 11
  secondary, 132
  selectivity, 345
  specific, 89, 135, 195, 345
  spot, 136
  therapeutic, 48, 49
Antibody–antigen binding, 84
Antibody–antigen interactions, 129
Antibody–antigen pair, 136
Antibody-based immunoassays, 155
Antigen–antibody pairs, 161
Antigens, 9, 11, 20, 84, 90, 110, 119,
      121, 132, 136, 152, 161, 166,
      167
  abundance, 20
  accessibility, 20
  binding of, 131, 132
  cognate, 152
  concentration of, 18
  cross-reactive, 111
  detection of, 191
  parallel generation, 11
  performance of, 138
  recombinant, 9, 11, 12
  spot, 128, 136
  valence, 18
Antisense RNA, 292
Apoptosis, 291, 305
Applied Biosystems, 187, 339, 345
Aptamers, 9, 10, 84, 122
Arabidopsis thaliana, 3
Arabinose (araBAD), 182, 184
Arabinose anhydrotetracyclin, 182
Archaebacteria, 305
Arg209, 317
ArgC, 305, 306
ArQule, 342

Array-based
  assays, 46
  nucleic acid microarrays, 114
  protein assay, 83
  proteomics, 2, 8, 10, 22, 23, 26, 50, 52,
    53, 66, 162
  technologies, 20
Arrayed colonies, 135
Arrayer piezoelectric, 159
Arrayer
  DNA, 151
  ring, 159
  mechanical, 113
  piezo, 159
  pin, 159
  ring and pin, 159
Arraying, protein, 23
Arrays, 11–13, 55, 62, 89, 113, 115, 121,
    122, 129, 133, 152, 191, 193–195,
    215, 326
  antibody, 2, 15, 19, 20, 30, 55, 66, 113,
    120, 122, 128, 175
  antigen, 155
  cDNA, 18
  cellular, 20
  chromatography affinity capture, 20
  construction, 121
  DNA, 20, 154
    high-density, 218
  ESTs human, 267
  formats, 160
  genechip, 326
  generation, 193
  high-density, 17, 83, 115, 150
  immunogens, 15
  lipid, 55
  macroscopic, 155
  microspot, 82
  nucleic acid, 83, 156
  oligonucleotide, 17, 18, 120
  peptide, 2, 17, 20, 128, 132, 138
  protein, 2, 11, 12, 19, 20, 30, 55, 66,
    121, 128, 175
  reverse, 19, 20
  spotted proteins, 128

[Arrays]
  surface, 159, 160, 220
  manufacturing, 129
  peptide, 17
  protein, 19
  tissue, 46
Artifacts protein-labeling, 121
Artificial sites recombination, 258
Aspartic acid, 245
AspN, 305, 306
Assay, 89
  activity, 319
  ambient analyte, 100
  bead-based, 149
  β-galactosidase, 264
  binding, 343
  colorimetric, 209
  competitive, 89, 90, 92–94, 100, 101,
    111, 120
  data, 87
  design, 85, 87, 89, 92
  enzymatic, 148, 343
    high-throughput, 215
  fluorescent, 283, 285, 286
  format, 118, 119
  functional, 156, 260
  immunometric, 104
  in-solution, 56
  interaction, 56
  kinase, 225
  ligand, 82, 83, 90
    ambient analyte, 112
  microspot, 97, 98, 101
    kinetics, 107
    radiometric, 100
  miniaturization, 112
  multianalyte, 115, 120, 190
  multiplexed, 53
  noncompetitive, 84, 89, 90, 92, 93, 94,
    100, 101, 104, 111, 120
  nucleic acid, 119
  protein-binding, 84, 91
  protein fragment complementation, 278
  resin-based, 149
  sandwich, 112, 120, 154, 155, 195

[Assay]
   sensitivity, 106
   solid phase, 114
   stringency, 279
   survival, 285
      selection, 283, 285, 287
Association rapamycin-enhanced serum-
      indiced, 291
Association rate, 108, 109
Asymmetric, 311
   unit, 319
AT/GC content, 43
Atomic force microscopy (AFM), 29, 54,
      121, 133
Atomic structure, 279
Atoms, 311, 322
   backbone, 312
   heavy, 307
   side-chain, 312
ATP, 177, 195, 225
ATPase, 5
Atrazine, 164, 167
Attachment, 129, 131, 149, 150
   chemistries, 129
   methods
      adsorption, 132
      affinity binding, 132
      covalent, 132
      multiple-point, 149, 150
      random, 149
      site, 150
AttB site, 258
AttL site, 258
AttP site, 258
AttR site, 258
Attrition
   affinity enrichment, 39
   amplification, 39
   cloning, 39
   transcription, 39
   translation, 39
Autoantigens, 132, 138
Autofluorescence, 177
Autofluorescent protein (AFP), 177
Autogen, 176

Autoimmune diagnostics, 155
Autoimmune disease, 138
Autoimmunity, 19, 155
Automated data collection, 300
Automated devices, 175
Automated DNA sequencing, 304, 326
Automated high-throughput approach,
      145, 176
Automated protein expression, 193
Automated protein purification, 193
Automatic devices, 188
Automation, 8, 36, 66, 187, 205, 215,
      237, 262, 344
Autophosphorylation, 290
Autoradiography, 218
Autoreactive sites, 139
Avogardro's number, 98
Average molar mass, 163
Avidin, 131, 187, 190, 225, 226
   beads, 225
   biotin, 224
α,ω-amino functionalized PEG, 163
AxCell Biosciences, 341, 342, 345


B factor refinement, 320
B-amyloid, 226, 227
B cells, 11, 12, 15
B-cell lymphoma, 65
B-GAL, 176, 177, 179, 193
B-galactosidase (B-GAL), 176–178
   activity, 264
*Bacillus subtilis*, 173, 304, 330, 331
BACE, 226
BACE1, 226
BACE2, 226
Backbone, 320
   atoms, 312
Background, 131, 162
Background fluorescence, 116, 195, 288
Background noise, 103, 113
Background signal, 103
Bacteria, 1, 6, 119, 173, 183, 184, 292,
      307, 316, 327, 316
Bacterial cells, 181, 185, 186, 292

Bacterial colonies, 19, 135
Bacterial expression, 183
  system, 205
Bacterial release protein (BRP), 186
Bacterial species, 307
Bacterial two-hybrid system, 261
Bacterial vectors, 182
Bacteriolytic properties, 186
Bacteriophage, 155
Bacteriophage lambda, 258
  integration, 258
Bacteriophage particles, 155
Baculovirus, 183, 185, 203, 205, 208,
    211, 214
  -based protein production, 206
  -based strategy, 203
  expression, 215
  infection, 27, 206
  production, 205, 206, 210
  -recombinant, 206, 209, 210, 214
  replication, 205
Bait, 156, 238, 239, 258, 261, 262–264,
    266, 268, 269
  cognate, 268
  collection, 266
  dual, 264
  multiple, 262
  protein, 156, 238
  –prey interacting pair, 262–264, 267
  swapping, 263
Barrier, 164
Bead, 156, 226, 238
  avidin, 225
  based assays, 149
  iodacetic acid-linked, 225
  magnetic, 187, 189
  paramagnetic, 189
    core, 189
Beam line, 319
Becton Dickinson, 176
Beetles, 177
Benzonase, 187, 190
β-amyloid, 226, 227
  amyloidogenic forms, 226

β-barrel, 311, 314
β-galactosidase activity, 264
β-glucuronidase, 209, 212, 213
β-secretase, 226
β-strands, 311
Betain, 132
Biacore, 149, 163, 345, 346
Binders, 9, 11–13, 18, 20, 29, 47, 48,
    113, 135, 136
  high-affinity, 10, 12, 47
  high specificity, 12
Binding
  affinity, 91, 332
  agent, 82, 83, 87, 91, 92, 95, 98, 100,
    102, 105–109
    structurally specific, 91
  conditions, 220
  covalent, 129
  electrostatic, 129
  environments, 312
  events
    persistent, 342
    transient, 342
  interactions, antibody–antigen, 129
  linearity, 344
  noncovalent, 192
  partner, 128
  proteins, 344
  reaction, 96, 97, 107, 111, 122
    kinetics, 108, 109
    rate, 110
  reversible, 84
  sites, 48, 97, 98, 100
    concentration, 97, 104
    DNA-metal, 138
    occupancy, 100
    sensor antibody, 105
  -specific, 81, 130, 131
  strengths, 128
  uniformity, 344
Bioassays, 46
  cell-based, 46
Biochemical analysis, 173
Biochemical genomics, 30

Biochemical inactivation, 316
Biochemical machinery, 281
Biochemical model, 331
Biochemical networks, 282, 294, 279, 281, 292, 294
  mapping, 282
Biochemical organization, 277
Biochemical pathways, 282, 279, 280
  mapping, 277
Biochemical processes, 238, 281
Biochemical response, 280
Biochemistry, 301, 303
  preparative, 304
Biochip, 2, 5, 8, 17, 19, 25, 28, 30, 35, 57–59, 65, 81, 218
  architecture, 148
  cDNA, 19
  design, 155
  manufacture, 25
  material, 148
  readers, 59
  surface, 150, 151
    coating, 148
  technology, 148, 151
Biocompatibility, 23
  layer, 166
Biocompatible polymer, 163
Biocompatible surfaces, 163
Biocomputing, 56, 59
Bioinformatic analysis, 173, 303
Bioinformatic platform, 327, 328
Bioinformatic tools, 326
  individualized, 343
Bioinformatics, 145, 173, 228, 303, 343
Bioinformatics market, 344
BioInsights, 338, 341, 342, 343, 346, 347
Biological complexity, 347–349
Biological function, 260, 266
Biological information, 327
Biological macromolecules, structures of, 300
Biological module, 268
Biological networks, 327
Biological processes, 248, 145, 234, 256, 268, 294

Biological relevance, 278
Biological structures, 145
Biological system, 326
  complexity, 334
Biomarkers, 138, 221, 222
  discovery, 139, 218, 221, 222
  identification, 138
Biomaterials, 150
  engineering, 150
Biomolecules, 145
  active state, 160
  interactions, 46
Biopanning, 9, 11, 135
Biophysical analysis, 299
Biophysical criteria, 304
Biorchard, 345
Bioreactors, 206
Biorobotics, 176
Biosensors, 160, 162, 163
  surfaces, 166
Biosynthesis, 307
Biotech analysts, 29
Biotech companies, 29, 82
Biotin, 29, 131, 167, 168, 169, 189
  hydrazide-derivatized, 132
Biotin-streptavidin, 29, 169
Biotinylated capture, 130, 131
Biotinylated proteins, 131, 187, 225
Biotinylation, 167, 169, 225, 226
Biotrace, 193
Biotynylation, 167
Biphasic LC column, 244
Bladder disease, inflammatory, 222
Blocking, 21, 23, 25, 26, 113, 129, 130, 160, 162, 187, 290
  agents, 25, 130
Blood, 23, 81, 154
  sera, 132
Blood serum, 138
Blotting, 149
Blue/white screening, 176, 178, 190
Body fluids, 19, 85
Boehringer Mannheim, 106, 112, 115, 116, 121, 122

Bond, 322
Bottlenecks, 148, 174, 301
Bovine serum albumin (BSA), 23, 131,
    160
Brandeis University Detector Group, 345
Breast cancer, 61, 138
    markers, 138
5-Bromo-4-chloro-3-indolyl phosphate/
    nitro blue tetrazolium (BCIP/
    NBT), 193
5-Bromo-6-chloro-3-indolyl-β-D-
    galactopyranoside (5B6C3IG),
    177, 193
Brownian motion, 305
BSA, 23, 25, 131, 160, 162, 165, 167,
    169, 170
Buffer reservoirs, 189
Bulk solvent correction, 320

c-erb-2, 138
C-terminal site, 286
C-terminus, 287, 311
Caenorhabditis elegans, 3, 173, 240, 257,
    258, 259, 262, 265–268, 304, 305,
    307, 316, 318
C18, 242
calcium dynamics, 330
Calcium-binding protein, 227
Calyculin, 288, 291
Calyculin A, 289
    sensitivity, 291
Cambridge Antibody Technologies, 345
Cancer, 63, 65, 218, 221
    cell, 291
        lines, 291
    patients, 139
Capillary, 240
    -based spotters, 151
    column, 243
    effects, 192
    electrophoresis (CE), 1, 190, 223, 234
    glass, 113
Capture agents, 100, 104, 106, 107, 121,
    155, 155, 344, 345
Capture antibody, 92, 111, 131
Capture probes, 159

Capture reagent, 101
Capture-agent affinity constant, 107
Captured antigen, 132
Carbohydrate, 132
Carbon backbone, 224
Carbon dioxide, 205
Carbon source, 280
Carboxy groups, 23
Carboxy-substituted polymers, 164
Carboxylic acid groups, 149
Carboxylic anhydride groups, 165
Carosel Magnetic Levitation Stirrer, 210,
    211
Cartesian robot, 21
Catalysis, 316, 317
Catalytic activity, 226
Catalytic site, 285
    PP2A, 291
Catalyzes, 307
Cathepsin D, 226
Cation-exchange resin, 242
Caveats, 55, 237
Cavities, 311
Ca$^+$ chelating, 186
CCD-based devices, 116
CCD camera, 176, 178, 179
ccdb, 178
Cdc2p, 245
Cdc42, 291, 292
cDNA, 8, 18, 34, 39, 59, 63–66, 148,
    174, 176–181, 193, 203–207, 209,
    210, 213, 214, 257, 265, 266, 278,
    301
    biochip, 2, 19, 56, 61
    chips, 35
    equimolar, 267
    expressed sequence tags, 257
    inserts, 175–177, 179, 180, 183, 266
        short, 180
    libraries, 33, 135, 258, 262, 265, 266,
        267, 278
        screening, 278
    microarrays, 129, 135, 233
        platforms, 129

CE, 234
coupled Fourier transform ion-cyclotron resonance mass spectrometry, 223
Cell, 5, 6, 7, 15, 20, 46, 54, 135, 146, 147, 175, 176, 177, 182, 185, 186, 277, 278, 281, 327
architecture, 330
bacterial, 182
biology, 301, 303
culture insect, 206
cycle, 256, 279
regulation, 245
death, 182
density, 211
eukaryotic, 181
fluorescein-activated, 280
fluorescence emitting, 179
geometry, 330
growth, 206, 212, 247
heterogeneity, 46
insect, 207, 209
adherent, 210
culture, 211, 212
lines
bacterial, 32
baculovirus, 32
mammalian, 32
yeast, 32
lysates, 46, 154, 207, 221, 226, 235, 237, 242, 243, 248, 260
lysis, 184, 246
mammalian, 181, 182, 280
map proteomics, 233
negative-stained, 179
nonexpressing, 177
pathways, 330
physiology, 325
processes, 330
proliferation, 263
reactions, 330
recombinant, 177
RFP expressing, 177
sorting, 9
staininng, 176

[Cell]
survival, 177, 279, 285
type, 256
viability, 206, 211, 212
wall, 5, 186
Cellular arrays, 20
Cellular component, 294
Cellular differentiation, 265
Cellular processes, 245
Cellular function, 265
Cellular location, 287, 283
Cellular processes, 234, 266, 282, 326
Cellular response, 281
Cellulose, 113, 128
supports, 128
Cellulose-binding domain (CBD), 183
Center for Proteomics Research and Gene-Product Mapping, 21
Centrifugation, 186, 211, 319, 212
Chaperones, 203
Characterization, 235, 245
biochemical, 301
biophysical, 301
high-throughput, 300
microchemical, 328
Charge, 242
Charge states, 241, 305
Chelating agent, 186
Chemical lysis, 186
Chemical modification, 192
Chemical pathways, 277
Chemiluminescence, 102
enhanced, 52
Chemistry
of nucleic acids, 128
orthogonal, 150
protein, 130
Chip, 10, 62, 115, 116, 152
DNA, 337
images, 328
prefabricated, 346
surface, 151
Chitin, 207
binding protein, 304
Chitosan, 164

Chloroamphenical acetyltransferase
    (CAT), 177
4-Chloro-1-naphtol (4C1N), 193
6-Chloro-3-indolyl-β-D-galactopyranoside
    (6C3IG), 193
CHO cells, 285–287
Chromatographic beads, 188
Chromatographic matrices, 187, 189
Chromatographic methods, 146
Chromatographic techniques, 146
Chromatography, 1, 131, 146, 191, 237,
    260
  affinity, 163, 187, 207, 221
    capture arrays, 20
  anion exchange, 220, 227
  cation exchange, 220
  column, 207, 209
  high-pressure, 242
  high-resolution, 242
  interaction, 239
  ion-exchange, 219, 221
  liquid, 187, 223, 224, 234, 239, 242,
      243, 260, 306
    biphasic, 244
    high-performance reverse-phase, 223
    high-pressure, 234
    multiplexed, 339
    reverse-phase, 235
    single-channel, 187
  metal affinity, 220, 225
  metal-ion-affinity, 183
  microcapillary, 242
  multidimensional, 7, 243
  nanoscale, 234
  normal-phase, 220
  retentate, 220
  reverse-phase, 220
  two-dimensional, 239, 243
Chromosome, 258
Chymotrypsin, 304
Cibacron blue, 219
CIP, 188, 189
Ciphergen, 220, 345, 346
Cleaning in place (CIP), 183, 188, 189

Cleavage
  sites, 306
  ubiquitin-mediated, 263
Cleft, 316, 317
Clinic, 58
Clinical samples, 221
Clinical specimens, 223
Clinical trials, 47
Clogging, 207
Clone, 45, 174, 175, 176, 178, 180, 181,
    182, 184, 204, 267, 278
  cDNA, 208
  destination expression, 258, 259
  DNA fragment, 258
  entry, 258, 259
  EST, 268
  expression, 175, 176, 259
  fluorescent, 177
  full-length, 259, 265, 267, 268
  human genes, 268
  IMAGE, 203, 207
  in frame, 204
  low abundance, 268
  nonexpressing, 175, 176
  nonrecombinant, 178
  ORF, 268
  overlapping, 181
  phage display, 128
  pooled, 174
  positive, 283
  prey, 266, 267
  read-through, 180
  RFP, 178
  scFv, 136
  sequencing, 262
  survival, 176
  truncated, 267
  unidirectional, 258
  unstable, 45
Cloned fragment, 258
Cloned genes, 303
Cloning, 36, 205, 255, 301, 303, 318
  cDNA library, 257
  directional, 180
  domains, 181

[Cloning]
  expression, 277, 278
  functional, 278
  Gateway, 258
  genome-scale, 257
  large-scale, 195, 277
  limitations, 257
  ORFeome, 257, 258, 269
  recombinant, 135, 258, 265, 268, 269,
    177
  sites, multiple, 206, 209
Clontech, 209
Clusters
  hypothetical, 268, 269
  protein interactions, 268
Coat proteins, 155
Coated plates, 190
Coating oligomers, 166
Coating polymers, 166
Cocoating surfaces
  anionic proteins, 130
  denatured proteins, 130
Code, 61
Coding sequence, 318
Codon, 184, 318
  Adaptation Index, 43, 219, 244
  bias, 332
  rare, 43, 184
  stop, 209
  termination, 318
Coefficient of variation, 128
Cofactor, 305
  flavin, 309
  organic enzyme, 307
Cognate antigen, 152
Collision
  cell gas-phase, 234
  -induced dissociation, 234, 235
Colonies
  blue/white, 176
  lifts, 19
  lysis, 19
  phage display, 139
  picking, 33, 35
Colorimetric gold nanoparticle sensors, 54

Colorimetric resonant reflection, 54
Colorimetric-staining, 195
Colorimetry, 195, 279
Column, 188, 189, 319
  biphasic, 242
  capillary, 243
  cleaning, 188
  microcapillary, 242
  positive pressure, 189
  regeneration, 188, 189, 191
CombiMatrix, 345
Combinatorial chemistry, 82
Combinatorial libraries, 9, 10, 17
Common denominator, 14, 18
  domain, 66
Comparative fluorescence, 133, 136
Comparative models, 318
Compartmentalization cellular, 55
Competent cells, 181
Competitive, 93
Competitive assay, 94
Complementary binding partner, 128
Complementation, 278
Completeness, 321
Complex
  antibody–antigen, 52
  enzyme-substrate, 263
  functional, 260
  mixtures, 242
  molecular, 266
  multienzyme, 281, 284
  multiprotein, 281
  protein, 239, 282
  Rac1-Cdc42-p70S6K, 291
Complexity, 327, 347
  analysis, 58
  biological, 348
Compounds, 341
Compugen, 344
Computational tools, 326, 330
Computation recursive, 332
Computer cluster, 241
Computer multiprocessor, 241
Computing
  high-speed, 300
  infrastructure, 56

Con A lectin, 225
Concentration, 29, 37, 46, 53, 85, 87, 88,
    91, 95, 107, 112, 161, 169, 211,
    305, 306
  ambient analyte, 97, 109
  analyte, 86, 97, 98, 103, 136
    minimum detectable, 105
  antibody, 88, 137
  binding agent, 104
  binding reagent, 98
  binding site, 104, 96, 97
  capture antibody, 111
  detergent, 130, 131
  free ribosome, 333
  gradient, 109
    analyte, 110
  metabolite, 334
  mRNA, 333, 334
    normalized, 333
  protein, 219, 334
    normalized, 333
  ribosomal, 331
  salt, 242
  sensing, 96
  time-dependent, 330
Conductivity electrochemical, 195
Confocal laser scanner, 116
Conformation active, 344
Conformational flexibility, 304
Conjugation, 134
Connections
  intercluster, 268
  intracluster, 268
Contact printing, 45, 151, 193
Contact sites, 138
Contact-free dispensing process, 152
Content chip, 66
Contrast, 241
Control, 214, 222, 285
Conventional antibody production
    methods, 136
Conventional proteomics, 159
Coomassie blue, 39, 213, 218, 219, 227,
    235, 319
Coomassie gels, 235

Copolymers, 169
Core β-barrel, 311
Coregulation, 234
Correlation coefficient, 320
Costs, 182, 189, 190
Cotransformation, 262
Cotranslational modifications, 2, 33, 56
Coupling, 148
  agent, 149
Covalent attachment, 131–133, 149, 150,
    160, 162, 164, 207
  antibodies, 223
  strategy, 131
Covalent binding, 129, 130, 131
  protein, 150
Covalent bonding, 23, 150
Covalent coupling, 162
Covalent interaction, 152
Covalent structure, 304
Coverage
  genomic, 267
  minimal, 267
Cowen, 337
Creator vector system, 35
Cross reactivity, 12, 13, 33, 51, 52
Cross-linking, 130, 131
  gluteraldehyde, 131
Cross-reactants concentrations, 112
Cross-reactivity, 111
Cryptates rare-earth, 114
Crystal
  freezing, 300
  growth, 300
  quality
    diffraction limits, 305
    maximizing size, 305
    morphological appearance, 305
  structure, 300, 303, 307, 309, 311, 314,
    318
    determination, 300
Crystallization, 174, 181, 301, 304, 305,
    319
  conditions, 301
  high-throughput, 300
  serial, 304
  trials, 304, 305

Crystallography, 299, 300, 319
    analysis, 319
    model, 311
    X-ray, 215
Crystals, 301, 303, 305, 307, 319
    diffraction-quality, 301
C-terminal fusions, 303
Culture medium, 185
Culture volume, 185
Curagen, 341
Current revenues matrix, 339
Cy3, 64, 120, 121, 133, 136
Cy5, 65, 120, 121, 133, 136
Cy5-FKBP12, 160
Cyanuric chloride activation, 163
CyBio, 188
Cysteine, 226, 340
Cysteine
    oxidized, 225
    residues, 147, 246
    thiols, 149
Cysteinyl residues, 224
Cystic Fibrosis Foundation, 342
Cytomation, 176
Cytosol, 286, 287, 289, 292, 293
Cytosolic, 288, 290

D. melanogaster (see Drosophila
    melanogaster)
DALI, 311, 314
Dam 1p, 244
Data, 61, 65, 326, 327, 59
    analysis, 326, 328
        platforms, 334
    automation, 235
    biochemical, 330
    biochip, 59
    biological, 326, 328
    collection, 145, 319, 320
        automated, 300
    communication, 326
    dissemination, 301
    DNA sequence, 277, 325, 326, 334
    electrophysiological, 330
    filtering, 241
    gene expression, 343

[Data]
    genechip, 325
    genomic, 145
    handling, 328
    independent, 65
    integration, 328, 329, 330, 331, 334,
        343, 327
    integration mRNA-DNA, 327
    interpretation, 235, 327
    interrogation, 58, 59
    MAD, 301
    management, 59
    manipulation, 343
    mass-spectral, 242
    mass spectrometric, 240
        tandem, 240
    mRNA expression, 326
    peptide mass fingerprint, 240
    processing, 65
    protein structure, 343
    proteomic, 325s, 337s, 344s, 346s,
        347s, 349
        high-throughput, 347
    quality, 240
    storage, 59, 326
    structural, 307
    transfer, 327
    types, 327
    variance, 59
    visualization, 65
Databanks, 326, 328, 330
    DNA sequence, 326, 328, 330
Database, 208, 209, 214, 221, 237, 239,
        240, 299, 318, 330
    2D gel electrophoresis, 328
    expresses sequence tag, 240
    gene ontology, 294
    genome, 240
    mining, 40
        tools, 40
    nonredundant, 318
    nucleotide, 235, 239
    peptides, 241, 260
    protein, 235, 239, 240, 260
    search, 236, 235

Datasets, 56, 57, 59, 62, 319, 326–329, 334
  2D gel electrophoresis, 327
  derivative, 319
  genome-wide, 329
  relationships, 328
De Novo Pharmaceuticals, 342
Deactivation cycle, 293
Decarboxylases, 307
Deep-well plate, 212
Defects, 150
  surface, 25
Deglycosylation, 226, 228
Dehydration, 207
Deinococcus radiodurans, 6
Delta-lac mutation of B-GAL, 177
Denaturation, 132, 148, 151, 226
Denaturing conditions, 183
Densities, 307
  side-chain, 307
Density gradient, 237, 238
  separations, 237
DENZO package, 319
Dephosphorylatet/deactivated kinases, 292
Dephosphorylation, 290, 293
  p70S6K, 289
Derivatized microscope slides, 129
Descriptors, 54
  files, 35
Desolvation, 240
Destination plasmids, 258
Destination vector, 303
Desulfovibrio vulgaris, 314, 315
Detection, 22, 52, 87, 107, 119, 129, 132, 139, 148, 149, 154, 169, 177, 219, 279, 344
  antigen, 191
  ccd-based, 116
  direct, 132
  electrochemical, 54
  electromagnetic, 195
  fluorescent, 132
  label-independent, 148
  labeled, 194
  limits, 86, 88, 104, 107, 128, 136, 161, 162

[Detection]
  lower limit, 88
  methods, 132, 147, 148
  multiple wavelength, 133
  nonlabeled, 52, 53, 67, 148, 154, 194
  optical, 177
  PCA, 294
  physics, 145
  radioisotopic, 132
  secondary, 132
  signal, 8, 149
  strategies, 20, 52
  technologies, 129, 194, 195
Detectors, 187, 234, 238, 319
  technology, 160
  UV, 190
Detergent, 130, 131, 189
  concentration, 131
  extraction, 228
  nonionic, 186
Deuterium, 224
Deutsche Bank Alex. Brown, 339
Deviation, 311, 322
Devices, 176, 186, 189
Dextran, 163–167
  amino-modified, 164
  -hapten, 167
    conjugate, 167
DHFR, 279, 285, 286
  fluorescence assay, 286
  fragments N- and C-terminus, 285
  PCA, 279, 285
Diagnosis, 19, 54
  precocious, 54, 66
Diagnostics, 55, 66, 83, 122, 139, 150, 195, 222, 337
  assays in, 149, 155
  autoimmune, 155
  microarray-based, 115
  markers in, 47
Dialysis, 225
Difference isomorphous, 320
Differential display, 233
Differential profiling, 20
Differential splicing, 2

Differential transcriptional analysis, 19
Differentially expressed proteins, 127
Diffraction, 319
 limits, 305
Diffusion, 107, 108, 111, 163, 319
 analyte, 98
 coefficient, 109, 305
 hanging-drop, 307
 variation, 117
Digestion, 234, 236, 245, 247
 ArgC, 306
 gel, 236
 multiproteolytic, 245
 proteolytic, 242
 restriction, 257
 tryptic, 218, 221, 225
Dihydrofolate reductase (DHFR), 183
 murine, 279
 reassembly, 278
Dimer, 310, 315, 317, 319
 interface, 311, 312, 314, 316, 317
 structure, 315
Dimerization, 311, 317
 constitutive, 288
 of PNPO, 317
Diphosphate, 284
Diploid yeast cells, 262
Direct detection, 132, 133
Direct labeling, 133
Discovery, 19
 chain, 6
 science, 147
Disease, 2, 5, 8, 9, 30, 47, 54–56, 58,
  65–67, 127, 154, 337
 diagnosis, 67
 marker, discovery of, 139
 mechanisms, 234
 outcome, 54
 predisposition, 59
 process, 349
 progression, 154
  monitoring of, 55, 58
 status, 154
 tissue, 139
Disk storage robots, 59

Dispenser protein-compatible, 148
Display libraries, 136, 139
Display particles, 135
Dissemination, 301
 results, 304
Dissociation
 collision-induced, 235
 constants, 29
 rate, 108, 263
Distance, 314
Distribution, 8
 cytosolic, 287
Disulfide bond, 205
Disulfide bridges, 184
Disulfide cross-linking, 205
Dithio-bis succinimidylundecanoate, 149
DM, 320
DNA, 2, 8, 14, 37, 48, 56, 81–83, 119,
  120, 148, 156, 174, 186, 257, 258,
  264, 301, 318, 327
 arrayers, 151
 arrays, 154
 baculovirus, linear, 206, 210
 -binding domain of yeast, 261
 biochips, 145, 147
 damage response, 266, 269
  pathways, 265
 fragments, 258
 genomic, 206
 hybridization, 19, 147
  kinetics, 344
  thermodynamics, 344
 metal binding sites, 138
 microarrays, 127, 129, 133, 146–148,
  279, 299, 303
  performance, 128
 peptide constructs, 20
 polymerase, 318
 primer, 133
 recombinant, 300
 sequence, 46, 145, 173, 264, 325, 326,
  328, 329, 330, 331
  data, 326
 sequencing, 145, 147
  automated, 326

[DNA]
  strand, 133, 135
  template, 258
Domain, 9, 50, 181, 265, 291, 300, 318,
    342, 282, 305
  activation, 267
    loop, 286
  boundries, 305
  catalytic, 181
  cellulose-binding, 183
  chitin-binding, 207
  DNA-binding, 261, 266
  FKBP12-rapamycin binding, 160
  functional, 209
  hydrophobic, 225
  mapping, 305
  PH, 293
  PH N-terminal, 292
  screening, 19
  structure, 50
  transmembrane, 262
Dominant selection, 279, 285
Dominant-negative forms, 292
Donor plasmid, 258
Dose zero, 87
Dose-response curve, 87, 88
  slope, 85
Dot blot, 14, 19, 128, 192
  hybridization, 14
Droplet formation, 151
Drosophila, 305, 318
Drosophila melanogaster, 173, 240, 318
Drugs, 30, 47, 48, 50, 55, 66, 81, 119,
    174, 282, 283
  binding sites, 10, 48
  design, 233
  development, 48, 49, 173, 218, 343
  discovery, 48, 156, 173, 337, 349
  inhibition, 286
  response, 337
    monitoring, 55
  small-molecule, 48, 49
  targets, 47, 337
DSU, 149
DTASelect, 241, 243

Dual-affinity enrichment, 36, 37
Dual-affinity purification, 34
Dual-tag bacterial expression, 193
Dual-vector expression, 186
Dual-wavelength surface plasmon
    resonance (SPR), 28
Dyax, 345
Dynamic flow, 187
Dynamic light scattering (DLS), 305
Dynamic range, 18, 21, 53, 159, 219,
    220, 221, 223, 224, 226, 227, 237,
    238, 242, 248, 260
Dynatech Microfluor, 113
Dynatech microtiter-well polystyrene
    strips, 113

E-values, 318
E. coli, 39, 51, 167, 169, 173, 178, 179,
    181, 186, 203, 208, 210, 244, 258,
    304, 307, 311, 314, 316, 318, 319,
    329, 341
E. coli-based recombinant expression,
    203, 207
Echo cloning, 181
EDTA, 186, 187
EGTA, 186
Elastase, 245
Electrically neutral surface, 165
Electrochemical conductivity, 195
Electrochemical detection, 54
Electromagnetic detection, 195
Electron densities, 312
Electron density map, 312, 313, 320
Electronic detection, 52
Electronic magnets, 189
Electronic oligonucleotide-localization,
    121
Electrophoresis, 237
  gel, 212, 239, 319
Electrophoretic separation, 147
Electroporation, 182
Electrospray ionization (ESI), 234
  source, 240
Electrospray ionization tandem mass
    spectrometry (ESI-MS/MS), 240

Electrospray ionization-mass
spectrometry, 304
Electrospray micro-, 242
Electrostatic, 129
Electrostatic absorption, 167
Electrostatic binding, 129
Electrostatic interactions, 161
Electrotransfer, 15, 22
ELISA, 16, 18, 21, 28, 52, 66, 132, 133,
149, 152, 155, 160, 162, 177, 191,
227
sandwich, 28, 132
Ellipsometry, 54, 154
Elution, 188, 191, 225
EM Separations Technology, 319
Emerging technologies, 147
Endocrinology, 119
Endonuclease activity, 332
Endoplasmic reticulum, 5
Endoprotease, 226
Endoproteinase, 223, 226
Endosomes, 238
Engineered materials, 148
Enhanced chemiluminescence, 52
Enolase, 5, 46
Enrichment, 34, 235, 244
Golgi, 227
sample, 237
Entrapment, 28
Entry clone, 37
Entry vector, 35, 41
Environmental changes, 281
Environmental monitoring, 81
Environmental perturbation, 329, 331
Environment, chemical, 282
Enzymatic activity, 310, 314
reconstitution of, 279
Enzymatic assays, 148
Enzymatic pathway, 303
Enzymatic processes, 238
Enzymatic reactions, 167
Enzymatic tag, 133
Enzyme, 133, 166, 174, 177, 195, 224,
226, 263, 278, 282, 285, 286, 290,
292, 307, 329

[Enzyme]
active, 285
activity, 263, 286, 312
reconstitution, 279
biochips, 156
cleavage, 56
Enzyme conserved domain, 226
degradative, 206
digestion multiple, 63
function, 19, 30, 56, 62, 193, 329
insulin-activated, 284
Enzyme-linked immunosorbent assay
(ELISA), 18, 132, 149, 160, 177,
227
optimal pH, 56
precursor activation, 56
proteolytic, 206
purification, 304
reassembly, 278
restriction, 209
restriction digest, 214
substrate, 56, 122, 193, 194, 226
specificity, 330
stable, 263
turnover rate, 263
Enzymology, 307
Epimerases, 307
Epitopes, 9, 10, 18, 28, 48, 110, 111, 122,
138, 267
cellular, 20
conformational, 8, 10, 32
glycosylated, 32
linear, 8, 9, 32, 35
mapping, 17, 20, 132
phosphorylated, 32
spreading, 139
succinylated, 32
tag, 135, 183, 207, 209, 212
N-terminal, 209
Epoxide, 167
EPR Labautomation, 188
Equilibrium, 98, 103, 108, 110
final, 96
thermodynamic, 107
Equimolar, 268

Equipment, 129
Equivalence point, 18
Erembodegem-Aalst, 176
Error, 94
    random, 87, 93
*Escherichia coli* (*see E. coli*)
ESI-MS, 304–306
ESI-MS-MS sequencing, 37
Esters, 23
    activated, 23
ESTs, 240, 257, 267, 268
    human, 267
    mouse, 208
Etch technology, 152
Ethanol, 319
Ethanol-amine, 23
Eukaryotic expression systems, 204
Eukaryotic expression vectors, 258
Evaporation, 132, 187
Evolutionary insights, 299
Evolutionary relationships, 277
Excision, 258
Exogenous expression, 257
Exogenous systems, 255
Exon-intron structure, 267
Exonuclease activity, 332
Experimental design, 58, 59, 217
Express, 303
Expressed proteins, 128
    antibody specificity, 135
    expression, 135
    folding, 135
Expressed sequence tag (EST), 208, 240, 257
Expression, 174, 181, 182, 301, 304, 318
    basal level, 182
    cassette, 179, 180, 181, 184
        multifunctional, 180
    clone, 37, 174, 278
    dual-tag, 193
    heterologous, 260
    high-throughput, 184, 300
    levels, 182, 233
        mRNA, 154

[Expression]
    libraries, 174, 188, 190, 191
    patterns, 145
    profiling, 155, 175, 195, 346, 347, 348
        mRNA, 327
    protein, 338
    repression, 182
    switched, 182
    systems, 185, 193
        bacterial, 39
        eukaryotic, 204, 207
        insect, 39
        mammalian, 39
        prokaryotic, 207
        yeast, 39
    temporal, 46
    testing, 303
    ubiquitous, 305
    vector, 35, 39, 178, 182, 301, 303, 304
        host, 32, 33, 38, 43, 56, 183, 204, 263
Extraction, 236
    solid-phase, 188

Fabrication, 8
FACS, 176–179, 280
Factors structure, 322
False negative rates, 264, 268
False positives, 156, 160–162, 180, 261, 265, 278, 341
    interactions, 278, 286
    rate, 161, 264, 268
FastRed, 194
Fast Red/Naphtol AS-TR phosphate, 193
Fc region, 131, 132
Features of microarrays
    quantitative accuracy, 128
    sensitivity, 128
    speed, 128
Femtomoles, 234
Fetal calf serum, 161
Fiber optics, 54
Fibrinogen, 163, 164, 169
Film polystyrene, 155
Filter, 193, 195, 207, 319
    -based screening, 190

[Filter]
  membranes, 155
  plates, 189, 207
Filtration, 187, 189, 221
  devices, 189
  gel, 319
  vacuum, 212
Five-stranded, 311
FKBP-rapamycin-associated protein, 160
FKBP12, 284, 290
FKBP12-rapamycin-associating protein
    (see FRAP)
FKBP12-rapamycin binding domain
    (FRB), 160
FKBP12-rapamycin complex, 284
FLAG-Tag, 183
Flattening, 320
Flavin cofactor, 309
Flavin mononucleotide (FMN), 307
Flexibility, conformational, 304
Flow rates, 183, 187, 223
Fluorescein, 114
Fluorescein di-βD-glucopyranoside, 177
Fluorescein-conjugated methotrexate
    (fMTX), 279, 285
Fluorescence, 102, 132, 133, 146, 154,
    176, 177, 195, 218, 234, 279, 280,
    288, 114
  -activated cell sorters (FACS), 176
  assays, 285
  background, 195
  -based detection systems, 146
  Cy3, 64, 120, 121, 133, 136
  Cy5, 65, 120, 121, 133, 136
  intensity, 283
  intrinsic, 113
  microscopy, 286
  ratios Cy5 : Cy3, 137
  resonance energy transfer, 195
  two-color, 134, 139, 140
Fluorescent background, 102, 116
Fluorescent dyes, 12, 133
  detection, 129
Fluorescent intensity, 12
Fluorescent labels, 102, 195

Fluorescent microspheres, 114, 117, 121
Fluorescent rare-earth chelate labels, 94
Fluorescent sensitivity, 102
Fluorescent signals, 121
Fluorescent substrate, 179
Fluorescent tag, 133
Fluorescent time-resolving, 102
Fluorescently labeled oligonucleotides,
    135
Fluorescin-di-βD-glucopyranoside
    (FDGlu), 177
5-Fluoro-orotic acid, 264
  resistance, 264
Fluorophor, 114, 133
  conjugation, 133
    of reactive groups, 133
Fluorophosphonate, 226
Fluoroscein, 114
FMN, 317, 320, 322
  -binding, 314
    protein, 314
    sites, 310–312, 314
  cofactor, 317, 319
  ligand, 312–315
  molecule, 316
Foaming, 189
Fold, 300, 309, 311, 314
  assignment, 300
  -axis, 311
Foldases, 204
Folding, 285
  interaction-induced, 278
Food industry, 81
Forensic investigation, 81
Forward proteomics, 255, 256, 259
Fourier method, 320
  cross-difference, 320
Fourier transform ion cyclotron resonance
    mass spectrometry, 6
Fourier wavelet decomposition, 65
  tools, 65
Fraction
  collection, 187, 188
  insoluble, 243
  soluble, 243
  washed, 243

Fractional occupancy, 91, 96–98, 100, 104, 105, 107
Fractionation, 173, 227, 239, 242, 219, 222
  sample, 237
  subcellular, 227, 237
  two-dimensional, 242
Fractions, 188, 190
Fractogel column, 319
Fragmentary antibodies, 155
Fragments, 279, 306
  complementary, 285, 288
  heavy-chain, 9
  ions, 234
  light-chain, 9
  pattern, 234, 241
  probe, 279
  swapping, 285
Frame, 179
FRAP (FKBP12-rapamycin-associating protein), 284, 289, 290, 292, 293
  autophosphorylates, 290
  homodimerization, 290
  pathway, 286
  wortmannin-sensitive association, 288
Freeze-thawing, 186, 190
  cycles, 186, 190
French press, 186
FRET, 195
Frits, 188, 242
Frozen crystals, 319
FTICR-MS, 6
Fugu, 208
Full-length cDNAs, 268
Full-length clones, 267
Function, 173, 290
  biochemical, 174
  chemical, 260
  enzymatic, 19
  known, 269
  organismal, 260
  protein, 318
  putative, 260
  suggested, 269
  surface, 318

Functional analysis, 174
Functional annotation, 294, 301
Functional assemblies, 278
Functional characterization, 173
Functional cloning, 278
Functional genomics, 5, 257, 277
Functional inferences, 301
Functional mapping, 292
Functional validation, 282
  profile, 283
Functionalized surfaces, 150, 159, 160, 167
Fusion in-frame, 174, 177
Fusion points 3', 180
Fusion points 5', 180
Fusion protein, 183
  tag, 34
Fusion proteins, 33, 206, 207
Fusions, N- or C-terminal, 303
F[1,2] fragment, 278
F[3] fragment, 278

GAL pathway, 281
Gal-4, 261
GAL4 protein, 261
Galactose, 280, 281
Gallium, 245
γ-secretase, 226
Gas-phase collision cell, 234
Gastight hood, 185
Gateway, 135
Gateway recombinational cloning, 265
  system, 135
Gateway system, 35, 181, 257, 258, 264, 268
Gateway(tm), 257, 258, 264, 265, 268
Gateway vectors, 33
Gel, 340
  electrophoresis, 212
  filtration, 221, 307
  images, 328
  matrix, 192
  pads, 166
    array, 167
  replicate, 341
  ultrazoom, 219

Gene, 2, 14, 19, 40, 45, 48, 61, 63–65,
119, 147, 173–178, 181, 184, 256,
265, 277, 278, 299, 326, 328, 332
  background suppressor, 178
  -by-gene approach, 277
  candidate, 328
  cloning, 257, 260
  deletion, 281
  discovery, 203
  duplication, 45
  existence, 257
  expression, 182, 233, 257, 263, 279,
    281, 325, 326, 327, 329, 333, 337,
    257
    genome-wide, 330
    patterns, 145
    profile, 331
  finding, 337
  function, 238, 248, 277, 278, 326, 327,
    328
    annotation, 292
    strategy, 292
  functional characterization, 173
  human, 44, 208
  knockout, 279
  microarray, 282
  mouse, 208
  network, 326, 277, 327
  novel, 278
  number, 267
  ontologies, 294
  ontology, 294
    database, 294
  predicted, 257
  products, 2, 46, 277, 282, 303
  regulation, 59, 327–329
  reporter, 264
  sequences, 277, 301, 326
  structure, 245, 257
  target, 264
  toxic, 45, 177, 178, 263
  unknown function, 257
  yeast, 267
Genechip, 326, 328–331, 333
  experiments, 328

GeneFormatics, 344
GeneMachines, 185
Genesis RSP 150 (Tecan), 210
Genetic lysis, 186
Genetic modifications, 185
Genetic perturbations, 331
Genetic programs, 279, 280
Genetic reporters, 177
Genetic selection, 176
Genetic switch, 176
Genetix, 176
Genome, 30, 146, 173, 255, 256, 267,
277, 299, 300, 326, 327, 330
  complete, 329
  coverage, 267
  database, 303
  entire, 279
  human, 14
  microbial, 240
  phage, 258
  projects, 300
  sequence, 279, 299, 327, 329
  whole, 299
  -wide, 325–327, 329, 330, 334
    biochemical processes, 282
    effects, 330
    mapping, 269
      of biochemical pathways using
      PCA, 282
    screens, 265, 267
Genomic data, 195
Genomic DNA, 5, 34, 318
Genomic sequence, 9, 39, 260
Genomics, 5, 47, 217, 277, 318, 337
  biochemical, 301, 256
  comparative, 257
  functional, 5, 257
  industry, 338
  pipeline, 301, 302
  structural, 30, 39, 256, 300
Genomics Institute Novartis Research
  Foundation, 342
Genosensor Consortium, 82
Genotype–phenotype relationship, 331

Geometry, 307, 320
GFMN-binding sites, 309
GFP, 30, 177
Glass, 23, 149, 155
  capillaries, 113, 151
  microscope slides, 23
  silynated, 131
  slides, 25, 113, 155, 161, 193, 204
Glaucus Proteomics, 59, 169, 170
Glu-Glu tag, 135
Glucose, 165, 280, 281
Glucosylation, 35
Glutamic acid, 245
Glutaraldehyde, 113, 130
Glutathione-S-transferase (GST), 30, 33,
    183, 303, 304
Gluteraldehyde cross-linking, 131
Glycan moeity, 225, 226
Glycerol, 132
3-Glycidoxypropyltrimethoxysilane, 162
Glycopeptides, 226
Glycoproteins, 225, 225
Glycoproteome, 226
Glycosidase, 226
Glycosylation, 32, 35, 204, 205, 224, 226
  N-linked, 205
  O-linked, 205
Gold surface, 149
Golgi, 227
Golgi subproteome, 227
GPC-Biotech, 195
GPI-anchored proteins, 228
GPI-linked protein, 228
Gradient
  organic, 242
  reverse-phase, 242
  sucrose, 228
Graft copolymers, 169
Grafting, 162, 163, 166
Graphic user interface, 41, 45
Grating coupled surface plasmon
    resonance, 54
Greek-key β-barrel, 311, 315
Green fluorescent protein (GFP), 30, 31,
    175, 177, 136

Gridding, protein, 23
Grooves, 312, 314
Group
  aldehyde, 130, 131, 167
  amine, 130, 131, 149
  amino, 148, 152
  carbohydrate, 130, 131
  carboxylic acid, 149
  carboxylic anhydride, 165
  carboxymethyl, 164
  hydrazide, 167
  hydrophobic, 165
  lipid, 228
  succinimide groups, 131
  sulfhydryl, 132
  terminal hydroxy group, 166
  thiol, 148
Growth, 184, 281
  factors, 283, 284
  medium minimal, 262
  -phase logarithmical, 185
  potential matrix, 339
  rate, 347
    protein structure analysis, 342
Growth-factor-receptor, 284
  tyrosine kinase (RTK), 284
GST, 30, 33, 183, 204, 239, 304
  -fusion protein, 30, 204
GTPases, 289, 291
Gus, 212–214
  construct, 209

*H. influenza*, 330
*H. pylori*, 127, 172, 173, 266
*Haemophilus influenzae*, 329
Hampton Research, 304
Hanging-drop diffusion, 307, 319
Haploid strain, 262
Hapten (atrazine), 159, 164, 167
Haptoglobulin, 5
Hartman interferometry, 54
Harvard Institute of Proteomics, 345
Health care, 58
  delivery, 58
Heat shock proteins, 46

Heavy atoms, 307
  derivatives, 301, 320
  sites, 320
Heavy reagent, 147
HEK library, 179
*Helicobacter pylori*, 127, 172, 173, 266
Helper plasmid, pLysS/E, 186
Hepatitis B, 119
Hepatitus C, 119
HEPES, 319
Heterogeneity conformational, 305
Heterogenous molecular composition, 122
Heterologous fusion protein, 183
Heterologous protein, 183, 203, 204
  expression, 184
  induction, 184
Heterologous systems, 255
Hexa-histidine, 303
Hierarchical organization, 283, 286
High-affinity single chain antibody (scFv)
  libraries, 139
High-density array of proteins, 145
High-detergent conditions, 130
High-dimensional space, 56, 57, 65
High-field nuclear magnetic resonance
  spectrometers, 300
High-pressure, 187
High-performance liquid chromatography
  (HPLC), 21, 234
High-salt conditions, 130
High-specific-activity labels, 105
High-speed computing, 300
High-speed shaker, 185
High spot density, 159
High-stringency binding conditions, 131
High-throughput, 304
  antibody screening, 135
  characterization, 300
  crystallization, 300
  crystallography, 299
  expression, 184, 185, 300
  genome sequencing, 299
  instruments, 326
  methods,structural determination, 300
  protein analysis, 127, 148

[High-throughput]
  protein expression, 135, 174
    systems, 260, 266
  protein purification, 135, 174
  purification, 183, 300
  robotic-based methods, 257
  screening, 139, 156, 279
  structural biology, 301, 304
  transformation, 181
  yeast two-hybrid technology, 127
His tag, 182, 183, 187, 189
Histidine
  conserved, 317
  residue, 313, 314, 316
Histochemistry, 9
Histogram matching, 320
History, 82, 83
HIV, 15
Hoffmann-La Roche, 115
Homeobox protein PAL-1, 265
*Homo sapiens*, 173
Homodimer, 269, 309, 311, 316
Homodimerization, serum-induced, 290
Homogeneity, 23, 305, 306
Homologous recombination, 181, 205
  cloning, 214
Homologous site, 287
Homologs
  domain, 38
  sequence, 38
  tertiary structure, 38
Homology, 173, 284
  high, 309
  modeling, 300
  search, 337
  sequence, 316
  structural, 316
    three-dimensional, 343
Hormones, 81, 83, 85, 119, 282
  thyroid-stimulating, 115
Horseradish perioxidase (POD), 177
Host, 32
  bacteria, 182
  cell, 183, 185, 263
  organism, 184
  system, 40

Housekeeping gene, 46
HPLC, 187–190, 234, 242, 339
  robot, 190
  two-dimensional, 239
HPP, 256
hsp70, 328
HTS, 37
HTS Biosystems, 345
Human embryonic kidney, 179
Human fetal brain, 204
  cDNA expression library, 204
Human genome, 2, 8, 13, 14, 47, 48, 145,
    326, 337, 66
  blueprint, 5
  sequence, 145, 255
  working draft, 5
Human Genome Project, 52, 66, 82, 84,
    147, 203, 208, 255, 342
Human proteome, 2, 5, 14, 30, 35, 51, 55,
    56, 66, 203
Human Proteome Project (HPP), 2, 8, 14,
    56, 255, 256
Humans, 240, 300, 305, 316, 330
Human serum albumin, 169
Humidity, 132
Hybridization, 147
Hybridomas, 11, 15
Hybrigenics, 341, 342
Hydra 96, 210
Hydration, 132
Hydrazide-activated surface, 131
Hydrazide-derivatized, 132
Hydrazide groups, 167
Hydrazine, 150
Hydrogel, 23–25, 28, 48, 56, 163, 165,
    166, 169, 170
  coating, 23, 28
  matrices, 163
Hydrogen, 224
  bond, 312, 314, 317
    acceptor surface, 165
    donor, 165, 166
    interactions, 311, 312
  peroxide, 23
Hydrophilic, 165, 166, 312

Hydrophilic polymer, 162
Hydrophobic, 219, 312
  core, 318
  force, 130
  hapten, 164
  interactions, 311
  side, 312
  silanes, 130
Hydrophobicity, 181, 235, 242
Hydroxyl groups, 166, 312
Hydroxylapatite, 146
  chromatography, 146
Hypophosphorylation, 291, 293
  p70S6K, 291
Hypothesis testing, 217, 283
Hypothesis-driven science, 147
Hypothetical cluster, 268
Hypothetical protein interaction network,
    269

IBM, 342
ICAT, 146, 147, 223, 247, 340
Identification, 237
  of proteins, 339
Identity, 316
IgG, 114, 131, 132
IMAC, 183, 225, 227, 245
Image, 150
  analysis, 327
  real, 330
  simulation, 330
IMAGE (Integrate Molecular Analysis of
    Genomes and their Expression),
    203, 207–209, 267
IMAGE Consortium, 203, 208
Imager phospho-, 225
Imaging ellipsometry, 154
Imidazole, 264
Immobilization, 20, 23, 28, 56, 148, 150,
    152, 153, 160, 169, 189, 192, 193,
    195, 207, 226, 238, 344
  antibodies, 152
  covalent, 149, 150, 160
  density, 160, 163
  DNA, 148
  protein, 150, 152

[Immobilization]
site-specific, 150
strategies, 160
Immobilized, 236
antibodies, 132, 152, 154
colonies, 19
functional groups, 131
metal-affinity chromatography (IMAC), 245
ProteinChip, 225
proteins, 161
Immortalized B-cells, 11
Immune precipitation, 18
Immune response, 138
Immune system, 10
Immunization, 11, 12, 84, 193
parallel, 12, 13
Immuno-dot-blot asssay, 128
Immunoaffinity matrix, 212
Immunoaffinity purification, 238
Immunoaffinity tag, 209
Immunoassays, 84, 85, 89–91, 106, 114, 115, 128, 133, 148, 154, 155, 167
competitive, 91, 93
microarray-based, 122
multiplexed, 223
noncompetitive, 91, 93
parallel approach, 128
Immunobilized resins, 238
Immunoblotting, 218, 224, 236
Immunochemistry, 195
Immunodiagnosis, 15, 83, 94
Immunogenicity enhancement, 11
Immunogens, 11
Immunoglobulin, 5, 14, 132
Immunoglobulin G (IgG), 160
Immunohistochemistry, 138
Immunological reactivity, 138
Immunometric assay, 104
Immunoprecipiation, 239
Immunoradiometric, 90
Immunoradiometric assay (IRMA), 89
Immunosensors, 131
Immunosorbent, 90
Implant integration, 150

Improperly folded proteins, 181
In silico, 39, 42, 337
analysis, 257
In vitro biopanning, 11
In vitro interaction assay, 181
In vitro recombination, 33, 258
In vitro transcription, 183
In vitro translation, 183
In vivo immunological biopanning, 11
In-frame fusion of cDNA, 175, 176
In-frame fusion, resistance markers, 177
In-frame read-through, 180
In-gel digestion, 236
Inclusion bodies, 33, 34, 45, 182, 184, 203, 207
cause of, 203
Independent variables, 58
Indicator, 92
Individualized medicine, 55, 59
Inducer, 182, 185
anhydrotetracyclin, 182
arabinose, 182
cost, 182
Induction, 184, 185, 205, 286
of early genes (IEG), 292
insulin, 286
serum, 286
time, 45
wortmannin, 290
Industry value, 342
Infectious disease, 119
Inflammatory bladder disease, 222
Information
biological, 327
content, 40
density, 127, 128
DNA sequence, 331
genome sequence, 329
integration, 329
levels, 327
posttranslational modifications, 327
quantitative, 327
spatiotemporal, 327
technology, 327, 328, 329
Infrastructure computing, 60, 61

InFusion PCR, 35
Inhibition, 290
   P13K, 291
   PKB, 291
   PP2A-specific, 291
   selective, 265
Inhibitors, 175, 191, 279, 282, 283, 288,
      341
   libraries, 156
   pathway-specific, 285
   specific, 292
Inhibitory, 291
Inhibitory effect, 290
Initiation, 318
Injection, 187
Injection-molded plastic wells, 116
Ink-jet printing, 116, 118
Ink-jetting, 155
Inorganic surfaces, 149
Insect cells, 205, 206
   baculovirus-infected, 185
Insoluble particulates, 212
Insoluble target, 303
Institute for Genome Research, 5
Institute for Systems Biology, 348
Instruments, 186
   liquid-handling, 187
Insulin, 283, 284, 285, 288, 290, 291
   inducible, 290
   receptors, 286
   stimulation, 285
Integration, 234, 258
Intein, 175
Intensity, 327
   flourescent, 283
Interacting partners, 279
Interactions, 128,146, 193, 280, 290, 291
   4EBP1-FRAP, 290
   affinity, 160, 239
   analysis, 346, 348
   antibody-antigen, 129
   bait-prey, 262–265, 267, 269
   biological relevance, 278
   biomolecular, 67
   compound-protein, 159

[Interactions]
   cross-cluster, 268
   electrostatic, 161
   enzyme-substrate, 122, 138, 263
   false-positive, 286
   gene products, 278
   hydrogen-bond, 311, 312
   hydrophobic, 311
   improbable, 285
   inhibitors, 279
   intermolecular, 36
   kinase-dead, 285
   ligand-target, 26
   macromolecular, 255
   mapping, 19
   maps, genome-wide, 265
   novel, 287
   p70S6K, 289
   p70S6K-4EBP1, 287
   partners, 156
   PDK1-PKB, 287
   potential, 266
   profiling, 175, 190, 195
   protein, 278, 279, 338, 342
      clusters, 268
      –DNA, 281
      –drug, 19
      networks, 260, 269
      –nucleic acid, 19, 122, 138, 145
      –phospholipid, 30
      –protein, 19, 25, 30, 122, 138, 145,
         155, 156, 159, 162, 218, 233, 234,
         238, 239, 244, 256, 260–264, 266,
         268, 269, 278, 280–282, 285–288,
         338, 341, 342, 344, 348
      high-throughput, 344
      interactions, 280
      maps, 127, 255, 257
      proteome-wide, 266
   proteome-wide, 257
   small molecule, 19, 138, 145, 156,
      155
   surface, 148
      optimization, 148
      Rac1-Cdc42-p70S6K, 291

[Interactions]
    screening of, 283
    subcellular, 282
  rapamycin induced, 289
  receptor-analyte, 169
  serum/insulin-insensitive, 289
  serum/insulin/rapamycin-induced, 289
  single molecule, 52
  stacking, 317
  van der Waals, 311
  wild-type, 285
INTERACTIVA Biotechnologie, 345
Interactive proteomics, 238
Interactome, 269
Interactor potential, 264
Interation maps proteome-wide, 269
Interface, 162
  liquid interface, 151
  microcolumn-microelectrospray, 242
  solid-liquid, 148
Interference, 195
Interferometry Hartman, 54
Interlayer structured, 165
Intermediates reactive, 282
Intermolecular docking, 28
Intervention strategies, 66
Intrachromasomal recombination, 258
Intuitive reasoning, 65
Invitrogen, 181, 182
Iodacetic acid-linked glass beads, 225
Ion etching, 152
Ion exchange, 307
Ion traps, 240, 234
Ion-exchange chromatography, 219, 221
Ionization, 223, 234, 238
  soft, 240
  source, 238
Ions
  charged, 240
  fragment, 241
  gas-phase, 240
  predicted, 241
Iosoforms, 122
IPTG, 19, 178, 182, 184
IRMA, 89, 90, 91

Iron, 245
Isoalloxanthine ring, 317
Isoalloxazine, 313
  group, 312, 314
Isoelectric focusing, 218
Isoelectric point (pI), 242
Isoforms, 2, 20, 122, 255
  acetylated, 2
  amidated, 2
  glycosylated, 2
  myristolated, 2
  palmitoylatedd, 2
  phosphorylated, 2
Isopropyl-β-D-thiogaloctropyranoside
    (IPTG), 178
Isotope, 246
  heavy, 223
  ratio, 223
  tag, 224
Isotope-coded affinity tag (ICAT), 146,
    147, 223, 340
Isotope-encoded affinity tag, 330
Isotopic label, 146, 147

Jain PharmaBiotech, 341
Junctions, 242

Kcat, 309
Kd, 29
Keto group, 150
Kinase, 174, 181, 195, 285, 292
  assays, 162
  cyclin-dependent, 245
  -dead, 285
  FRAP activity, 290
  p70S6K, 287
  protein, 284
  radioactive ATP, 195
  serine, 284, 286
  substrates, 225, 285
  threonine, 284, 286
Kinetic parameters, 309
Kinetics, 107–109
  binding, 28
Kinetochore, 244
  complex, 246

Kits, phage display, 139
Km, 309
Knob, 317
Knockouts, 279
Kyte-Doolittle hydrophobicity plots, 42

Label, 83, 89, 105, 114, 154
  binding agent, 100
  chemiluminiscent, 102
  dual fluorescent, 102
  enzymatic, 102
  fluorescent, 102, 114, 152, 155
    rare-earth chelate, 94
  -free detection, 194
  high specific activity, 104, 105
  -independent imaging elliposometry,
    154
  isotopic, 146
  luminescent, 129
  metabolic, 247
  nitrogen 14, 15, 246
  nonisotopic, 94
  peroxidase, 184
  radioactive, 129, 181
  radioisotope, 162
  -specific activity, 102, 104
Labeled detection, 194
Labeled fluorescent, 135
Labeled secondary antibodies, 132
Labeling, 134
  chemistries, 129
  dyes, 133
  fluorescent, 138
  metabolic, 146
  site, 180, 181
  two-color, 134
Laboratory Information Management
    System (LIMS), 35, 38
Laboratory Systems, 113
LacZ alpha peptide, 180, 190
Large Scale Biology, 340
Large-scale protein production, 304
Large-scale protein structure analysis, 300
Laser, 102
Laser-based scanners, 112

Laser capture microdissection (LCM),
    221
Laser etching, 25
Laser pulse, 240
Laser scanner confocal, 116
Laser-scanning confocal microscope, 114
Lawrence Livermore National Laboratory,
    345
Layers, 163
  biocompatible, 166
LC/CE/MS/MS, 63
LC coupled MS/MS, 244
LC/MS, 247
LC-MS/MS, 223
Lead
  compounds, 343
  molecules, 48, 50
  optimization, 38, 47–50, 53, 56, 66
  selectivity, 343
  specificity, 343
  validation, 38, 50, 53, 343
Learning set, 55
Lehman Brothers, 337
Leucine zipper, 288
Library(ies), 11, 17, 19, 48, 155,
    174–176, 178, 179, 181, 184, 185,
    264
  baits, 266
  cDNA, 174, 204, 257, 258, 262, 267,
    278
    array, 208
    cell type, 267
    developmental stages, 267
    tissue, 267
  chemical, 48
  combinatorial, 9, 17
  complexity, 268
  construction, 341
  conventional, 268
  display, 136
  enzyme inhibitors, 156
  enzyme substrate, 156
  equimolar, 267
    metazoan, 268
  expression, 174, 191, 190
  generation, 178

[Library(ies)]
  normalized, 267
  oligo-dT-primed, 174, 180
  peptide, 9
  phage display, 135
  plated, 176
  prey, 262, 264, 266, 268
  ribosome display, 135
  scFc, 139
  screening, 278
  size, 179
  standard, 267
  yeast display, 135
Life technologies, 178
Ligands, 9, 10, 12–14, 53, 83, 88, 89,
      189, 191, 195, 311, 312
  affinity, 12, 14
  assay, 82, 83–87, 89–91, 93–96, 99,
      101–104, 111, 112
    antibody-based, 110
    nucleic acid, 110
    ultrasensitive, 94
  binding, 191
  cross-reactivity, 11, 13
  FMN, 314
  generation, antibody screening, 11
  high-affinity, 11
  high-specificity, 11, 38, 39
  interacting, 191
  –target interactions, 26, 29
Ligase, 213
Ligation, 213, 214, 257, 259
Light reagent, 147
Limit of detection, 128, 138
LIMS, 35, 38
Linear gradient, 319
Linear signal response, 128
Linker chemistry, 164
Linker layers, 163, 165
  chemical composition, 163
  density, 163
  thickness, 163
Liquid chromatography (LC), 21, 187,
      223, 224, 234, 306
Liquid handling, 21, 187, 262
  robots, 188, 190

Liquid-separation techniques, 234
Lithium hydroxide, 225
Local ambient analyte assay, 94
Local ambient analyte concentration
      (LAAC), 18
Localization, 146, 233, 284
  membrane, 286
  studies, 184
  subcellular, 244, 262
Location, 279
  cellular, 286
  compartments, 282
  cytosolic, 291
  subcellular, 234, 282, 292, 294
Logarithmic growth phase, 185
Long-chain ω-functionalized alkyl thiols,
      163
Loop, 312, 313
  activation, 287
Low-amplitude shaker, 185
Low-complexity, 300
Low-density arrays, 156
Lower limit, 87
Luciferase, 177
Luminescence, 176, 195
Luminescent label, 129
Luminescent product, 133
Lysates, 190, 207, 236
Lysine, 131
  residues, 131
Lysis, 185
  buffer, 190
  chemical, 185
  genetic, 186
  host cells, 184
  physical, 185
Lysosomes, 238
Lysozyme, 186, 187, 190
  expression, 186

m/z ratio, 238, 240
Mach-Zehnder, 54
Macroarrays, 95
Macrocomplexes
  constitutive, 282
  transient, 282

Macromolecular complexes, 260, 303
Macromolecules
  monodisperse, 305
  structure, 300
Macrospot, 95
MAD, 300, 301
  data, 301
Magnetic beads, 187, 189
Magnetic manifold, 189
Magnetic stirring, 185
  bars, 185
Magnetic tweezers, 54
MALDI-MS, 223, 226, 305
MALDI MS/MS, 225
MALDI-TOF MS, 2, 20, 21, 38
  peptide mass fingerprinting, 37
  total mass spectrometry, 37
Maleimide, 132
Maltose-binding protein (MBP), 183, 303,
    304
Mammalian cells, 183, 290, 292
Mammalian Gene Collection, 208
Mammalian two-hybrid system, 261
Mammals, 307, 316
Manufacture, 100, 148
Mapping, genome-wide, 282
Markers
  antibiotic resistance, 175
  diagnostic, 47
  fluorescent, 154
  gene, 176
  Golgi, 227
  metabolic, 177
  resistance, 177
  selectable, 262, 264
  tumor, 128
Market, 343
  growth, 342
  position MS, 339
  size, 342
Mass, 239, 304
  action, 67, 107
    laws, 96
  analyzers, 234, 238, 240
  fingerprint, 240

[Mass]
  resolution, 241
  sensing, 96
  spectometry, 133, 260
  spectra, 235, 241, 327, 328
  -to-charge ratio, 238, 240
Massive parallel testing, 83
Mass spectrometry, 1, 5, 6, 8, 21, 22, 54,
    63, 133, 146, 147, 156, 195, 223,
    226, 234–239, 241, 242, 244, 245,
    247, 256, 260, 304, 306, 327, 328,
    338–341, 344, 345
  -based analysis, 260
  CE-coupled, 223
  detection, 54
  ESI, 239
  ESI-MS, 306
  ESI-MS-MS, 37, 39, 239, 240
  FTICR, 6, 223
  HPLC, 21
  hybrid LC/CE/MS/MS, 63
  inlet, 240
  ion trap, 234
  LC, 241
  LC/MS/MS, 223
  MALDI/MS/MS, 225
  MALDI-TOF, 20, 21, 37, 38, 220, 223,
    226, 239, 305
  MALDI or SELDI, 226
  quadrupole TOF, 234, 240
  SELDI, 155, 220, 221, 226
  SELDI/MS/MS, 225
  sensitivity, 340
  tandem, 218, 221, 223–225, 234, 235,
    237, 239, 241, 243, 245
  triple quadrupole, 234, 240
Mass spectrum, 235, 237, 239
Material science, 145, 147
Materials, porous, 149
Mathematical framework, 331
Mathematical models, 331, 334
Mating type yeast, 262
Matrix, 7, 156, 169, 187, 188
  cleaning-in-place, 183
  effects, 160, 169

[Matrix]
    gel, 192
    immunoaffinity, 212
    metal-chelating, 53, 207
    pairwise, 269
    porosity, 28
    regeneration, 183
Matrix-assisted laser desorption ionization
    mass spectrometry, 305
    time-of-flight, 240
MBP, 304
McKinsey & Co., 337
Mechanism compensatory, 291
Media, 147
Medical devices, 59
Medium, 19
Membrane, 14, 15, 19, 20, 22, 81, 113,
        128–130, 135, 155, 186, 191, 192,
        193, 195, 263, 291
    association, 263
    cell, 81
    cellular, 291
    localization, 291
    nitrocellulose, 19
    nylon, 19
    PVDF, 21
    plasma, 227
    proteins, 223, 300
        porin, 311
    receptors, 228
    -spanning regions, 300
    translocation, 291
Mendelian inheritance, 1
Metabolic labeling, 146, 147, 248
Metabolic pathways, 329, 330
Metabolic/signal transduction pathways,
        281
Metabolism, secondary, 307
Metabolites, 329, 331
    concentration, 334
Metal, 193
    binding, 305
        proteins, 183
    chelate, 187
        chromatography, 187, 189
    noble, 149

Metallohydrolases, 226
Metazoans, 5
Methionines, 307
Methotrexate, 285
Methotrexate-resistance mutation, 285
Methyl esters, 245
Methylation, 246
MIAME, 35
Mice, 11
Microarray, 2, 81, 83, 85, 89, 94–96,
        102, 103, 111, 112, 115, 117, 119,
        120, 128, 132, 160, 162, 167, 169,
        174, 208, 235, 279, 282
    antibody, 136
    applications, 169
    cDNA, 233
    DNA, 127–129, 146, 147, 235, 279,
        299, 303
    format, 127, 128
    high-density, 96
    manufactured, 129
    methods, 101
    nucleic acid, 260
    panel, 119
    printing robots, 129
    production, 100, 113
    readers, 133
    technology, 120–122
    tissue, 20
Microarray-based assay, 95, 102
Microarray-based diagnostics, 115
Microarray-based immunoassays, 122
Microarray-based ligand assay, 94
Microbial cells, 186
Microbiology, 218
Microcalorimetry, 54
Microcapillary liquid chromatography,
        242
Microcolumn-microelectrospray interface,
        242
Microdevice, 145
    technology, 145
Microdissection, 46
Microdissection laser capture, 221
Microdrops, 167

Microelectrospray interface, 242
Microenvironment, 108
Microfluidics, 340
  system, 156
Microheterogeneity, 225
Micro-organisms, 6, 8, 19
Microprocessors, 59
Microscale spots, 127
Microscope images, 330
Microscope slide, 45, 129
Microscopy, 330
  confocal laser-scanning, 114
  confocal time-resolved, 114
  fluorescence, 134, 280, 283, 286, 288
  scanning tunneling, 150
Microsphere sensors, 155
Microspheres, fluorescent, 114, 117
Microspot, 15, 18, 81–83, 89, 94, 95,
    98–102, 104, 106–109, 112–116,
    117, 121, 149, 155, 193, 194
  area, 104
  array, 82
  assay, 97, 104, 106, 107, 111, 112, 115
  kinetics, 107
  fluorescent, 118
  format, 107
  ligand assay, 82, 101, 104
  plates, 113, 129, 303
    coated, 187, 189, 190
  radius, 109
  visibility, 104, 105
Microwells, 19, 155
Migration, 108
Military defense, 81
Milk powder, 23, 160
Milk proteins, 130
Milteny, 189
Mimotopes, 8, 17, 42
  technology, 17
Miniaturization, 8, 20, 64–66, 112, 145,
    147, 195, 205, 206, 209–211, 213,
    214
Miniaturized assays, 2
  technology, 81
Miniaturized DNA biochips, 145

Minimal growth media, 262
MIR, 320
  map, skeletonized, 320
  phases, 320
Misonix, 186
Mitochondria, 5, 225, 238
Mitochondrial subproteome, 227
MLPHARE, 320
MODBASE, 318
Model, 301, 318, 329, 330, 334, 348
  biochemical, 331
  comparative, 318
  crystallographic, 311
  mathematical, 330
  organisms, 173, 300
  parameters, 334
  polyserine, 307
  predictive, 347
  protein synthesis, 331
  refined, 311
  side-chain, 307
Modeling, 299, 300, 307, 318
  comparative, 318
  programs, 303
  structural, 318
  predictions, 334
Modifications, 233
  cotranslational, 2, 33, 421, 56
  posttranslational, 2, 8, 20, 32, 33, 35,
    38, 42, 56, 203, 205
Modular maps, 268
Modular screens, 266
Module, 268
Moiety
  biofunctional, 169
  glycan, 225
  phosphate, 225
  protein-resistant, 150
Molecular density, 149
Molecular interactions, 279
Molecular mass, 53, 238, 241, 242
Molecular probes, 177, 223
Molecular scanner, 340
Molecular structure, 110
  3D, 110

Molecular weight, 236
Molecular-recognition, 83
Monoclonal antibodies, 10–13, 17, 152
  production, 84
Monodispersity, 305
Monolayers, 25, 28, 95, 149, 150, 165,
  166
  mixed, 150
  multifunctionality, 345
Monomeric, 315
Monomeric protein, 305
Monomers, 162, 310, 314
Monospecific polyclonal antibodies, 9
Monte Carlo simulation, 332
Motifs, 48
Mouse, 240, 305
  liver, 222
  lung, 4
Mr, 37
mRNA, 2, 5, 8, 45, 83, 146, 233, 267,
  281, 293, 332
  abundance, 146, 233, 328, 329
  binding site, 332
  degradation, 331
  distribution nonuniform, 267
  editing, 217
  expression, 146, 325–327, 329, 331,
    332
    genome-wide, 325, 326
    large-scale, 281
    levels, 154
    profile, 135, 331, 332
  length, 332
  level, 326, 328, 329
  perturbation, 330
  processing, 328
  sequence, 332
  stability, 332
MS, 6, 7, 63
MS-Tag website, 218
MS/MS, 242
  spectrum, 245
mTOR (mammalian target of rapamycin),
  284
MTP, 186, 188, 189, 190

MudPIT (Multidimensional protein
  identification technology), 239,
  242–256, 248
Multianalyte assay, 120, 190
Multianalyte testing, 101
Multidimensional chromatography, 7
Multidimensional protein identification
  technology (MudPIT), 239,
  242–256, 248
Multidomain, 305
Multigenic phenomena, 1, 57
Multipin synthesis, 128
Multiple expression vector hosts, 38
Multiple-isomorphous replacement (MIR),
  307, 320
Multiplexed assays, 53
Multiplex immunoassay, 223
Multiwave anomalous dispersion (MAD)
  phasing, 300
Murie dihydrofolate reductase (DHFR),
  278
Mutant, 19
  kinase-dead, 285
  substrate-trapping, 263
Mutation
  complementary, 262
  delta-lac, 177
*Mycobacterium tuberculosis*, 119, 316,
  318
*Mycoplasma genitalium*, 1, 5, 300
Myriad Genetics, 341, 342
Myristylation, 35

N-ethyl-N(3-dimethylaminopropyl)-
  carbodiimide (EDC), 149
N-glycosidase F, 228
N-hydroxysuccinimide, 133, 164
  conjugation, 134
N-terminal, 287
  fusions, 303
  sequencing, 236
N14, 146, 246
N15, 146, 246
Nanoarrays, 121
Nanocantilevers, 54
Nanogen, 345

Nanoscale liquid chromatography, 234
Nanospots, 121
Nanotechnologies, 46
Nanotopography, 150
Narrow-pH-gradient-range gels, 219
National Innovation Centre, 21
National Institutes of Health, 345
National Synchrotron Light Source, 307, 319
Native signal sequence, 205
Native stop signal, 180
NC, 193
NCS
  constraints, 320
  map, 320
Near target space, 38, 49
Negative feedback circuit, 291
Negative selection, 177
Negative staining, 179
Network, 233, 282
  biochemical, 279, 281–283
  biological, 327
  hydrogen-bonding, 312
  mapping, 285
  reaction, 330
  RTK-FRAP, 289, 292, 287
  signaling, 285
Neuroblastoma cells, 330
Neuropsychiatry, 218
New England Biolabs, 139
New York Structural Genomics Research
  Consortium, 300
Nitrocellulose, 113, 130, 148, 149, 152, 192
  membrane, 194
Nitrogen 14, 15, 146, 246
  isotopes, 223
NMR, 55, 215
  spectroscopy, 304
  structure, 314
Noble metals, 23, 149, 163, 165
  layers, 163
  substrates, 166
  surfaces, 163, 165, 166
Noise, 86
  background, 103

Noncandidate space, 38, 48, 49
Noncompetitive assay, 94
Noncontact printing, 46
Noncovalent binding, 192
Noncovalent interactions, 150
Noncrystallographic, 311
  two-fold, 320
Nonequilibrium pH-gradient
  electrophoresis, 3
Nonidet NP40, 186
Nonionic detergents, 186
Nonisotopic label, 94, 106
Nonlabeled detection, 52
Nonlabeled technologies, 53
Nonparametric calculation, 61
Nonpolynomially complete problems, 57
Nonspecific adsorption, 160–167, 169, 170
  suppression, 164
Nonspecific binding (NSB), 10, 21, 23, 25, 27, 67, 94, 117, 129–131, 138, 148, 150, 160, 162–164, 166, 189, 190, 207
  plastic, 190
Nonspecificity, 46
  target binding, 47
Nonspecific protein adsorption, 131
Nonspecific protein binding, 150
Normalization, 133
Normalized libraries, 267
Novagen, 139, 182, 186, 318
Novel gene, 278
NSB, 23, 25
Nuclear magnetic resonance (NMR), 42, 55, 215
  spectrometry, high-field, 300
Nuclear targeting, 262
Nuclei, 238
Nucleic acids, 20, 48, 120, 128, 129, 145, 187, 191
  arrays, 119, 156
    analysis based on, 83
  biochemistry, 147
Nucleic assay, 114
Nucleotides, 148, 256, 285

Nucleus, 5, 263
Nunc, 113
Nutrients, 282, 284

Occupancy, 92
  rate, 109
  ribosomal, 332
Oligo-dT-primed cDNA library, 174
Oligo-dT-primed libraries, 180
Oligoethylene glycol, 166
Oligomer, 162, 166
  layers, 162, 163
Oligomeric assembly, 205
Oligomeric compounds, 165
Oligomerization, 205
  state, 304, 305
Oligonucleotide, 81, 84, 92, 114, 121,
      148
  array, 17, 18, 120
  microspots, 114
One-dimensional gel electrophoresis, 15,
      226, 227
On-line detector, 188
Open reading frames (see ORF)
Optical dark fibers, 59
Optical density, 176
Optical detection, 177
Optical devices, 176
Optical reporter, 178
Optical tweezers, 54
Optimization, 138, 148, 344
Ordered arrays, 129
Orders of magnitude, 128, 136
ORF, 2, 8, 30, 35, 39, 43, 45, 48, 56, 66,
      173–175, 184, 190, 217, 255, 258,
      257, 262, 267, 268
  candidate, 262
  cloning, 258
  exon-rich, 45
  full length, 268
  mispredicted, 257
  novel, 173
  predicted, 260
  small, 45
ORFeomes, 255–259, 266, 268
  entire, 257

Organelles, 224, 234, 236, 238, 244
  purification, 227
Organic thin films, 149
Organisms, 1, 5, 7, 56, 85, 146, 147, 156,
      159, 184, 277, 299, 303, 326, 334
  marine, 177
Orientation, 132, 285
  antibody, 132
  fusions, 287
  random, 149, 150
  specificity, 285
Orthogonal approaches, 55
Orthogonal chemistry, 150
Orthologs, 303–306, 316
Ovalbumin, 164, 165, 169
Ovarian cancer, 65
Oxford Glycosciences, 345
Oxidase, 311
  activity, 309
  family, 306, 314
Oxidation, 246, 309
  of carbohydrate groups, 131
Oxidization, 132
Oxygen, 185, 312
Oxygenation, 206
Oxygen carbonyl, 312

P005, 303
P008, 301, 303, 306, 309, 316, 318, 319
P088, 305
p70S6K, 286, 287, 289–293
p70S6K-PP2A, 291
Packard Biosciences, 133, 345
Packing density, 106
PAL-1, 265
Palmitoylation, 35
Panning, 136
Paradigm, 325, 329, 331, 334
Parallelization, 8, 20, 66
Parallel quantitation of proteins, 127
Parameters
  kinetic, 334
  model, 334
Paratope, 110
Partners, interacting, 260
Patents, 17, 23

Pathogens, 299

Pathways, 139, 208, 238, 282–284, 286, 292, 328, 329, 337, 348
- amino acid biosynthesis, 329, 330
- biochemical, 30, 279–281, 282, 334
- biological, 330
- biosynthetic, 227
- cell-death, 227
- cellular, 234, 330, 342
- component, 330
- convergence, 287
- cross-talk, 290
- DNA damage response, 265
- enzymatic, 303
- excretion, 45
- FRAP, 284, 290
  - mapping, 290
- functional, 349
- GAL, 281
- glycolysis, 348
- growth-factor-mediated, 292
- interconnected, 292
- mapping, 277, 279, 281
- maps, high-resolution, 342
- metabolic, 227, 329, 334
- parallel, 292
- perturbation, 280, 292
- protein, 341
- rapamycin-sensitive, 287, 288
- Ras, 266
- Rb, 266
- RTK, 286, 290, 292
- serum/insulin-stimulated, 288
- signaling, 266, 290
- signal transduction, 156, 265, 282, 283, 292, 286
- wortmannin, 287, 288

Patients, 7, 54, 64, 128, 139
- cohorts, 1, 5, 66

Pattern, 140, 145, 288
- 2D gel electrophoresis, 328
- fragmentation, 234, 241
- recognition, 59, 65

Patterson map, 320

PBS, 193

PCA, 278–283, 285, 292
- -based analysis, 281
- strategy, 283, 292

PCR, 8, 33, 37, 43, 83, 92, 204, 205, 207, 208, 210, 213, 217, 304, 318, 209
- gene-specific, 268
- high-fidelity, 209
- nested, 33
- primers, 318
  - ORF-specific, 258
- product, 209, 259

PDB, 314

PDK1, 286, 287, 290–293
- activated, 292

PDK2, 286

PE Biosystems, 190

Peak collection, 188

Peak ratios, 247

PEG, 131, 150, 164–166, 169, 319
- α,ω-amino functionalized, 163
- layers, 163

PEG-coated surface, 131

PEG-modified poly-L-(lysine), 168

PEG poly(L-lysine)-graft, 167

Peptide alpha, 177

Peptides, 8, 17, 42, 63, 127, 128, 146, 159, 160, 167, 181, 226, 235–237, 241, 242, 245, 265, 285, 224, 316
- arrays, 48, 128, 132, 138
- backbone, 234
- chain release, 332
- closure, 41
- competition, 212
- conserved, 316
- coverage, 5, 245
- C-terminal, 317
- digestion, 239
- -DNA constructs, 20
- elution, 236
- enzyme-specific, 292
- fingerprint, 221
- flexible, 285
- fragments, 218, 226, 239, 242, 306
- heavy, 147
- internal sequencing, 236

[Peptides]
ions, 234
libraries, 9, 10
light, 147
linker, 285
mass fingerprinting, 37, 43, 218, 219,
    221, 239, 240
mass mapping, 239, 240
mixtures, 235, 242, 243
multiply charged, 241
overlapping, 245
parent, 241
phosphorylated, 241
random expression, 349
regions, 316
sequence, 181, 245
sequencing, 234, 239
    synthetic, 9
    tryptic, 221, 223
signal, 204
signature, 260
spurious, 267
string, 40–42
synthesis, 332
tag, 183
tryptic, 218
Performance, analytical, 161
Periplasmic space, 184, 244
Perkin Elmer, 133
Perturbation, 281, 282, 279, 282
P-γ-ATP, 162
PET, 163
vectors, 186
PH, 9, 23, 242, 286
domains, 291, 292
gradients, 218, 219, 236
Phage
display, 84, 136, 139, 140, 155, 174,
    263
    clones, 128
    -generated scFv antibodies, 135
    libraries, 135, 139
    methods, 84
    technology, 9, 12, 136
filamentous, 17
particle, 135

Phagmids, 9
Pharmaceutical companies, 55, 343
Pharmaceutical therapies, 173
Pharmacological perturbations, 292
Pharmacological profiles, 282, 283,
    285–288, 291, 292, 294
Pharmacological responses, 283, 286
Pharmacology, 337
Phases, 301, 307, 320
    combination, 307, 320
    reflections, 301
Phasing, 320, 321
Phenotypes, 264, 327, 329, 349
    mutant, 328
Phenyl sepharose resin, 227
Phosphatase, 174
    inhibitor, 289
    PP2A, 293
    serine, 284
    serine-threonine phosphatase, 290
    subunit regulatory, 289
    threonine, 284
Phosphate buffered saline, 193
5'-Phosphate group, 312
5'-Phosphate oxidase, 301
Phosphatidyl inositol-3-kinases PI3K, 284
Phosphatidyl inositol (3,4,5) triphosphate
    (PIP3), 284
Phosphatidyl inositol (4,5) diphosphate
    (PIP2), 284
Phosphatidyl inositol triphosphate, 284
Phosphoimager, 225
Phosphoinositides, 291
Phospholipase C phosphatidylinositol-
    specific, 228
Phosphopeptides, 225, 241, 245
Phosphoproteins, 224
Phosphoproteome, 224
Phosphorous oxychloride, 319
Phosphorylate, 290
Phosphorylation, 32, 35, 162, 204, 205,
    224, 241, 245, 246, 284, 286, 289,
    290, 292, 293
    of eIF-4E-binding protein (4EBP1),
        284

[Phosphorylation]
  growth-factor-mediated, 284
  PDK1, 287
  rapamycin/FKBP12, 290
Phosphotases, 174, 181
Photoaptamers, 344
Photolinker, 167
Photolithography, 155, 326
Photometric assays, 177
Photopolymerization, 166
Phycoerythrin, 169
Phylogenetic relationships, 277
Phylogenic relationships, 277
Phylos, 344, 345
Physical lysis, 185
Physical methods, 186
Physiochemical properties, 85, 108, 326, 327
Physiological relevance, 342
Physiological response, 325
Physiology, cell, 325
Physisorption, 148, 152
PI, 218, 223, 236, 244
  extremes, 219
PI3K, 291
Pichia, 181
Pichia/*E. coli* shuttle vector, 181
Picking robots, 176, 178
Pierce, 186, 319
Piezo-based ink-jet technology, 116
Piezo printer, 16, 118, 167
Pin-based methods, 17
Pinhead, 193
Pins, 17
PIP3, 286, 292, 293
Pipeline, 303
Pipetting robots, 187, 188
Pipetting tips, 189
Piranha treatment, 23
Pitfalls, 349
PKB, 286, 287, 290–293
  association, 286
  kinase, 286
Planar surfaces, 156
Plant cells, 292

Plaque-forming units, 211
Plaque purification, 213
Plasma, 17, 219
  membrane, 238, 263, 284, 286, 288, 289, 291–293
  polymerization, 163
  treatment, 23
Plasmids, 189, 205, 214
  bait, 262
  destination, 258
  DNA, 186
  donor, 258
  expression, 264
  helper, 186
  prey, 262
  purification, 207
  vector, 258
Plastics, 23, 84, 193
Plate-washing devices, 190
Platinum Pfu, 318
Pleckstrin, 284
Pleckstrin homology (PH) domain, 284
PLP, 319
Pluronic F-68, 206, 211
PMP, 309
PNOP, 301
PNP, 309
PNP oxidase (PNPO), 306–311, 314–319, 321, 322
  activity, 314
  family, 306, 314
  structure, 310, 312
Pocket, 312, 314
POD, 193
Point-of-care diagnostics, 299
Polar residues, 312
Polarization, 195
Polyacrylamide, 131
  -co-acrylic acid), 164
  gel patches, 155
Polyalanine, 307
  model, 307
  trace, 320
Polydisperse systems, 305

Polyethylene, 128
    glycol (PEG), 131, 150, 163, 319
        layers, 163
    supports, 128
    terephthalate (PET), 163
Poly(ethyleneimime), 164
Polyhedrin promoter, 205, 206
Polyhistidine, 207
Poly-histidine, 135, 304
Poly-L-lysine, 130
    coated glass, 130
    -graft PEG, 167, 169
Polylysine, 148, 149, 161, 344
    materials, 149
    supports, 148
Polymer, 162, 165, 166, 169
    amino, 164
    biocompatible chemical composition,
        163
    carboxy-substituted, 164
    chains, 162, 163
    diffusion, 163
    layer, 23
Polymerase chain reaction (PCR), 7, 8,
        204, 205, 209, 213, 258, 304
    nested, 33
    reverse transcriptase, 8
Polymerase T7, 186
    RNA, 319
Polymerization, 162
Polymyxin, 186
Polynucleotide, 82, 84, 92, 110, 122
    arrays, 120
Polypeptide, 8, 9, 82, 314
    chain
        multifunctional, 179
        segments, 305
    nonsensical, 267
Polypropylene, 113
Polysaccharides, 163
Polyserine, 307
    model, 307, 320
Polysome, 332
Polystyrene, 129, 130, 149
    clear, 113

[Polystyrene]
    film, 155
    plates, 113
        black, 113
        clear, 113
        white, 113
Polyvinylidene difluoride, 192
Polyvinylidene fluoride (PVDF), 130, 192
Pombe, 305
Ponceau S, 193
Population
    doubling, 211
    outliers, 65
    variance, 2
Pores, 186, 188
Porin membrane proteins, 311
POROS, 187, 189
Porosity, 28
Porous materials, 149
Positive clones, 179
Positive controls, 46, 48
Posts, 152, 154
Posttranslational isoforms, 255
Posttranslational modifications (PMTs), 2,
        33, 56, 122, 146, 181, 183,
        203–205, 217, 224, 235, 245, 246,
        248, 255, 262, 262, 263, 327, 328,
        341, 344
    characterization, 245
    protein, 338
Potency, 111
Potter-Elvehjem tissue homogenizers, 186
PP2A, 284, 289–292
    activity, 292
PP2A-p70S6K, 291
PP2A-PKB, 291
Precipitation, 89, 319
Precision, 90, 104
Preclearing, 184
Prefactionation, 178, 219, 221, 227, 237
    nonspecific, 221
    specific, 221
Pressure, 183
Prey, 261–264, 266, 267, 269
    interactions, 265
    library, 268
    swapping, 263

Primary
  alcohol, 307
  antibody, 132
  sequence, 306, 307
Primate, 208
Primer
  design, 39
  melting point, 43
Printer, 151
  capillary-based, 151
  piezoelectric, 116, 118, 159
  pin, 159
  protein-compatible, 148
  ring, 159
Printing
  contact, 151, 193
  ink-jet, 116, 118, 155
  noncontact, 46
  robotics, 129
Probe fragments, 279
Procheck, 320
ProDom, 303, 305, 307
Profiles, quantititive, 146
Profiling experiments, 195
Prokaryotic expression
  host, 183
  systems, 181
  vectors, 258
Promega, 189, 192
Promoter, 180–181, 184, 205
  araB, 182
  araBAD, 182, 184
  bacterial *Lac*I, 182
  constitutively active, 182
  *Lac*I, 184
  p10, 206
  polyhedrin, 205, 206
  strong, 264
  T7, 182, 184
  TET, 184
  weak, 261
ProSite, 42
Prostate cancer, 138
  marker, 138
Prostate specific antigen (PSA), 138, 222

PROT@GEN, 345
Protease, 174, 184, 226, 236, 242
  amino recognition sequences, 180
  carboxyl-terminus recognition
    sequences, 180
  cleavage, 180
    sites, 181, 183, 304
  inhibitors, 187, 190
  purification, 304
  Xa, 180
Proteasome, 266
Protein(s), 82, 127, 145, 150, 159, 160,
    167, 169, 174, 181, 185, 277, 278,
    280, 305, 314, 322
  A, 113, 114, 131, 132, 169, 304
  A and B interacting, 280
  abundance, 5, 46, 147, 159, 217, 219,
    237, 299, 326, 328
    profiles, 159
  accessibility, 46, 151
  acidic, 244
  activity, 30, 145–148, 326, 338
  activities, 191
  adsorption, 150
  analysis, 234
    high-throughput, 148
  analytes, 133
  anionic, 130
  arrays, 30, 35, 42, 45, 54, 55, 121, 153,
    138, 190, 192, 195
    challenges, 129
    coated with agarose, 192
    universal, 174
  assay, 118
  attachment, 129
    methods, adsorption, 129
    methods, affinity binding, 129
    methods, covalent binding, 129
  autofluorescent, 177
  bacterial, 184
  basic, 7
  binding, 220
    agents, 84
    assay, 91
    capacity, 167

[Protein(s)]
  biochips, 14, 38, 46, 48, 56, 145, 147,
    148, 151, 152, 154, 155, 260, 327
  biomolecules, 48, 49
  biophysical analysis, 299
  biotynylated, 131
  -bound, 129, 150
  calcium-binding, 227
  cancer-related, 304
  capture, 129
    reagents, 135
  cellular, 341
  characeterization, 1, 67, 173, 238
  charges, 129, 146
  chemistries, 128
  chips, 2, 12, 32, 35, 40, 54, 66, 327,
    338, 340, 341, 344, 346
    alternative technologies, 346
    applications, 346
    competitive environment, 346
    customer demand, 346
    market, 344, 346, 348
    market growth rate, 347
    technology, 346
    prefabricated, 346
    researcher-assembled, 346
  class, 244
  cleavage, 2, 4, 7, 15, 19, 21, 22, 45
  cluster, 268
  complexes, 55, 156, 234, 235, 238,
    244, 245, 256, 260
    putative, 260
  conformation, 56, 148
  coregulation, 234
  crystallization trials, 305
  crystals, 319
  degradation, 6, 184, 265
    machinery, 265
  delivery, 340
  denaturation, 130, 132, 148, 242
  density, 152
  detection, 149, 152
  differentially expressed, 127
  discovery, 136, 155
  disease-related, 127

[Protein(s)]
  dispensers, 148, 151
  display libraries, phage display, 135
  distribution, 152
  domains, 282
  dots, 192
  encoded, 260
  engineering, 207
  entrapment, 150
  environment, 233
  eukaryotic, 204
  expression, 6, 63, 128, 135, 145, 173,
    203, 217, 233, 255, 260, 281, 300,
    301, 325–329, 331, 332, 338, 343,
    344
    eukaryotic, 205
    exogenous, 255
    heterologous, 255
    high-throughput, 215, 260, 345
    large-scale, 281
    levels, 234
    patterns, 145, 139
    profile, 237, 331, 332
    proteomics, 233
    quantification, 340
    systems, 300
    vector, 33
  extraction, 7
  family, 303
  FMN-binding, 314
  folding, 56, 135, 181, 203, 279, 314
    interaction-induced, 278
  fractions, 212
  fragments, 181, 278, 349
    complementation, 278
  full length, 267, 268, 286
  function, 22, 30, 127, 139, 224, 245,
    260, 281, 337, 342, 348
  functional, 244
  fusion, 33, 155, 183, 206, 207, 265,
    280, 285
    C-terminal, 258
    N-terminal, 258
  G, 132, 160, 187
    matrices, 187

[Protein(s)]
  glycosylated, 225
  glycosyl phosphatatidylinositol-linked, 228
  GPI-linked, 228
  GST-fusion, 204
  heavy, 246
  heterologous, 182, 184, 203, 262
    fusion, 183
  high abundance, 5, 20, 26, 47
  higher-molar mass, 167
  highly basic, 223
  human, 35
  hydrophobic, 7, 43, 219, 227, 237
  identification, 219, 221, 235, 236, 238–243, 282
    multidimensional, 241
  immobilization, 149
  inaccessible, 148
  inactivation, 150, 151
  inactive, 148
  insoluble, 43, 55, 212
    cellular, 81
  integral membrane, 263
  interactions, 146, 234, 268, 283, 288, 338, 347
    clusters, 268
    -induced folding, 278
    maps, 265, 268
    network, hypothetical, 269
    networks, 260
    protein–DNA, 281
    protein–nucleic acid, 145
    protein–protein, 145
    protein–small-molecule, 145, 155
    protein–surface, 148
  intracellular, 264
  isoforms, 8, 20, 35
  isotopically labeled, 146, 147
  kinases, 225
  known, 269
  labeling, 301
  large, 167, 221, 237
  level, 328
  light, 246

[Protein(s)]
  load, 219
  localization, 146
  location, 46
  low-abundance, 30, 34, 46, 146, 235, 237, 244
  low complexity, 43
  matrix, 169
  membrane, 205, 223, 237, 244, 262, 300
    anchoring, 291
    -associated, 20, 55
    metal-binding, 183
  microarrays, 48, 127–129, 131–133, 135, 136, 138–140, 149, 159, 160, 166, 174
    generation, 174
    methods, 128
    technology, 127
  microchips, 167
  misfolding, 203
  mixtures, 154, 235, 236, 242, 243
  modification, 224
  molecular weight, 146
  –mRNA correlation, 233, 325, 331, 333
  native, 181, 319
  optically active, 177
  pairs, 286
  pathways, 344
  perturbation, 330
  pharmaceuticals, 299
  phosphorylated, 224, 225
  plasma membrane, 228
  posttranslational modifications, 338
  production, 184, 185, 207, 215
    baculovirus, 206
    high-throughput, 205, 206
    Pluronic F-68, 206
  profiling, 155, 217, 218, 221, 223
    activity-based, 226
  –protein, 122
    contact sites, 138
    interactions, 25, 155, 156, 260, 264, 266, 268, 269, 278–282, 285, 286

[Protein(s)]
    maps, 127, 255, 257
   purification, 173, 301, 307, 314, 343
    high-throughput, 345
   purity, 46, 212, 305
   quantification, 152, 340
   -reactive surfaces, 149
   recognition, 122
   recombinant, 30, 32, 36, 37, 48, 55, 66,
      176–178, 180–182, 187, 203, 204,
      207, 212, 213, 238
    dissemination, 301
    expression, 19
    fusion, 32
    high-throughput, 208, 209
    production, 209
    soluble, 193
   refolding, 182, 207
   regulation, 338
   relative abundance, 223
   separation, 340
   sequence, 128, 236, 245, 338
   sets, 135
   signaling, 234
   size, 129, 242
   small, 160, 219
   soluble, 205, 212, 214, 318
   spots, 21, 236, 237, 327, 328
   stability, 305
   "sticky," 161, 265
   structural status, 147
   structure, 145, 148, 195, 218, 301, 342,
      299, 338
    analysis, 300
    determination, 155
   synthesis, 284, 329, 332
    model, 331
    rate, 332
   targets, 136, 234, 343
   therapeutic, 193
   titer, 210
   toxic, 182, 184
    denatured, 9
    dispensing, 151
    highly basic, 7

[Protein(s)]
   hydrophobic, 7, 43
   hydrophobicities, 129
   insoluble, 43
   mammalian, 183
   nonuniform nature of, 191
   polarities, 129
   recombinant, 19, 24, 174, 304
   structures, 128, 129
  unknown function, 281
  variants, 129
  visualization, 228
  yeast single domain, 303
ProteinChip™, 220, 221, 222, 222, 223,
    224, 226–228
  profiling, 221
  reader, 227
Proteolysis, 187, 206, 239, 304–306
  ln-gel, 239
Proteome, 1, 2, 5–9, 21, 22, 28, 30, 32,
    41, 146, 147, 159, 184, 217, 226,
    246, 255, 256, 267, 326, 327, 330,
    334, 347
  analysis, 146, 234, 235, 341
  cellular, 217, 233
  complete, 256
  complexity, 237, 348
  diversity, 237
  entire, 32, 224, 227
  expected, 8
  expressed, 256
  expression, 331
  gene-based direction, 259
  Golgi, 227
  human, 14
  identification, 259
  instantaneous, 260
  mitochondria, 227
  predicted, 266
  protein-based direction, 259
  small, 184
  systems, 345
  visualization, 219, 227
Proteometrics, 344
Proteomewide screens, 266

Proteomic analysis, 236
Proteomic coverage, 5, 7
Proteomics, 1, 5, 8, 10, 30, 47, 55, 56,
    63, 83, 111, 122, 145, 146, 150,
    167, 175, 217, 228, 248, 277, 325,
    329, 333
  array-based, 8, 22, 26, 50, 52, 162
  cell map, 233, 234
  challenges, 341
  current technologies, 339
  data, 347, 349
  discovery, 4, 159
  expression, 247
  forward, 255, 256, 259, 260
  high-throughput, 248, 299
  interactive, 238
  market, 337–339, 341, 347, 349
    growth rate, 341
  protein expression, 233
  quantitative, 246, 247
  reverse, 255, 256, 259, 260
  sensitive, 248
  shotgun, 233, 235, 241
  structural, 299, 342, 343
  technologies, 338, 347
  tools, 128, 348
  traditional, 1, 18, 20, 26, 224, 226
  two-dimensional gel-based, 241
Proteomic technologies, 146
Proteosome 26S subunit, 244
Protoplate, 209
Prototrophic growth, 262
PSA, 138, 222
PSI-BLAST, 318
pSPORT, 178
PTMs, 8, 20, 32, 35, 38
Pufferfish, 240
Pumping station, 189
Pumps, 187
Purdue University, 345
Purification, 173–175, 181–184,
    186–190, 206, 207, 213, 221, 260,
    304, 318
  affinity, 147, 206, 224, 244
  biochemical, 256

[Purification]
  denaturing conditions, 183
  enzyme, 304
  filtration gel, 307
  genomic scale, 215
  high-throughput, 300
  immunoaffinity, 207, 238
  ion exchange, 307
  organelle, 227
  plaque, 213
  RNA, 267
  strategy, 185
  Strep-Tag, 187
  vacuum, 189, 212
  vector, 205
Purity, 301
Putative function, 260
PVDF, 130, 191, 192
  membranes, 21
Pyridoxal-5′-phosphate (PLP), 307
Pyridoxamine, 301, 319
Pyridoxamime-5′-phosphate (PMP), 307
Pyridoxamine-5′-phosphate oxidase, 301,
    306
Pyridozal-5′-phosphate (PMP), 307
Pyrosequencing, 189

Qiagen, 183, 189, 209
Quadrupole time-of-flight (TOF)
    instrument, 234, 240
Quality assurance, 28, 36, 44
Quality control, 35, 38, 100, 116
Quantative proteomic analysis, 248
Quantification, 340
Quantitation, 136, 237, 246, 248, 127,
    327
  high-throughput, 248
Quantitative analysis, 147
Quartz crystal microbalance, 54
Quartz optical fibers, 113
Quasiequilibrium, 111
Quenching detection, 223

Rabbit liver, 309, 314
Rac1, 291, 292
  activated, 292

Rac1-Cdc42, 293
Radioactive label, 129
Radioactivity, 195
Radioimmunoassay (RIA), 87, 89, 106,
    149
Radioisotope, 29, 89, 90, 94, 114, 133
    detection, 132
    labeling, 102, 133
Radiolabeled small molecules, 50
Radiolabeling procedures, 50
Raft1, 284
Ramachandran plot, 307
Random error, 87, 93
Random orientation, 149, 150
Rapamycin, 160, 284–291, 293
    -associating protein, 284
    -induced complex PP2A-p70S6K, 290,
        291
    insensitive, 290
    resistance, 288, 290
Ras, 265, 266
    -based signaling cascade, 263
    recruitment, 264
Rat, 208, 316, 318
    liver, 227
Ratiometric microspot assay, 100
Rational drug design, 343
RAVE package, 320
Rb, 265, 266
RCA, 133, 135
Reaction
    network, 330
    site, 163
    velocities, 107
Reactivation, 291
Reactive intermediates, 282
Read-through, 180
Reading frame, 33, 34, 37, 174, 265, 267
    correct, 267
    incorrect, 267
Readout, 154
Reagent
    costs, 128
    protein microarray, 135
Rearraying, 176, 208

Reassembly, 279, 285
Receiver operator characteristic, 221
Receptins, 9
Receptors, 17, 19, 92, 159, 160, 227, 284,
        292
    membrane, 228
    soluble, 284
    tyrosine kinase, 181, 292
Recessive selection strategy, 285
Recognition
    profiles, 48
    sequence, 180
Recombinant antigens, 9, 11, 12
Recombinant baculoviruses, 135
Recombinant clones, 176
Recombinant DNA technology, 300
Recombinant expression, mammalian, 204
Recombinant proteins, 13, 19, 20, 23, 30,
        32–36, 38, 39, 43, 45–48, 50, 56,
        174, 177, 178, 181–185, 187, 190,
        193, 304
    expression, 36
    high-throughput generation of, 30
    radioactively labeled, 181
Recombination, 205
    artificial site, 258
    homologous, 181, 205, 214
    in vitro, 258, 259
    intrachromosomal, 258
    site, 258
Recombinational cloning, 257, 258, 268,
        269
Recrystallization, 319
Red fluorescence, 136
Red fluorescent protein (RFP), 177
Red-to-green ratio, 136
Reduction, 226, 242
Reference mixture, 136
Refinement, 307, 322
    annealing, 320
    cycles, 311
    individual B factor, 320
    simulated annealing, 320
    structural, 320
Reflections, 301, 321, 322
    phases, 301

Reflectometric interference spectroscopy, 54
Refolding, 182, 303
Refraction, 176, 195
Regeneration, 191
Regulation
  allosteric, 282
  FRAP, 289
  PKB, 289
  positive/negative, 289
  protein, 338
Regulatory networks, 233
Relative intensity of fluorescent signal, 133
Repeat sequences, 181
Rephosphorylation, 291
Replicates, 62, 63, 65, 66
  on-array, 50, 57
Replicator pins, 185
Reporter, 30
  construct carboxyl-terminus, 178
  constructs, 176
  gene, 179, 180, 261, 264
    activators, 263
  optical, 178
  protein, 181
  systems, 176
Repression, 182
Reproducibility, 2, 7, 8, 20, 23, 28, 46, 57, 148, 152, 219, 341
Residues, 289, 311, 312, 318
  active site, 317
  conserved, 316, 318
  nonphosphorylated, 245
  polar, 312
  terminal, 312
Resin, 238
  -based assays, 149
  cation-exchange, 242, 243
  reverse-phase resin, 243
Resistance
  3-aminotriazole, 264
  5-fluoro-orotic acid, 264
Resolubilization, 203
Resolution, 146, 193, 220, 224, 307, 320–322

Resonance light scattering, 52
Resonant mirrors, 54
Responder mouse, 12
Response
  minimum, 86
  stimuli/inhibitor, 287
  -stimulus ratio, 85, 87
  variable, 85
Response-dose curve, 85–88
Response-dose ratio, 87
Restriction, 259
  -based enzymatic cloning, 257
  digestion, 257
  enzyme, 206, 257
    digest, 214
    rare-cutter, 206
Retentate chromatography, 220
Retroviruses, 349
Revenue, 338, 340
Reverse arrays, 19, 20
Reverse-phase-high performance liquid chromatographs (RP-HPLC), 223
Reverse-phase liquid chromatography, 235
Reverse-phase resin, 242
Reverse proteomics, 255, 256, 259, 260
Reverse transcriptase-PCR (RT-PCR), 8
RFP, 177
Rheumatoid arthritis, 138
Rho family, 291
  GTPases Rac1 and Cdc 42, 291
RIA, 89, 91, 149
Ribbon diagrams, 315
Ribityl group, 312
Ribosomal binding site, 180
Ribosomal complex, 244
Ribosomal proteins, 5, 46, 284
  S6, 293
Ribosome, 331, 332
  binding affinity, 331, 333
  binding constant, 333
  binding site, 332
  display libraries, 135
  free, 332, 333
  -mRNA binding, 332
  number, 333

Rifampicin, 119
  resistance, 119
Rigel Pharmaceuticals, 348
Ring, 313
Rms deviation, 314
RNA, 82
  antisense, 292
  polymerase, 119
  profiling, 191
  purification, 267
  -splicing machinery, 266
Robbins Scientific, 210
Robot, 189
Robotics, 22, 35, 59, 129, 190, 208, 209,
    257, 262, 269, 326
  -based screening systems, 264
  colony-picking, 176
  infrastructure, 176
  pipetting multichannel, 188
  platforms, 257
  printer, 129
  screening, 264
  system, 12
  units, HPLC, 187
Robots, 188, 189, 208, 209
ROC curve, 221
Roche, 190
Rolling circle amplification (RCA), 52,
    133
Root mean square (rms), 311
Rotamer conformation, 307
Rotring drawing pens, 113
RP-HPLC, 223
RTK, 290
  pathway, 286
RTK-FRAP, 293
  pathways, 286
  signaling network, 285, 286
Rubella, 119
R.W. Johnson Pharmaceutical Research
    Institute, 342

S6 protein, 293
*Saccharomyces cerevisiae* 30, 173, 204,
    237, 240, 242–244, 246, 248, 266,
    278, 279, 303, 318, 341

Saccharomyces Genome Database, 303
*Saccharomyces pombe*, 244, 245, 246,
    316, 318
SAGE (serial analysis of gene
    expression), 146, 233
Saliva, 95
Salt, 19, 130, 131
  calcium, 319
  concentration, 242
  pulses, 242
SAM, 163, 164, 192
Sample
  area, 108
  complexity, 242
  consumption, 128, 190
  enrichment, 221, 237
  fractionation, 237
  homogeneity, 304, 305
  identity, 304
  injection, 188
  load, 219
  loss, 237
  preparation, 8, 22
  size, 121, 220, 221
  volume, 97, 98, 108, 220, 95
SAMs, 149
Sandwich, 91
  assay, 112, 120, 154, 195
  ELISA, 132
  layers, 195
  -type assays, 154, 155
SARA supercomputing facility, 59
Scaffolding, 317
Scanning, 112, 129, 193
  time, 116
  tunneling microscopy, 150
Scatchard analysis, 107
Scattering mass, 307
scFv clones, 136
*Schizosaccharomyces pombe*, 318
Screening, 176
  antibody high-throughput, 135
  blue/white colonies, 176, 190
  cDNA libraries, 278
  enzyme inhibitor, 156

[Screening]
enzyme substrate inhibitor, 156
factorial, 204
filter-based, 190
genome-wide, 265, 269, 278
genomic, 279
high-throughput, 26, 37, 279, 343
interactions, 283
library, 278
modular, 266
parallel, 12
parallel nonlabeled, 48
proteome-wide, 262, 266
robotics, 264
small-molecule binding, 215
strategies, 262, 278
ultrahigh-throughput, 190
Y2H, 258, 262, 264, 265, 267–269,
    280, 303
    genome-wide, 267
yeast two-hybrid, 303
SDS, 186
sample buffer, 184
SDS-PAGE, 184, 218, 223, 239, 306, 319
Secondary detection, 132
antibody, 133
structure, 119, 320
Secretion system, 181
Segments, polypeptide chain, 305
Seiving gel, 15
SELDI, 155, 220, 221
mass spectrometry, 155
MS/MS, 225
Selection, 175, 261
dominant, 279, 285
genetic, 176
medium, 177
metabolic, 177
positive, 176, 179
size, 180
Selective medium, 176, 262
Selectivity, 342
Seleno-methionine, 301
Self-assembled monolayer (SAM),
    163–165

Self-activators, 265
Self-assembled monolayers, SAMs, 149
Self-proteins, 138
Semiconductors, 23
Sensitivity, 8, 21, 28, 30, 52, 82, 85, 86,
    88–90, 93, 94, 96, 101, 103,
    104–108, 114, 116, 117, 119, 121,
    127, 128, 146, 148, 162, 219,
    221–224, 234, 239, 279, 285, 299,
    341, 348
relative, 88
Separation, 235
multidimensional, 243
orthogonal, 220
sciences, 1, 7, 66
technologies, 1
Sepharose resin, 227
Sequence, 300, 318
alignment, 316
amino acid, 235, 246
coding, 301
conserved, 316
homology, 38, 314
identity, 303, 316, 318
motif, 314
orthologs, 257
primary, 306, 314
primary gene, 277
protein, 338
searches, 329, 330
similarity, 318
translated, 180, 299
Sequencing, 30, 35, 301, 326
amino acid, 328
DNA, 304
N-terminal, 236
SEQUEST, 235, 241, 243
SEQUEST-PHOS, 241
Sequestering, 95
Sera, 15
SEREX, 190
Serial analysis of gene expression
    (SAGE), 146, 233
Serial crystallization, 304
Serine hydrolase, 226

Serine protease, 311
Serological analysis serological, 190
Serum, 5, 23, 46, 47, 154, 161, 169, 219,
  222, 288, 291, 290
  inducible, 290
  protein expression, 139
  stimulation, 285
Serum/insulin-stimulated profile, 288
Serum/ovalbulim, 169
SETOR, 310
Shaker, 185, 211
Shaker or spinner flasks, 206
Shelf-life, 23, 190
Shielding steric, 160
Short alpha-helices, 314
Shotgun, 242
  proteomics, 235, 241
Shuttle vector, 181
Side chain, 314, 317
  atoms, 312
  densities, 307, 320
Side effects, 48, 52
Signal, 86
  amplification, 8, 30, 52
    strategies, 8
  -background ratio, 103, 104, 108
  detection, 8, 149
  intensity, 44
  measurement, 104
  -to-noise ratio, 25, 27, 28, 53, 86, 102,
    103, 109, 113, 113, 114
  peptide recognition, 204
  peptide sequences, 204
  termination, 205
  transduction, 245, 282, 283, 285
    cascades, 227
    networks, 283
    pathways, 38, 265, 282, 292
Signaling
  cascades, 225
    Ras-based, 263
  networks, 292, 293
  pathways, 135, 294
Signature peptides, 8, 35, 41, 42
Silane
  linkers, 162
  monolayers, 155

Silicon, 192
  substrate, 152
  wafer, 192
Silinated glass, 130
Silver staining, 36, 218, 219, 223, 234
Silynated slides, 131
Similarity, 311
Single cells, 176
  sorting, 178
Single-channel LC system, 187–189
Single-domain structure, 311, 314
Single-domain yeast proteins, 303
Single-point mutations, 119
Site(s), 321
  active, 281
  antibody-binding, 90
  autoreactive, 139
  binding, 93
  mercury, 320
  occupancy, 23, 28, 53, 67, 91, 100,
    106, 107
    effective, 28
  phosphorylation, 286
  -specific immobilization, 150
  -specific recombination, 258
  unoccupied, 91, 93, 100
Six-stranded Greek-key β-barrel, 309,
  310, 311, 314
Size, tractable, 300
Slope, 86, 88, 188
Slotted-pin spotters, 132
Small compounds, 191
Small molecules, 28, 29, 53, 175, 191,
  193
Small-molecule therapeutics, 53
Sodium dodecyl sulfate, 184
  -polyacrylamide gel electrophoresis
    (SDS-PAGE), 218
Sodium ethylmercurithiosalicylate
  (EMTS), 320, 321
Software, 56, 129, 240, 243, 330
  packages, 299
Solid-phase, 131
  extractions (SPE), 188
  immunoassays, 133

Solid substrates, 192
Solid support, 81, 92, 95, 97, 98, 102, 103, 108, 113, 114, 117, 148, 174, 190
Solid–liquid interface, 148
Solid-phase assay, 114
Solid-phase extraction, 188
Solubilization, 182, 227, 228
Soluble protein, 303
Solvent, 312, 314
  boundaries, 320
  correction, 320
SomaLogic, 84, 344, 345
Sonics, 186
Sorting device, 177
Sound waves, 186
Space group trigonal, 319
Spatial distribution, 282
Spatio-temporal, 327
SPE, 188
  cartridges, 189
Specific activity, 102
Specific stimuli, 279
Specificity, 13, 48, 90, 96, 109, 112, 119, 121, 122, 135, 136, 221, 264
  structural, 91
Spectroscopy, 280
  absorbance, 213
  fluorescence, 280
Spectrum quality, 245
Speed, 96, 128
Spin columns, 221, 227
Spindel integrity, 244
Splice isoform, low-abundance, 267
Spliceosome, 244
Splice variants, 8, 38, 66
Splicing, 217, 265
  alternate, 255, 265, 267
Spots, 61, 98, 103, 104, 108, 128, 129, 132, 133, 135, 146, 147, 159–161, 169, 326
  area, 102, 103
  change, 328
  density, 115, 159, 192
  intensity, 220, 237, 327

[Spots]
  location, 328
  protein, 11, 18, 22, 57, 340
  size, 18, 104, 108, 110
  synthesis, 128
  volume, 100
Spotted array technology, 326
Spotted proteins, 131
  drying, 131
Spotters, capillary-based, 151
Spotting, 129, 131
  density, 192, 193
  procedures, 113
  solution, 132
SPR, 29, 149, 165, 169, 195
Spreading, 192, 195
Squamous cell carninoma, 127
β-Sheet, 311
  eight-stranded antiparallel, 309
Stability, 23, 192
Stabilization, 138, 311
Stable core, 306
  fragment, 306
Stable isotopes, 64, 223, 246
  dilution, 223
Stacking interactions, 317
S-Tag, 183
Stainless-steel rods, 113
Standard deviation, 87
Standard errors, 288
Standard international, 35
Standardization, 344
Standard operating procedure, 35
Standards, 326
  data communication, 326
  data storage, 326
Standing waves, 117
Staphylococcus aureus, 169
Statistical analysis, 87
Statistical confidence, 7, 57
Statistical methodologies, 61
Statistical methods, 65
Statistical variation, 104
Statistics, 228, 320
  data collection, 321
  refinement, 320, 322

Stereoview, 313
Steric hinderance, 23, 28, 162, 163
Steric shielding, 160
Stimulants, 288
Stimulation, 291
    cycle, 293
Stimulators, pathway specific, 283
Stimuli, 279, 282
Stirrer magnetic levitation, 211
Stop codon, 255, 258
Stop signal, 180
Strands, 311
Stratagene, 182, 184, 318
Strategies, 303
Strep-Tag, 183, 184, 187, 190
    purification, 187
Streptacin, 184, 187
Streptavidin, 29, 167, 169
Stress
    cellular, 256
    metabolic, 256
Stringency, 279
Structural Bioinformatics, 342
Structural biology, 300, 301
    high-throughput, 301, 304
Structural coordinates, 301
Structural crystallization, 174
Structural genomics, 30, 299–301
    biophysical/biochemical analysis, 300
    pipeline, 302
Structural GenomiX, 342
Structural homology, 173
Structural information, 300
Structural modeling, 303, 318
Structural predictions, 344
Structural refinement, 320
Structural status, 147
Structural studies, 195
Structure, 285, 301
    –activity relationships, 175
    analysis, 183, 338, 343
    covalent, 304
    crystal, 309, 314
    determination, 155, 193, 318
    derivative, 321

[Structure]
    native, 321
    nucleic acids, 128
    partial, 320
    protein, 338
    solution, 300
    solving, 300
    three-dimensional homology, 343
    x-ray crystal, 301
Structured interlayers, 165
Structure–function, 260
    relationships, 342
Student's $t$-test, 61
Subcellular compartments, 227, 235, 282
Subcellular fractionation, 227, 237
Subcellular locations, 292, 294
Subcloning, 264, 266, 214
    elimination vector, 208
Subproteomes, 217, 224, 227
    cellular, 228
    membrane protein, 228
Substrates, 17, 20, 23, 26, 28, 162, 163,
        175, 191, 193, 291, 293, 317
    4EBP1, 289
    bind, 314
    chemical composition, 163
    chemiluminescent, 194
    dissociation rate, 263
    enzymatic, 193
    fluorescent, 177
    fluorogenic, 194
    glass, 48
    kinase, 225
    libraries, 156
    morphology, 163
    noble metal, 166
    phosphorylated, 225
    silicon, 152
    sites p70S6K, 291
Subtilisin, 245
Subunits, 310–312, 316, 317, 320
Succinic anhydride, 164
Succinimides, 131
Succinylation, 32
Sucrose gradient, 228

Sulfhydryl group, 132
Sulfhydryls, 225
Sulfuric acid, 23
Supercomputing, 59, 61
Superdex 75 gel filtration column, 319
Supernatant, 181, 212, 319
Superposition, 314, 315
Suppressor gene, 179
Surface, 23, 25, 127, 129–131, 148, 160,
    161, 163, 166, 167, 130
  acoustic waves, 54
  activation, 149, 166
  adsorptive, 192
  area, 207
  attachment, 128
  biocompatible, 163
  bioreactive, 149
  cavities, 150
  -charged negative, 167
  chemistry, 19–23, 26, 28, 53, 67, 162,
    344
    multilayer, 23
    optimization, 28
  coated, 164, 189
  compatibility, 23
  defects, 23, 25, 150
  density, 98, 100, 103–109, 113
  derivatization, 131
  display, 174
  electrically-neutral, 165
  engineering, 148
  -enchanced laser desorption/ionization
    (SELDI), 155, 220
  filamentous phage, 17
  flatness, 150
    atomic, 150
  functionalization, 149, 150, 159, 160,
    163
  gold, 164
  hydrazide-activated, 131
  hydrazide-derivatized, 132
  hydrogen-bond acceptor, 165
  hydrophilic, 165, 166
  irregularities, 150
  modification, 162, 163, 170

[Surface]
  multilayer, 25
  noncharged, 166
  optimization, 150
  PEG, 165
    -coated surface, 131
  PET, 163
  plasmon resonance (SPR), 24, 25, 29,
    54, 149, 154, 163, 165, 169, 345
    dual wavelength, 28, 29
    grating coupled, 54
  poly(ethylene terephthalate), 163
  polymer-modified, 164
  polystyrene, 129
  preactivated, 227
  properties, 162
  protein-reactive, 149
  topography, 150
Survival
  assay, 285
  screen, 285
  -selection assay, 285
Suspension cultures, 206
SwellGel 20, 207
SwissProt, 303
Symmetry, 311
Synchrotron
  data, 307
  radiation sources tunable, 300
Synexpression, 64
Synthesis, 286
Synthetic binder libraries, 11
Synthetic peptides, 20
SYPRO Ruby, 223
Syrrx, 342
Systemic lupus erythematosus, 138
Systems approach, 326
Systems biology, 277, 330, 334

T7 polymerase, 186
Tag, 150, 183, 206
  affinity, 182, 206, 238, 330
    isotope-coded, 146, 223
  cysteine, 246
  epitope (*see* Epitopes, tag)
  expressed sequence, 208, 240, 257

[Tag]
fusion, 150
Glu-Glu, 135
glutathione-S-transferase, 303
heavy, 224
hexa-histidine, 303
isotopic, 224
light, 224
maltose binding protein, 303
on-rate, 183
poly-histidine , 135
removal, 304
selection, 204
Tagged image format files (TIF), 328
Tandem mass spectrometry (MS/MS),
218, 221, 223, 224, 225, 234, 235,
237, 239, 242, 243
peptide sequencing, 239
TAP, 245, 246
-purified protein complexes, 246
tag, 245
Targets, 8, 11–13, 28–30, 35, 38, 42, 46,
48, 50, 53, 63, 66, 111, 112, 135,
150, 173, 284, 300, 301, 303–305
accessibility, 46
analytes, 81, 89, 90, 95, 109, 121 , 86
amplification, 83
binders, 9, 48
discovery, 66
gene expression, 265
Y2H, 262
genes, 264
ligands, 111, 122
lists, 301
dissemination, 301
molecules, 159
pharmacological, 173
proteins, 150, 304
recognition, 12, 30
selection, 300, 301, 303
strategies, 303, 318
selectivity, 11, 42, 46–50, 66
sequences, 318
specificity, 66
strategy, 318

[Targets]
therapeutic, 55, 66
validation, 66, 341
Tecan, 210
Technological improvements, 339
Technologies, 326
Teflon, 113
TeleChem International, 345
Temperature, 205, 281
Terminal amines, 131
Termination rate, 332
Terminus amino, 207
Terminus carboxy, 207
Tertiary homologs, 38
Tertiary structure, 41
Tertiary structural homologs, 38
Testing
clinical, 48, 55
preclinical, 48, 55
toxicological, 48
Tetracycline (TET), 182
Texas Red, 114
TF-1 tumor cells, 305
TFAR19, 305
Therapeutic molecules, 48
Therapeutic proteins, 193
Therapeutics, 59, 66, 119, 195, 299
novel, 66, 342
Thermal lens microscopy, 52, 54
Thermo Electron, 339
Thermodynamic equilibrium, 107
Thin films, 149, 152
Thiol groups, 23, 148, 165
Thioredoxin (THX), 33, 183
Threading algorithms, 38
Three-dimensional hydrogel matrix, 23,
48, 56
Three-dimensional polyacrylamide gel
patches, 155
Three-dimensional structure, 316
Throughput, 145, 146, 299
Time, 29, 46, 107–109, 282, 327
incubation, 82, 90
Time-of-flight mass spectrometer, 220
Time-resolution, 94

Tissue, 6, 7, 54, 147, 175, 235
  arrays, 9, 46
  extracts, 19
  heterogeneity, 46
  homogenizers Potter-Elvehjem, 186
  microarrays, 20, 47
  slices, 46
Titin gene, 2
TOF instrument, 240
Topography
  chip surface, 150
Toxic genes, 177, 178, 182
  products, 182
Toxic proteins, 182, 184
Toxicological testing, 47
Toxicology, 337
Tracers, 90, 92, 100
Traditional proteomics, 1, 2, 5, 7, 20, 23, 26
Traditional vectors, 257
Trails crystallization, 305
Transactivator, 263
Transcript profiling, 208
Transcription, 5, 263, 281, 318, 265
  factor, 244, 261, 263
  in vitro, 183
  profiles, 217
Transcriptional activator, 263
Transcriptional activation domain (AD), 261
Transcriptional output, 290
Transcriptional profiling, 265
Transcriptome, 326, 327, 330, 334
Transduction metabolic, 281
Transfection, 210, 213, 214, 209
  insect cells, 206
Transfer vector, 206, 208, 210
Transferrin, 5
Transformants, 267
Transformation, 182
  efficiency, 213
  high-throughput, 181
Transitional cell carcinoma (TCC), bladder, 222
Translated sequences, 299

Translation, 183, 281, 318, 331, 332
  completion, 332
  initiation, 282, 284
    amino acid-activated, 284
  rate, 332
Translational diffusion coefficient, 305
Translocation, 293
Transmembrane regions, 181
Transmembrane-spanning region, 9
Transporters drug development, 227
Triple quadrupole, 240
  instrument, 234
Tris-HC1, 187
Triton X-114, 227
tRNA, 184
Truncation, 306
TRX, 33
Trypsin, 146, 225, 234, 242, 245, 304, 306
Tryptic digestion, 221
Tryptic peptides, 225, 244
  fingerprints, 218
Tuberculosis, rifampicin-resistant, 119
Tumor
  antigens, 190
  cells, 155, 305
  markers, 128
Tunable synchrotron radiation sources, 300
Tween 20, 186, 187
Two-color comparative fluorescence, 139
Two-color detection system, 120
Two-dimensional (2D) electrophoresis, 127, 236–328
  database, 328
  gels, 15, 19, 21, 22, 57, 63, 128
  reproducibility, 236
  resolution, 236
  sensitivity, 236
Two-dimensional fractionation, 242
Two-dimensional gel electrophoresis (2DGE), 1, 3, 5–7, 47, 63, 127, 146, 156, 175, 191, 218–221, 223–228, 237, 242–244, 260, 327, 328, 338, 340, 347

Two-dimensional gels, 2, 46, 223, 327
Two-dimensional polyacrylamide gel
  electrophoresis (2D PAGE), 146,
  236, 237, 242–244
  equipment, 341
  market, 341
  reagents, 341
  resolution, 146, 147
  sensitivity, 146, 147
  throughput, 146, 147, 146
Two-dimensional protein gels, 328
Twofold noncrystallographic symmetry,
  312
Two-gene network oscillatory behavior,
  331
Two-hybrid screening, 280, 303
Two-hybrid system, 263, 265

Ubiquitin split, 264
UBS Warburg, 338, 339
Ultra-high-throughput
  analysis, 154
  screens, 190
Ultrasensitive ligand assay, 82
Ultrasound, 185, 186
  waterbath, 186
Ultra Turrax, 186
Ultrazoom gels, 219
Unit, 311
  asymmetric, 320
Universal donor vector, 303
Untranslated regions, 37, 46, 180, 209, 37
Upregulated, 329
Urea, 242
UTRs, 180
  3′, 37, 46
  5′, 37, 46
UV detector, 188, 190

V8 protease, 306
Vacuum, 21, 23, 339
  filtration, 207, 212
  manifold, 212
  purification, 189
Validation, 221, 278

Valuations, 338
Valve
  automatic, 189
  electronic, 189
  pneumatic, 189
van der Waals interaction, 311
Vanishingly small, 82, 98
Variability interarray, 120
Variance, 5, 7, 8
  background, 65
Variants, 285
Variation coefficient of, 128
Variomag, 185
Vector, 32, 37, 176–181, 210, 257, 258,
  264, 301, 304
  bacterial, 182
  design, 34, 37
  destination, 34, 258, 303
  donor, 259
  entry, 35
  expression, 34, 35, 182, 258, 301, 303,
    304
    bait, 266
    dissemination, 301
    prey, 266
  low-copy-number, 264
  multicopy, 264
  plasmid, 258
    single-copy, 261
    transfer, 205
  secretion, 181
  shuttle, 181
  transfer, 206, 214
  universal donor, 303
  Y2H, 258
Video, 330
Viral amplification, 214
Virtual Cell Project, 330
Viruses, 85
Visibility microspot, 104, 105
Visualization, 220, 227
  dynamic, 277
Vitamins, 85
Vm, 319

V & P Scientific, 210
Vulval development, 266

Wafers, 193
Wallac Oy, 94
Washes, 131
Washing, 113, 130, 188, 191, 193
    conditions, 220
Water, 322, 339
Wavelength, 133
Wells, 22
Well volume, 211
Western blot, 9, 15, 19, 21, 36, 39, 46,
        138, 184, 209, 212, 226
    blot analysis, 184
ω-functionalization, 149
ω-functionalized alkyl thiols, 163
Whatman, 185
Whole-cell lysate, 243
Whole-genome sequences, 299
Wilcoxon sign-ranked test, 61, 62
Wortmannin, 284–288, 290, 291
    inhibition, 290
    sensitive, 290


Xenopus, 208
X-gal, 177, 178
X-gluc, 214
X-PLOR, 320
X-ray area detectors, 300
X-ray beam line, 301
X-ray crystallography, 55, 215, 301, 304,
        343
X-ray crystal structure, 301
Xeroderma pigmentosum,
        complementation group C, 304

γ-secretase, 226
Y2H, 260, 261, 263–268
    screens, 258, 262, 268, 269
    vectors, 258
Yamanouchi Pharmaceuticals, 342
Yeast, 30, 127, 174, 177, 181, 183, 193,
        223, 233, 242, 246, 261, 262, 267,
        279–281, 300, 303, 305, 308–311,
        314–316, 318, 319, 322, 331, 348,
        290
    diploid, 262
    display libraries, 135
    expression, 181
    genes, 300
    genome, 244
    host, 262
    mating, 262
        strategy, 262
    ORFs, 193
    PNP oxidase, 311, 312, 314
    proteins single domain, 303
    proteome, 6
    two-hybrid, 46, 127, 156, 256, 257,
            258, 260–264, 268, 281, 341, 278
        approach, 46
        bacterial, 261
        high-throughput, 266
        mammalian, 261
        modular, 269
        screening, 303
        system (Y2H), 257, 278

Zebrafish, 208
Zero dose, 87
Zymark, 185
Zyomyx, 151, 345
    protein chip, 151, 343

9 780367 395070